FUNDAMENTALS OF FIRE AND EMERGENCY SERVICES

FUNDAMENTALS OF FIRE AND EMERGENCY SERVICES

Jason B. Loyd, MA Ed

Texas A&M Engineering Extension Service (TEEX)
Emergency Services Training Institute (ESTI)—Online Recruit Academy
Training Manager
Texas A&M University System

James (J.D.) Richardson, MA Ed

San Antonio College Protective Services Department
Assistant Professor of Fire Science
Coordinator of Emergency Management/Homeland Security
Administration (Retired)

PEARSON

Boston   Columbus   Indianapolis   New York   San Francisco   Upper Saddle River
Amsterdam   Cape Town   Dubai   London   Madrid   Milan   Munich   Paris   Montreal   Toronto
Delhi   Mexico City   São Paulo   Sydney   Hong Kong   Seoul   Singapore   Taipei   Tokyo

Publisher: Julie Levin Alexander
Publisher's Assistant: Regina Bruno
Editor-in-Chief: Marlene McHugh Pratt
Senior Acquisitions Editor: Stephen Smith
Program Manager: Monica Moosang
Development Editor: Eileen Clawson
Editorial Assistant: Mary Ellen Ruitenberg
Director of Marketing: David Gesell
Executive Marketing Manager: Brian Hoehl
Marketing Specialist: Michael Sirinides
Marketing Assistant: Crystal Gonzalez
Team Lead for Brady and Health Professions:
    Cynthia Zonneveld

Project Manager: Julie Boddorf
Full-Service Project Manager: Niraj Bhatt,
    Aptara®, Inc.
Manufacturing Manager: Vincent Scelta
Manufacturing Buyer: Nancy Maneri
Editorial Media Manager: Amy Peltier
Media Project Manager: Ellen Martino
Creative Director: Jayne Conte
Cover Designer: Karen Noferi
Cover Image: Jason B. Loyd
Composition: Aptara®, Inc.
Printer/Binder: R.R. Donnelley/Willard
Cover Printer: Lehigh/Phoenix Color Hagerstown

Credits and acknowledgments borrowed from other sources and reproduced, with permission, in this textbook appear on the appropriate pages within the text.

Library of Congress Cataloging-in-Publication Data
Loyd, Jason B.
    Fundamentals of fire and emergency services/Jason B. Loyd, MA Ed, Texas A&M Engineering Extension Service (TEEX), Emergency Services Training Institute (ESTI)—Online Recruit Academy Training Manager, Texas A&M University System, James (J.D.) Richardson, MA Ed, San Antonio College Protective Services Department, Assistant Professor of Fire Science, Coordinator of Emergency Management/Homeland Security Administration (Retired).—Second edition.
    pages cm
        ISBN-13: 978-0-13-341923-8
        ISBN-10: 0-13-341923-1
    1. Fire protection.   2. Fire extinction.   I. Richardson, James D.   II. Title.
TH9146.L69 2013
628.9′2—dc23   2013012088

10 9 8 7 6 5 4 3 2 1

ISBN 13: 978-0-13-341923-8
ISBN 10:   0-13-341923-1

CONTENTS

Foreword x
Preface xi
Acknowledgments xv
About the Authors xvii
16 Firefighter Life Safety Initiatives xix
Fire and Emergency Services Higher Education
  (FESHE) Grid xxii

Chapter 1 History of the Fire Service 1

Introduction 1

The Evolution of Firefighting Services 2

Early Firefighting Apparatus, Vehicles, and
  Extinguishers 6

Respiratory Protective Equipment or Self-
  Contained Breathing Apparatus 10

Fire Facilities 10

Fire Service Insignias and Traditions 11

      Red Fire Engine 11
      Badges 11
      Maltese Cross 11
      Dalmatian Dog 12

The Fire Dilemma in the United States 12

The Scope of Fire and Emergency
  Services Today 13

Summary 16

Review Questions 16

Suggested Reading 16

Endnotes 16

Chapter 2 Fire and Emergency Services Career Opportunities 18

Introduction 18

Firefighting Career Opportunities 19

      Fire Cadet/Recruit 19
      Firefighter 20
      Federal Firefighter 21
      Firefighter/EMT or Higher 21

      Wildland Firefighter 21
      Aircraft Rescue Firefighter 22
      Fire Apparatus Engineer 23
      Search and Rescue 23
      Industrial Firefighter 25
      Hazardous Materials Technician 25
      Fire Investigation Units/Arson Units 26

Nonfirefighting Careers in Fire Protection 26

      Dispatcher/Communications Operator 26
      Occupational Health and Safety
        Specialist 27
      Fire Equipment Salesperson 27
      Fire Extinguisher Service Technician 28
      Fire Marshal 28
      Public Information Officer 29
      Fire Prevention Specialist/Public Fire
        Education Specialist 29
      Fire Department Training Specialist 29
      Insurance Adjuster 29
      Fire Inspector 29
      Fire Investigator 30

Private Sector Fire Protection Careers 30

      Statistician 30
      Fire Protection Engineer 30
      Contract Firefighter Opportunities
        (Overseas) 30
      Inventor 31

Volunteer Opportunities 31

The Future of Fire and Emergency Careers 31

Summary 33

Review Questions 33

Suggested Reading 33

Endnotes 33

Chapter 3 The Selection Process 35

Introduction 35

Preparation for a Firefighter Position 36

      Prerequisites 36
      Written Test 36
      Physical Ability/Agility 38

Interviewing Skills 39

      First Interview or Chief's Interview 39

Personal History 41

    Medical Exam 41
    Background History 41

Hiring Conditions 42

    Eligibility List 42
    Probationary Period 42

Marketing yourself for the Firefighting Job 42

    EMT Status 42
    Firefighter I, Other Certifications,
      and Higher Education 42
    Volunteer Experience 43
    Bilingual Ability 43
    Clean Driving Record 43
    Excellent Physical Fitness 43
    Diversity 43
    Social Media 44

Summary 45

Review Questions 45

Suggested Reading 45

Endnotes 45

Chapter 4 Training and Higher Education 47

Introduction 47

Higher Education 48

Training 48

    Fire Department Training or Special
      Operations Bureaus 50
    The U.S. Fire Administration and
      the National Fire Academy 50

Advancement and Promotion 51

    Certification 52
    Specialty Certifications 53
    Training Instructor 54
    Training Officer 54

Post-Secondary Education 54

    The Fire Science Technology
      Curriculum 55
    Distance and Online Education 56

Setting Standards for Higher Education in the Fire Service 57

    FESHE Guidelines 58

International Fire Service Accreditation Congress 59

Pro Board Accreditation 60

Summary 62

Review Questions 62

Suggested Reading 62

Endnotes 62

Chapter 5 Fire Department Resources 64

Introduction 65

Fire Department Facilities 65

    Administration Buildings 65
    Training Facilities 66
    Fire Stations 67

Fire Apparatus 69

    Pumping Apparatus 70
    Wildland Firefighting Apparatus (Brush
      Trucks) 71
    Mobile Water Supply Apparatus 72
    Aerial Apparatus 73

Specialized Apparatus 76

    Ambulance Apparatus 77
    Rescue Apparatus 77
    Hazardous Materials Apparatus 79
    Aircraft Rescue and Firefighting Apparatus 80
    Fireboats 81
    Firefighting Aircraft 81

Personal Protective Equipment 82

    Clothing 82
    Self-Contained Breathing Apparatus 88
    Personal Alert Safety System 90

Tools and Other Equipment 91

    Thermal Imaging Cameras 91
    Mobile Data Terminal 92

Summary 94

Review Questions 94

Suggested Reading 94

Endnotes 94

Chapter 6 Fire Dynamics 96

Introduction 97

Atoms and Molecules 97

Matter, Energy, and the Combustion Process 97

    Solid Matter 98
    Liquid Matter 99
    Gaseous Matter 100

Heat Energy 102

 Fahrenheit, Rankine Scales, and British
  Thermal Units 102
 Specific Heat and Latent Heat 102
 Conduction, Convection,
  and Radiation 102

Chemical Energy 104

 Heat of Combustion or Caloric Value 104
 Spontaneous Heatings 105
 Heat of Decomposition 106
 Heat of Solution 106
 Heat of Reaction 107

Electrical Energy 107

 Forms of Electrical Heat Energy 107

Mechanical Energy 109

 Friction 110
 Compression 110
 Combustion Process 110
 Burning Process 111

Fire Classification 114

 Class A: Ordinary Combustibles 114
 Class B: Flammable Liquids 114
 Class C: Energized Electrical Fires 115
 Class D: Flammable Metals 115
 Class K: High-Temperature Cooking
  Oils 115

Summary 116

Review Questions 117

Suggested Reading 117

Endnotes 117

**Chapter 7 Building
     Construction 119**

Introduction 120

Loads and Forces 120

 Load Type 121
 Load Distribution 121
 Load Imposition 121
 Axial Loads 121
 Eccentric Loads 122
 Torsional Loads 122

Structural Components 123

 Foundations 123
 Beams 124
 Columns 125

 Walls 126
 Wood Trusses and I-Joists 128
 Ceiling Joists 130
 Roof Configurations 131

Construction Types 135

 Type I Construction 136
 Type II Construction 140
 Type III Construction 141
 Type IV Construction 144
 Type V Construction 146

Summary 149

Review Questions 149

Suggested Reading 149

Endnotes 150

**Chapter 8 Fire Prevention Codes
     and Ordinances 151**

Introduction 151

**Building Codes and Fire Prevention
Codes 153**

**History of Model Codes in the United
States 153**

 Building Officials and Code Administrators
  International 154
 International Conference of Building
  Officials 154
 Southern Building Code Congress
  International 154
 International Code Council 154
 National Fire Protection Association 155

**Historical Basis for the Legal Authority of Code
Enforcement 156**

 Dillon's Rule, the Cooley Doctrine, and
  Atkins v. Kansas 156
 Warrantless Entry 156
 Right of Entry 157

Types of Codes and Code Adoption 158

 Minimum Codes Versus Mini-Maxi
  Codes 158
 Model Codes 158

Code Administration and Enforcement 159

 Documentation 159
 Summons and Citations 160

Summary 161

Review Questions 161

Suggested Reading 161

Endnotes 162

Chapter 9 Fire Protection Systems 163

Introduction 164

Water Supplies and Distribution Systems 164

Sources of Water Supply 164
Water System Components 165
Pipe Size Requirements 168
Water Distribution Models 170

Fire Hydrants 171

The Base-Valve Dry-Barrel Hydrant 171
The Wet-Barrel (California) Hydrant 172
Dry Hydrants 174
Hydrant Maintenance and Flow Testing 174

Standpipe Systems 177

Class I System 178
Class II System 178
Class III System 179
Types of Standpipe Systems 179
Standpipe Flow Rate Requirements 179
Pressure-Regulating Devices 180

Automatic Sprinkler Systems 180

Automatic Sprinkler System
Components 182
Types of Automatic Sprinkler Systems 189

Signaling Systems 192

Types of Signaling Systems 193

Automatic Detection Systems 195

Heat Detectors 195
Smoke Detectors 197
Flame (Light) Detectors 197
Carbon Monoxide Detectors 198

Special Extinguishing Systems 198

Halon Systems 199
Carbon Dioxide Systems 200
Dry Chemical Systems 201
Wet Chemical Systems 202
Foam Systems 203

Summary 205

Review Questions 205

Suggested Reading 205

Endnotes 206

Chapter 10 Organizational Structure and Emergency Incident Management Systems 208

Introduction 209

Fire Department Basics 209

Types of Fire Departments 209

Rank Structure and Organization 211

Rank Structure 211
Organization 212

Basic Principles of Command 214

The History of the Incident Command Systems 215

FIRESCOPE 215
Fire Ground Command System 216
FIRESCOPE versus the FGC 216

The Incident Management System 216

Functional Areas of the IMS 216
Administration and Finance 218
National Interagency Incident Management System 219
Incident Command Systems—A Report Card 219
The National Incident Management System 220
Principles of NIMS 221

Summary 225

Review Questions 225

Suggested Reading 225

Endnotes 225

Chapter 11 Preincident Planning, Fire Strategy, and Tactics 227

Introduction 227

Preincident Planning 228

Building Construction 230
Occupancy 230
Apparatus and Personnel 231
Life-Safety Hazards 232
Water Supply 232
Auxiliary Appliances 234
Street Conditions 234

Weather 235
Area and Height (Topography) 236
Location and Extent 237
Time of Day 237
Hazardous Materials 238
Communications 239
Command Structure 239

Strategy 240

Tactics 242

Tactical Decision Making 242

Summary 244

Review Questions 244

Suggested Reading 244

Endnotes 245

Chapter 12 Emergency Medical Services and Emergency Management 246

Evolution of EMS 246

Level of Emergency Medical Care 248

Fire and Emergency Services Higher Education Guidelines 249

Postsecondary Education 250

Firefighters and EMTs 250

Emergency Management and Homeland Security 251

History of FEMA 251

Focus of Emergency Management 252

Preparedness 253
Response 254
Recovery 255
Mitigation 256

Postsecondary Education for Emergency Management 256

Summary 258

Review Questions 258

Suggested Reading 258

Endnotes 258

Appendix A   On Scene Suggested Answers 260

Appendix B   Answers to Review Question 263

Appendix C   Stop and Think Suggested Answers 270

Appendix D   Expand Your Learning Activities 276

Glossary 279

Index 289

I have written many pieces in my life—however, this one in particular is of greater significance to me. Writing this is an opportunity for me to publicly recognize and state my admiration of Jason Loyd and J.D. Richardson. I have known them both for more than 15 years. I saw them progress in their respective fields and admired their dedication and tenacity in the pursuit of their goals. With all this in mind, I am pleased to convey the following thoughts of my two good friends and colleagues.

In 1996, I met J.D. Richardson when I was hired by San Antonio College Fire Academy as an adjunct instructor. As instructed, I showed up for the first day of live fire training. There was heavy smoke billowing from the burn house, much like the stack on an old-fashioned locomotive, a figure appeared through the smoke. He walked towards a group of fire cadets who had just finished their live fire training evolution. He took off his self-contained breathing apparatus (SCBA) mask and the rest of his personal protective equipment (PPE)—a sweaty, smokey-smelling man with white hair started debriefing the cadets. I stood by and observed the look on the face of the cadets as he explained what went well and what could be improved. It was an inspiring sight. You could almost feel his knowledge and experience oozing out of every word he used. His love and dedication to his profession were evident not only on his face but in the faces of the students he was briefing. Through the years, J.D. Richardson continued teaching fire cadets and every time I saw him lecturing I would always observe the same look on his face and on the face of the fire cadets—simply awe inspiring.

In the spring of 1998, I was teaching for the San Antonio College Fire Academy when I met Jason Loyd. He was a young man wanting to become a firefighter. I remember him well, not because of his good grades, his insecure mannerisms, or his constant questions; it was his study habits that impressed me. Jason Loyd was the only cadet who would diligently stay every day after class to study. One day Jason approached me and asked if I could tutor him. He informed me that he was dyslexic and suffered from attention deficit hyperactivity disorder (ADHD). I agreed, setting up life-changing events for the two of us. Jason not only passed his firefighter exam, but he received the highest grade in his class. He went on and completed an associate's degree, bachelor's degree, and finally a master's degree. Among other achievements, he wrote this book. I worked with him overseas and I witnessed the positive influence he had over his colleagues. In my opinion, Jason Loyd exemplifies perseverance and dedication. His ability to overcome his deficiencies has inspired me to continue on my path and achieve my dreams.

This book, *Fundamentals of Fire and Emergency Services* by Jason B. Loyd and James D. Richardson, exemplifies their commitment for excellence and knowledge of the fire service. The book is well organized and easy to understand. Each chapter emphasizes Jason's and J.D.'s strengths / knowledge of the subjects. Chapter 4 focuses on training and higher education; this subject has been and continues to be Jason Loyd's flagship of knowledge. Chapter 9 concerns fire protection systems—I can almost see J.D. Richardson writing this chapter, his mind spinning at a high revolution as he explores the best ways to convey his message.

Fundamentals of Fire and Emergency Services is a must-have reference book in all firefighters' libraries and required reading for new fire recruits. It enhances the training experience through its well-organized thoughts, concepts, and presentation.

Martin D. Davila, Chairperson
San Antonio College First Responders Academy

PREFACE

Prior to 1970, the scope of fire service was narrow. The main focus was on fire prevention, arson investigation, and fire suppression. After 1970, the nation's fire service began adding duties to increase the level of public protection. Emergency medical services (EMS) was integrated into the fire service, followed by heavy rescue and hazardous materials response teams. The increased responsibilities required a change in applicants' requirements. Being able to perform the physical aspects of the fire service was no longer adequate. Education became an important prerequisite for employment. Men and women seeking careers in the fire service were now expected to bring a high level of critical thinking and problem-solving abilities. The demand for these skills increased greatly after the events of September 11, 2001, when the fire service incorporated emergency management into its list of responsibilities.

The modern fire service is a dynamic profession with ever-increasing challenges. This text is designed to introduce students to the firefighting profession as well as remind career firefighters that fire service is a profession requiring continuous learning.

WHAT'S NEW

- A new chapter was written on emergency medical services and emergency management to introduce the importance of these two functions in the fire service. The authors felt this was necessary due to the evolving role of the firefighter and the emergency medical service field. This chapter covers basic information in relation to emergency medical responses and emergency management procedures. It provides an introductory overview of the ever-changing emergency response occupation.
- Updated color design with color photos throughout.
- Updated key terms, references, examples and statistics.

Other chapter-specific updates are as follows:

Chapter 1
- Update of information on large-loss fire in the United States
- Update of information on fire facilities

Chapter 2
- Update of information on firefighter employment
- Introduction of statistician as a private sector fire protection career

Chapter 3
- Discussion on social media (i.e. Facebook, LinkedIn, Twitter) and how employers may use it in the hiring process

Chapter 4
- Lists the National Fire Protection Agency (NFPA) standard in the general professional certifications and specialty certifications tables
- Update of Fire and Emergency Services and Higher Education (FESHE) model curriculum
- Discusses FESHE initiative of the creation of the High School Pathway Program, which encourages the incorporation of fire service courses into the high school curriculum
- Overview of regional versus national accreditation

Chapter 5
- ■ Expands explanations for training facilities and work uniforms
- ■ NFPA 1971 recommendations for structural firefighting ensembles

Chapter 6
- ■ Adds fire flow computations after section on fire load computation and before rate of heat release

Chapter 7
- ■ Addition of formulas for soil and wind pressure computation
- ■ Addition of information and formulas on beams and columns
- ■ Description of roof designs

Chapter 8
- ■ 2012 NFPA Life Safety Code 101 occupancy classification list
- ■ International Code Council code development cycle

Chapter 9
- ■ Expands discussion on flow testing fire hydrants
- ■ Statistical information on sprinkler system reliability
- ■ Information on residential sprinklers

Chapter 10
- ■ Discussion on organizational structure
- ■ Basic principles of command
- ■ Emphasis on National Incident Management System (NIMS) and incident command system (ICS) training

Chapter 11
- ■ Update of gallons per minute (GPM) and human-power formulas
- ■ Information on wind-driven fires
- ■ Update of information on preincident planning and new technologies

Chapter 12 (New Chapter)
- ■ Evolution of emergency medical services and its importance in the fire service
- ■ Discussion of new national EMS education standards
- ■ EMS professional development model as noted by the USFA/NFA for Higher Education
- ■ Discussion of postsecondary education in EMS
- ■ Statistics and information with graphs depicting call volume in relation to fire and emergency service calls
- ■ Discussion of the process of acquiring a position as an emergency manager
- ■ Discussion of postsecondary education in emergency management

HOW TO USE THIS TEXT

Throughout the text, there are several features that enhance the material presented. Stop and Think questions ask the student to pause, consider what they have just learned, and apply the concept. General safety icons point out areas where extra precaution is required. The icon of the National Fallen Firefighter's Foundation is used appropriately to promote the 16 life-safety initiatives that ensure that "Everyone Goes Home." Each chapter concludes with an

"On Scene," bridging the gap between classroom learning and experience. Review questions reinforce the chapter content by providing a final check that the student has mastered the chapter content.

Chapter 1: History of the Fire Service

This chapter introduces students to the history of the fire service—how it began, where it is now, and what the future of the fire service profession may be. This chapter covers the ever-changing role and needs of the fire service. The instructor may utilize particular historical information pertaining to the audience to personalize the lecture and link the past to the future fire service.

Chapter 2: Fire and Emergency Services Career Opportunities

This chapter will acquaint new and experienced students with the variety of career opportunities that the fire service offers. The basic role of a fire service worker has evolved into many aspects and new careers over the ages. This chapter focuses on the many specialized opportunities available to a firefighter. It also notes that although a specific job may have various titles, depending on the area or locale, the job description remains the same. The text provides basic information that the instructor can expand on to make the information more relevant to the student's environment.

Chapter 3: The Selection Process

This chapter gives an overview of the procedure that is typically followed in the fire service. It is recommended that the instructor give real examples of this process and possibly carry out a mock interview to illustrate the significance of learning this process. This chapter emphasizes that, no matter what a person's background is, the selection process is a structured venue. The more prepared an applicant is for the job, the more likely he or she will get the job.

Chapter 4: Training and Higher Education

This chapter can be used to impress on new students that training and education does not end when they graduate from the fire academy. Experienced students can also benefit from this chapter by learning different avenues that will allow them to further their career in ways that may not have been available before. Continual training and education are paramount for a safe department and for individuals to move through the fire service ranks. To give relevance to the students' environment, instructors can describe firefighters they know who have used training and education to their advantage.

Chapter 5: Fire Department Resources

This chapter can be used either to introduce new students to a variety of apparatus, facilities, and equipment that are used in the fire service today or the instructor's focus on the resources that are available in the instructor's area. The text provides basic information that the instructor can expand on to make the information more relevant to the students' environment.

Chapter 6: Fire Dynamics

This chapter is an overview of the chemistry and physics of fire and the phenomena that occur during the combustion process. The instructor may utilize the entire chapter to give new students an overview of fire dynamics or focus on certain topics to emphasize areas of concern to the more experienced student. The instructor may choose to use demonstrations to enhance the lecture.

Chapter 7: Building Construction

The focus of this chapter is to enlighten the student on the hazards that buildings pose under fire conditions. The emphasis is on the dangers of a building collapse and how to "read" a building during fire operations. It is recommended that the instructor obtain samples of various forms of structural elements for the student to view and manipulate. The information presented is appropriate for new and experienced students.

Chapter 8: Fire Prevention Codes and Ordinances
This chapter is designed to introduce students to the codes and ordinances that regulate the actions of the fire service. Students will learn why the codes and ordinances were created, the historical events that inspired them, and the constitutional challenges that set judiciary precedents. It is recommended that the instructor utilize the entire chapter to maintain a smooth, chronological flow of information.

Chapter 9: Fire Protection Systems
This chapter introduces the various forms of water-based automatic extinguishing systems, public water supplies, special extinguishing systems, and fire detection and alarm systems. The instructor may choose the level of instruction based on the level of experience of the students.

Chapter 10: Organizational Structure and Emergency Incident Management Systems
Using this chapter, the instructor may choose to introduce new students to the fire service organizational and rank structure before detailing the history and functions of incident management systems. For more experienced students, the instructor may want to give a brief history of how the incident management system was developed before moving to the National Incident Management System (NIMS). The dynamic nature of this system should be emphasized.

Chapter 11: Preincident Planning, Fire Strategy, and Tactics
When covering this chapter, the instructor should continually emphasize the importance of preincident planning. This chapter will be best utilized with scenario-based problem-solving activities that are founded on the information from the text. This will challenge new students and maintain the interest of experienced students.

Chapter 12: Emergency Medical Services and Emergency Management
The focus of this chapter is to introduce EMS and emergency management, which play an important role in the fire service. The basic role of a fire service worker has evolved into many aspects and new careers over the ages. the chapter also notes that although a specific job may have various titles, depending on the area or locale, the job description remains the same. Emergency medical responses and emergency management procedures are major forces in today's private and public emergency services; therefore, it is essential that there is a framework for federal, state, and local communities to work together in our nation's security.

ACKNOWLEDGMENTS

The most important person I want to acknowledge is my loving wife, Melissa, for her unwavering support during this endeavor. This text would never have been possible without her. I would like to personally thank Stephen Smith from Brady for giving me the opportunity to write the second edition of this text. In addition, I would like to thank the entire Brady staff and Eileen Clawson, who have worked so hard to see this project through to completion. Many thanks are due to my coauthor, J.D. Richardson, for his encouragement to never stop learning and to achieve the highest education possible to prepare for the future. I would like to thank Chief Mike Wisby and Angela Israel at Brayton Fire Training Field for their assistance in providing photos. I am grateful for the support that Brazos County Precinct 3 Volunteer Fire Department has given me during the writing of this book. This has truly been the most rewarding project with which I have ever been involved.

I would like to thank the reviewers in helping shape the second edition to make this the best possible textbook you will read and learn from. I know I probably missed many other people who have given me support and information throughout this project, and to them—thank you.

Finally, this textbook is dedicated to Firefighter Dick Danes who lost his life during a training exercise May 17, 2012. Firefighter Danes was the first line of duty death (L.O.D.D.) at Brazos County Precinct 3 Volunteer Fire Department. Firefighter Danes and his family will always be a part of our family. Thank you, Betsy, for being so loving and caring. You will always be in my heart.

—Jason B. Loyd

I would like to thank my wife, Dianne, for her patience and understanding while I worked on this project. She has always been my rock and my foundation, and I could not have completed this book without her support. She is, and always will be, the love of my life.

I would also like to thank the following:

Jim Cox, Ansul Incorporated

Captain Willis Lamm (Retired), Water Supply Officer, Moraga-Orinda, California, Fire Department

The officers and firefighters of the Moraga-Orinda Fire District, Moraga-Orinda, California

Captain Gary Fox, Balcones Heights F.D. (Retired), current Director of Fire Science and Chairperson of the Protective Services Department, San Antonio College

Firefighter Travis Haveman and the officers and men of the Burnet Fire Department in Burnet, Texas

The officers and men of the Lampasas Fire Department, Lampasas, Texas

Mike Pickett, author of Pickett's Primers and former director of the San Antonio Fire Science Program

Assistant Chief of Training Joe Martinez and the United States Air Force, Randolph Air Force Base

Fire Chief Ross Wallace and the officers and firefighters of the Universal City (TX) Fire Department

Former Fire Chief Scott Lee and the officers and firefighters of the Selma (TX) Fire Department

Lieutenant Richard (Rick) Bachmeier (Retired), San Antonio (TX) Fire Department

Captain John De La Garza (Retired), San Antonio (TX) Fire Department

I would especially like to make a special acknowledgment to Captain Elmer Wayne (Red) Blevins (1932–2012), who served 52 years and 9 months with the Houston Fire Department. He was an inspiration to three generations of firefighters. God bless you, Red; you will be missed.

—James D. (J.D.) Richardson

REVIEWERS

We would also like to thank the following reviewers for their invaluable feedback in the preparation of this revision:

John M. Best
Lecturer
John Jay College of Criminal Justice
New York, NY

Scott R. Clarenbach
Adjunct Instructor
Lakes Region Community College
Laconia, NH

Kathleen M. Dayton, MA, NREMT-P
Professor
Montgomery College
Rockville, MD

Thomas L. "Tommy" Harper, MSDS
Assistant Professor
Jefferson College of Health Sciences
Roanoke, VA

George A. Hettenbach, MS
Retired Captain, Upper Darby Fire
*   Department*
Delaware County Community
*   College*
Havertown, PA

L. E. Hudson, MEd
Fire Instructor III Master, Hazmat
*   Technician, EOD Operator*
LBJ Fire Academy
Austin Independent School District
*   Career and Technology*
*   Education*
Austin, TX

Byron Mathews
Cheyenne Fire/Rescue
Cheyenne WY

Kevin Roy Nunn, BS EET/MS—
Emergency Management
Firefighter/EMT-IV
Adjunct Instructor
Walters State Community College
Morristown, TN

Dr. James Schaefer, BS Fire Science,
MA and PhD in History
Associate Professor of Fire Science
*   and History*
Lake Superior State University
Sault Ste. Marie, MI 49783

Professor Lee Silvi
Program Director
Fire Science & Emergency
*   Management Programs*
Lakeland Community College
Mentor, OH

Joshua M. Sunde
Engineer/Professor of Fire
*   Technology*
Nevada County Consolidated
*   Fire Protection District/Sierra*
*   Community College*
Grass Valley, CA

H. Scott Walker, MMOB
Faculty
Western IL University
Macomb, IL

Jason B. Loyd started his career in 1996 as a volunteer firefighter/EMT with Medina Valley in Castroville, Texas, while also attending San Antonio College. In 1998 he completed the San Antonio College Regional Fire Academy and was offered a job overseas as a contract firefighter on a U.S. military base. He was stationed in Hungary for two years and then promoted to Bosnia for two years. In Bosnia, he was responsible for training local national Bosnians to the Department of Defense firefighter standards for Fire Fighter I and II, Hazardous Materials—Awareness and Operations, Airport Fire Fighter, Driver/Operator-Pumper, and Driver/Operator-ARFF in addition to being a crew chief.

Mr. Loyd earned his associate of applied science degree from San Antonio College in Fire Science and his baccalaureate degree from Empire State College in Business Management and Economics with a concentration in Fire Service Administration. Mr. Loyd went through the National Fire Academy Distance Degree Program (DDP) to obtain his baccalaureate degree. After completing his BS degree in 2001, he was offered a full-time position at Weatherford College in Fire Science Technology. Mr. Loyd and his wife then relocated back to the United States from Bosnia. At Weatherford, he was responsible for the curricula of two associate of applied science degrees, the regional fire academy, and all fire service training courses offered at the college. He holds various certifications with the Department of Defense (DoD) and Texas Commission on Fire Protection (TCFP), including Master Fire Fighter, Aircraft Rescue Fire Fighter, Fire Officer IV, Fire Instructor III, Driver/Operator-Pumper, Fire Inspector II, Fire Investigator, and field examiner. Mr. Loyd was awarded the Texas Association of Fire Educators George Hughes Instructor of the Year Award for 2003 and was recognized in the Who's Who Among America's Teachers. In 2004, he was appointed to the Texas Commission on Fire Protection Curriculum and Testing Committee and in the same year appointed to the International Fire Service Accreditation Congress (IFSAC) Degree Assembly Board of Governors. As of 2013, Mr. Loyd still sits on the Texas Commission on Fire Protection Curriculum and Testing Committee. In 2005, he completed his master of arts degree in Education with a concentration in Higher Education. Furthermore, he remains a volunteer firefighter and EMS Training Coordinator with Brazos County Precinct 3 Volunteer Fire Department.

In May 2007, Jason was hired as an industrial fire instructor for the private sector group for the Emergency Services Training Institute (ESTI) Division of the Texas A&M Engineering Extension Service (TEEX), a part of the Texas A&M University System. In 2011, Mr. Loyd was selected as the subject matter expert (SME) to develop the new online fire academy program. In 2012, Mr. Loyd was promoted as the training manager to run the new Online Recruit Academy at Brayton Fire Training Field.

He has been married to his wife Melissa since 1993; she also works at Texas A&M University as an environmental services coordinator. They have two daughters, Keeley and Arabella.

James D. (J.D.) Richardson is a Vietnam Veteran who served four years with the United States Air Force as a munitions maintenance specialist. After returning from Vietnam in 1971, he met and married his beautiful wife, Dianne. Following his military service, he joined the Houston Fire Department. J.D. enrolled in the Fire Science Program at Houston Community College, but before completing his associate's degree, he transferred to the University of Houston and achieved a bachelor of science in Technology degree in 1977.

J.D. was promoted to chauffeur (engineer/operator) in 1979 and was assigned to an ambulance as an EMT. He was later reassigned to the newly created Hazardous Materials/Heavy Rescue Team in 1980. When the team was separated into two entities in 1981, J.D. was assigned as the engineer/operator for the Heavy Rescue Team. He remained with heavy rescue until 1989. From 1989 until 1991, he served as an engineer/operator in the suppression division. He was promoted to the rank of captain in 1991.

During his tenure as a captain in the Houston Fire Department, J.D. developed a hydraulic rescue tool training program that was used to train more than 500 Houston firefighters in the use of the Hurst rescue tool. He was temporarily assigned to the Houston Fire Department Training Division to work with the rescue teams in developing rescue training programs and was chosen to be a member of a hazardous materials training team with the International Association of Fire Fighters. As a member of this team, J.D. helped initiate the 24-hour hazardous materials awareness and operations pilot program now being mandated for the fire service.

Captain Richardson retired from the Houston Fire Department in 1997 and began a new career as an instructor of fire science with San Antonio College. During his tenure with the college, J.D. achieved his graduate degree in education, was instrumental in developing the Associate of Applied Science Degree in Emergency Medical Service in 1998, and in 2003 initiated the first Associate of Applied Science Degree in Emergency Management in the state of Texas. He modified the program to Emergency Management and Homeland Security Administration in 2006. He also developed and initiated a Certificate and Associate's of Applied Science Degree for the Department of Homeland Security, Transportation Security Agency. J.D. served as Codirector of the Fire Science Degree Program from 1999 to 2001 and chairperson of the Protective Services Department from 2001 until 2005. He achieved the rank of assistant professor before retiring in 2012. J.D. and his wonderful wife, Dianne, have been married 42 years. They have one son, R.J.; a lovely daughter-in-law, Kate; and a beautiful granddaughter, Cadwyn.

In March 2004, a Firefighter Life Safety Summit took place in order to address the need for reform within the fire service profession. As a result of this meeting, the 16 firefighter initiatives were developed. The National Fallen Firefighter's Foundation promotes programs and resources that facilitate in preventing line-of-duty deaths and injuries through the Everyone Goes Home website. These training programs like Firefighter Life Safety Resources Kit—Volume 1, 2, 3, and Courage to Be Safe, to name a few, can be downloaded off the website. Examples of the 16 Firefighter Life Safety Initiatives are stressed throughout this textbook in order to show how these initiatives can be applied in practice. Consider what other initiatives you can apply to practice as you read the chapters.

The following list includes each of the 16 initiatives and how they are applied in the chapters:

1. Define and advocate the need for a cultural change within the fire service relating to safety; incorporating leadership, management, supervision, accountability and personal responsibility.

   Chapter 4 discusses the challenges firefighters have had to adapt to in order to keep up with today's society while continuing to provide the best service to their community/customers.

2. Enhance the personal and organizational accountability for health and safety throughout the fire service.

   Chapter 3 illustrates the need to be in excellent physical fitness. This applies not only to individuals wanting to become firefighters but to current volunteer and career firefighters as well. For more information on this topic, refer to NFPA 1583, *Standard on Health-Related Fitness Programs for Fire Department Members, 2008 Edition.*

3. Focus greater attention on the integration of risk management with incident management at all levels, including strategic, tactical, and planning responsibilities.

   Chapter 11 discusses preincident planning, fire strategy, and tactics, which allows the instructor to facilitate talking points to students concerning greater attention on integration of risk management in the fire service today and in the future.

4. All firefighters must be empowered to stop unsafe practices.

   Chapter 10 discusses not only organizational structure but emergency incident management systems as well. Instructors should emphasize that everyone on the fire ground is empowered to stop an unsafe act. For more information on this topic, refer to NFPA 1521, *Standard for Fire Department Safety Officer, 2008 Edition.*

5. Develop and implement national standards for training, qualifications, and certification (including regular recertification) that are equally applicable to all firefighters based on the duties they are expected to perform.

   This initiative in Chapter 4 emphasizes the importance of training and higher education in fire and emergency services. Ultimately this initiative is paramount for national standardization for emergency personnel in all ranks and levels alike. Through lessons learned from September 11, 2001, plus Hurricanes Katrina and Rita, we have trained and educated our responders across the nation to be better prepared to mitigate these types of large-scale emergencies. It's important that our future fire and emergency services leaders have the knowledge required to perform to a certain level set by these national standards for the safety of our responders no matter where they're responding from (local or state).

6. Develop and implement national medical and physical fitness standards that are equally applicable to all firefighters, based on the duties they are expected to perform.

   This initiative in Chapter 3 discusses the selection process and the importance of national standards for medical and physical fitness for fire department members. For more information on this topic, refer to NFPA 1583, *Standard on Health-Related Fitness Programs for Fire Department Members, 2008 Edition.*

7. Create a national research agenda and data collection system that relates to the initiatives.

   Chapter 12 discusses support functions from different organizations. Instructors should introduce students to research, data collection systems, and the benefits to the department as a whole.

8. Utilize available technology wherever it can produce higher levels of health and safety.

   Chapter 2 discusses the career opportunities in occupational health and safety. An occupational health and safety specialist is an invaluable asset to the fire and emergency services industry. This career position might have a different title like health and safety officer (HSO) or incident safety officer (ISO). Regardless of the title, the key point is to promote higher levels of health and safety. For more information on this topic, refer to NFPA 1521, *Standard for Fire Department Safety Officer, 2008 Edition.*

9. Thoroughly investigate all firefighter fatalities, injuries, and near misses.

   This initiative in Chapter 2 discusses the career of fire investigators. This role can be different from locale to locale and even state to state. Regardless of the position title, it is of the utmost importance that every firefighter fatality, injury, or near miss be documented so others can learn from these events. Although this chapter may touch on basic information, the instructor should expand on this subject to make the information more relevant to the students' environment.

10. Grant programs should support the implementation of safe practices and/or mandate safe practices as an eligibility requirement.

    This initiative in Chapter 12 focuses on several support organizations and the vital role they play in the fire and emergency services industry. There are grant programs that assist fire departments with the essential funds to implement safe practices programs, although certain requirements may apply before a department is eligible for this type of assistance.

11. National standards for emergency response policies and procedures should be developed and championed.

    This initiative in Chapter 10 discusses the organizational structure and emergency incident management systems concerning policies and procedures for emergency response.

12. National protocols for response to violent incidents should be developed and championed.

    This initiative in Chapter 11 emphasizes the importance of preincident planning. It is vital that instructors stress to students about responding to violent incidents and the need to have national protocols in place before the incident occurs.

13. Firefighters and their families must have access to counseling and psychological support.

    This initiative in Chapter 2 discusses the many dangers firefighters face today. Because of the added responsibilities of the fire and emergency services personnel, it is important that firefighters have access to proper counseling and psychological care if needed.

14. Public education must receive more resources and be championed as a critical fire and life safety program.

Chapter 8, on fire prevention codes and ordinances, discusses the importance of life-safety education programs. It is recommended that instructors utilize the entire chapter so students will have a greater appreciation of fire prevention codes and ordinances, public education, and life-safety programs resources available.

15. Advocacy must be strengthened for the enforcement of codes and the installation of home fire sprinklers.

   This initiative in Chapter 8 discusses fire prevention codes and ordinances and why they were created. Instructors should emphasize to students the importance of becoming an advocate for enforcement of codes and the installation of home fire sprinklers.

16. Safety must be a primary consideration in the design of apparatus and equipment.

   This initiative in Chapter 2 discusses not only career opportunities in the fire and emergency services field but also new technology geared towards safety for responders. For example, new technology in apparatus and equipment has enhanced firefighter safety like Global Positioning Systems (GPS), Automatic Vehicle Location (AVL), and Geographical Information Systems (GIS), to name a few.

FIRE AND EMERGENCY SERVICES HIGHER EDUCATION (FESHE) GRID

The following grid outlines Principles of Emergency Services course requirements and where specific content can be located within this text:

COURSE REQUIREMENTS	1	2	3	4	5	6	7	8	9	10	11	12
Illustrate and explain the history and culture of the fire service.	X											
Analyze the basic components of fire as a chemical reaction, and the major phases of fire; and examine the main factors that influence fire spread and fire behavior.						X	X					
Differentiate between fire service training and education and explain the value of higher education to the professionalism of the fire service.				X								X
List and describe the major organizations that provide emergency response service and illustrate how they interrelate.										X		
Identify fire protection and emergency service careers in both the public and in the private sector.		X	X									X
Define the role of national, state, and local support organizations in fire and emergency services.										X		X
Discuss and describe the scope, purpose, and organizational structure of fire and emergency services.										X		
Describe the common types of fire and emergency services facilities, equipment, and apparatus.					X							
Compare and contrast effective management concepts for various emergency situations.										X		
Identify the primary responsibilities of fire prevention, including code enforcement, public information, and public and private protection systems.									X		X	
Recognize the components of career preparation and goal setting.			X	X								X
Describe the importance of wellness and fitness as it relates to emergency services.			X									

Courtesy of Jason Loyd.

KEY TERMS

apparatus Aldini, *p. 10*

conflagration, *p. 2*

couvre feu, *p. 3*

EMT, *p. 14*

familia publica, *p. 2*

FESHE, *p. 14*

fire science, *p. 2*

fire wardens, *p. 4*

firemark, *p. 3*

fireplug, *p. 5*

Great Fire of Boston, *p. 7*

Great Fire of London, *p. 3*

hand pumper, *p. 5*

Homeland Security, *p. 13*

Maltese cross, *p. 11*

NIMS, *p. 14*

NREMT, *p. 14*

siphon, *p. 6*

steam engine, *p. 7*

OBJECTIVES

After reading this chapter, you should be able to:

- Describe the evolution of firefighting in the United States and other countries.
- Explain the history of early firefighting organizations.
- Describe early fire vehicles.
- Describe early personal protective equipment.
- Understand fire service traditions.
- Describe the origins of fire facilities.
- Explain the fire dilemma in the United States and its causes.
- Discuss the changing role of the modern firefighter.

Introduction

Early humans experienced fire as a great and powerful mystery, a force to be feared. Gradually, as their knowledge of fire's properties grew, humans began to appreciate its potential as a tool for safety and protection against cold, darkness, and other dangers posed by harsh environments. The ability to control fire eventually was an important factor in the formation of stable communities of individual family dwellings and businesses. Fear slowly turned into respect for this great force. However, as humans became more and more accustomed to using fire in their everyday lives, their carelessness caused accidents that resulted in death and destruction. As communities grew in size, so did the need for protective measures against out-of-control fires. Unfortunately, most safeguards were not enacted until after a tragic loss of life and property made the need for prevention obvious.

Fire remains something of a mystery to this day, but over the centuries, the study of fire science has disclosed many of its secrets. The work of fighting fires has evolved into a highly respected occupation for many people, both full-time professionals and dedicated volunteers. The history of firefighting is a fascinating, inspiring story. It is the record of how our inventive and courageous forefathers fought to control one of nature's most useful and potentially most destructive forces. Their efforts lay the foundations on which our modern fire service is built and pioneered the field of study that we call **fire science**.

fire science ■ Study of the behavior, effects, and control of fire.

The term *field of study* is used because fire science is not a single discipline. Topics studied in a fire science curriculum include, but are not limited to, protective systems, building construction, fire codes and ordinances, hazardous materials, chemistry, strategy and tactics, and of course, fire behavior. Each of these areas has a history that explains its importance to the study of fire science. Each became part of this field of study because lives had been lost in the past, inspiring efforts to prevent similar losses in the future. The fire science student must learn not only the when and where of developments in the field but also the why. We will begin exploring the "why" by looking back at the evolutionary history of firefighting services and firefighting technology.

The Evolution of Firefighting Services

The earliest recorded effort to organize a firefighting service dates to the Roman Republic. In the first century BC, the crowded city of Rome was subject to repeated **conflagrations** that resulted in massive loss of life and property damage. A body of approximately 600 public slaves (the *familia publica*), strategically positioned at the gates of the city, were assigned the task of alerting the public to existing fires and fighting the conflagrations (Dio Cassius, 54.2–3). They were, however, notoriously ineffective and slow to react, perhaps in the interests of preserving their own safety.[1] After a particularly devastating fire in 6 AD, Emperor Augustus organized a corps of 7,000 freedmen, the *vigiles* (watchmen), who patrolled the streets at night. They began to combat fires using bucket brigades, as well as poles and hooks to tear down buildings in advance of the flames. It is generally thought that this is where the "hook" in "hook and ladder" originated.

conflagration ■ A fire that increases in size and spreads beyond human-made and natural barriers.

familia publica ■ "Servants of the commonwealth." These men were strategically positioned near the city gates to protect the city from fire.

Rome continued to suffer from serious fires, most notably the July 19, 64 AD, conflagration that started near the Circus Maximus and eventually destroyed two-thirds of the city. Emperor Nero was blamed; although it is unlikely that Nero actually ignited the blaze, some historians believe that he did permit it to burn unchecked. Evidence exists that at least one Roman citizen profited richly from this fire, buying properties in advance of the flames and using teams of slaves to defend his recent acquisitions from being consumed.[2] Whatever his faults, Nero was apparently a man of some vision and intelligence who recognized the dangers of the unregulated construction that was the norm in Rome before his reign. Much wealth and many resources had been expended on the construction of enduring public edifices, but almost all other structures were built of wood without care for safety standards. Nero can be credited with rebuilding Rome in accordance with sound principles of construction, sanitation, and utility. From

64 AD until the fall of the Roman Empire, both public and private building projects were closely regulated.

It was not until the 11th century that fire prevention was again addressed. William the Conqueror of England instituted one of the first recorded fire prevention regulations, which required that candles and all fires used for heating and cooking must be extinguished at nightfall. Every day about 8 PM, a bell was rung, which was the signal to "cover the fire" (*couvre feu* in William's native French). **Couvre feu** gradually was corrupted to the English word *curfew*. The curfew bell was utilized in parts of England and France for over 800 years, long after its original purpose had been forgotten. However, this simple precaution was not always adequate, and the archives of French towns and churches contain records of many disastrous fires and rebuilding works.

In 60 AD, four years before Nero's fire in Rome, a fire ignited by rebellious natives leveled the fledgling Roman town of Londinium. This was the first of many conflagrations that would destroy large sections of London (as the town would later be named). History preserves a record of major blazes in 675 and 1087, three in the course of the 12th century, and a massive blaze in 1212. These tragedies pale, however, in comparison to the blaze that has come to be known as the **Great Fire of London**. The Great Fire started on September 2, 1666, in the royal baker's shop on Pudding Lane. It burned relentlessly for four days, consuming about 430 acres (80% of the city's center), and destroying 13,000 homes, 89 churches, and 52 guild halls. Tens of thousands were left homeless and financially ruined. Much of central London today shows the results of the reconstruction ordered by King Charles II, which called for wider streets and brick buildings to replace the earlier structures built of highly flammable wood and pitch.

One result of the Great Fire was the formation of what may well have been the world's first insurance company. Dubbed "The Insurance Office," it was the first of a number of such companies that took on the responsibility not only of indemnifying the insured against losses but also of actually fighting fires. Each company formed its own private fire brigade that would battle the flames—but only in those buildings insured by the company. Insured buildings were marked with the company's badge, called a **firemark**, and any building without this identifying mark was left to burn. Clearly, this was not the best solution to the problem of protecting the public. (See Figure 1-1.)

couvre feu ■ A French name for a metal lid used to cover an open hearth. The English word *curfew* was derived from this name.

Great Fire of London ■ A devastating fire in central London that was considered one of the major events in the history of England.

firemark ■ Metal marker that used to be produced by insurance companies for identifying their policyholders' properties.

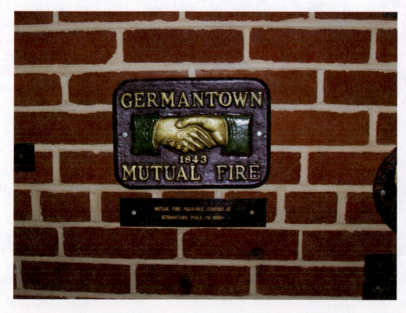

FIGURE 1-1 Firemark.
Courtesy of James D. Richardson.

FIGURE 1-2 The fire rattle was used to alert the town of an emergency. *Dorling Kindersley Media Library.*

fire wardens ■ Volunteers who patrolled the cities, inspected chimneys, and issued fines for noncompliance. Funds collected from fines were used to purchase new fire equipment.

The fire problem was not limited to European countries, of course. In 1631, a major fire broke out in Boston, which prompted the implementation of fire regulations banning thatched roofs and wooden chimneys. Colonial laws also required each house to have a bucket of water on its front stoop for the use of bucket brigades. In 1648, New Amsterdam (later renamed New York) created an organization of volunteer **fire wardens** who patrolled the city, inspected chimneys, and issued fines for noncompliance. Funds collected from fines were used to purchase new fire equipment. New Amsterdam also appointed a so-called Rattle Watch—men dressed in long capes who would patrol the city's streets by night. They carried wooden rattles to sound an alarm, rousing citizens to form bucket brigades whenever a fire broke out (see Figure 1-2).

Public interest in firefighting continued to increase. In 1679, Boston took the first step toward what has become today's professional fire department, initiating the efforts in establishing the first paid firefighting service. The firefighters were part-time employees of the city who were assisted by volunteer organizations known as mutual fire societies. However, it was not until 1736 that Benjamin Franklin organized what is known today as the first formal volunteer fire company. This volunteer assembly was named the Union Volunteer Fire Company of Philadelphia.[3] The amount of staffing and skill necessary for firefighting eventually prompted Franklin, like the citizens of Boston, to institute a paid fire company, but the concept of volunteer companies had caught on and was spreading across America (see Figure 1-3).

Men (this was before women joined the fire service) joined their local volunteer company for social reasons as well as to serve the public and to protect their homes and towns. Membership was considered prestigious, the mark of a public-spirited citizen, and would earn a man community respect. The result was long waiting lists of eager volunteers. The volunteer firefighter remains a proud American tradition to this day and is the heartbeat of any community public service.

FIGURE 1-3 American Founding Father Benjamin Franklin wears a fire helmet. *Courtesy of James Foley.*

Even after the formation of paid fire companies, there were still disagreements and fights over territory, because insurance money was awarded to the company that was first on the scene of the fire.[4] This was the beginning of a long-standing relationship between fire departments and fire insurance companies. The situation was complicated by the nature of water delivery in the towns. Public water systems consisted of mains constructed of wooden logs hollowed out to carry water down a street. Approximately every half block, wooden plugs were inserted in the logs. These plugs could be removed to access the water supply for firefighting, which has given us the term **fireplug**. Some fire companies would hire saboteurs to race ahead to the scene of the fire and hide the plug from competing fire companies. These miscreants were called "plug uglies." Even after fire hydrants replaced the fireplug, sabotaging a responding fire company's access to the water supply continued. New York City companies were famous for sending runners out to fires with large barrels to cover the hydrant closest to the fire in advance of the engines. Fights would often break out between the runners and the responding fire companies for the right to fight the fire, which continued to rage while they squabbled. Town parades were another occasion for spirited competition as each company marched down Main Street in its own elaborate, colorful uniforms, helmets, and hand-painted stovepipe-shaped "fire hats." Fancy fire axes and torches, lavishly painted fire buckets, and, if the company had one, a **hand pumper** decorated by a famous artist, added to the finery. Most striking of all, perhaps, were the engraved silver speaking trumpets that each company used to shout insults at its rivals.

The earliest steam pumpers were used in the 1800s. Volunteer firefighters viewed these as a threat to their existence as they needed less manpower to run and thus could diminish their ranks. Another change in equipment in the 1800s was the use of horses to pull the engines. Then, the motorized era began in the 1900s. The earliest motorized fire vehicles were used by the chiefs to drive to the scene of the fire and take control. After that came the motorized pumpers that pumped the water for the firefighters. The motorized pumpers worked much faster than did the steam or hand pumpers.

The profession of full-time paid firefighter was not a reality until around the middle of the 19th century. There were exceptions, but typically rural areas were served by volunteer firefighters, while paid organizations dominated urban areas. Today in the United States and around the world, fire and rescue remains a patchwork of both paid and volunteer responders. Of the approximately 26,350 fire departments in the United States, for instance, about 87% are totally or mostly volunteer (see Figure 1-4).

fireplug ■ A wooden plug inserted in logs that carried the town's water supply approximately every half block. These plugs could be removed to access the water supply for firefighting.

hand pumper ■ A fire engine that has long bars (also known as brakes or pumping arms) running parallel to the body that operate the pump. Firefighters physically use the pumping arms to make a full up and down motion to build up pressure, allowing water to spray out of a hose.

FIGURE 1-4 Early volunteer fire company with station and equipment in the late 1800s. *Courtesy of Andrea Franks/Franks Photography and Weatherford Fire Department.*

More recently, the U.S. Congress created the National Fallen Firefighters Foundation to lead a nationwide effort to remember America's fallen firefighters. This foundation is located in Emmitsburg, Maryland, and one of its tasks was to create a National Fallen Firefighters Memorial Park. This memorial was constructed in 1981 and lists the names of all members of the fire service who have died in the line of duty since 1981. Future plans include the creation of a special area in the park to honor pre-1981 fallen firefighters. The Memorial Park also features a brick Wall of Honor that connects the monument, the historic Fallen Firefighters Memorial Chapel, and memorial plazas. (See Figure 1-5.)

As mentioned earlier, fire departments and fire insurance companies have a long intertwined history. Today the Insurance Service Office, Inc. (ISO), exists, playing an integral part in how insurance rates are assessed for the community that the fire department serves. Special committees are usually formed with members consisting of firefighters and government officials. Their mission is to find ways to improve their ISO rating.

Early Firefighting Apparatus, Vehicles, and Extinguishers

The use of technology for firefighting dates back to the second century BC. The Greek engineer Ctesibius of Alexandria is credited with inventing a basic hand pump that operated on the principle of a **siphon** (Latin *siphona*) and could propel a jet of water toward a blaze.[5] For reasons unknown, this technology fell out of use, and we know of it today only because archaeologists have unearthed parts of the machinery.

siphon ▪ A large syringe used to deliver a stream of water.

The only apparatus available to fight the Great Fire of London in 1666 were buckets, hooks, axes, ladders, and hand-operated water syringes called "squirts." This equipment proved woefully ineffective as the fire continued to burn for four days.

Elsewhere in 17th-century Europe and in the American colonies, firefighting equipment was equally rudimentary. The ravages of London's Great Fire were a timely wake-up call, however, and the late 1600s saw a surge of technological advancements in firefighting equipment, especially in Europe. Hand-operated pumps were placed on wagons, greatly enhancing their mobility. A number of different designs were attempted, including one that required 28 men to work its levers and could shoot a stream of water 80 feet into

FIGURE 1-6 Old Silsby steam engine.
Courtesy of Andrea Franks/Franks Photography and Weatherford Fire Department.

the air. In 1672, the Dutch inventor Jan van der Heiden designed the first fire hose. It was constructed of flexible leather and coupled with brass fittings at intervals of 50 feet—the interval that remains the standard to this day.[6] These inventions changed the way fires were fought, enabling more sophisticated strategies and tactics.

In 1679, Boston imported America's first fire engine. Our nation would continue to rely on European technology until 1743, when Thomas Lote of New York produced the first American-built fire pump.

A steam-powered fire pump, capable of pumping 30 to 40 tons of water per hour, was constructed in London in 1829 by John Braithwaite. One of these new marvels was brought to America, but volunteer firefighters saw the invention as a threat to their social organizations because the **steam engine** would eliminate the need for large numbers of men to operate the hand pumps. Acceptance of the new technology was delayed for several years, but by the second half of the century, the steam-driven pump had become the norm (see Figure 1-6).

In 1841, a group of insurance companies in New York contracted the construction of a self-propelled steam fire engine that resembled a train engine. Again, firefighters resisted its use. Existing pump wagons required numbers of men to push and pull them through the streets, and a self-propelled model would put them out of work.

Steam-powered vehicles may have been out of favor, but the early 19th century did see the advent of horse-drawn fire wagons. Volunteers at the New York Mutual Hook and Ladder Company No. 1 voted in 1832 to purchase a horse to pull their engine. A yellow fever epidemic had reduced available manpower in the city, and it is suspected that this led to the purchase of the horse. Another theory is that the volunteers had simply tired of pulling heavy steam-driven pumpers through the streets. Whatever the reason, the idea caught on fast, and horses became an integral part of the fire service all over the world. They were often trained to the pitch of the bell that signaled an emergency. This pitch prompted them to get out of their stalls and stand at attention in front of the fire apparatus (see Figure 1-7).

In the United States, the end of the horse-drawn era was no doubt accelerated by the **Great Fire of Boston** in 1872. Tragically, the fire began at a time when the horses used to

steam engine ■ A heat engine that performs mechanical work using steam as its working fluid.

Great Fire of Boston ■ Boston's largest urban fire and one of the largest fires in American history.

FIGURE 1-7 Horses pulling a large hook and ladder truck. *Courtesy of Andrea Franks/Franks Photography and Weatherford Fire Department.*

pull the heavy steam pumpers were suffering from an epidemic of equine influenza. Without horses, firefighters again had to pull heavy steam pumpers through narrow, twisting streets to the scene of the fire. This, of course, greatly reduced their efficiency. The fire burned for 20 hours, consuming over 60 acres of downtown Boston (776 buildings). Thirty people lost their lives, and property losses were estimated at over $70 million (over one billion in today's dollars). A review of the problems faced during the Boston fire accelerated the search for alternative ways to transport firefighters and equipment to fire scenes, but Boston did not lead the way.

Earlier, about the middle of the century, the citizens of Cincinnati, Ohio, had become disgusted with the delayed responses and inefficient firefighting that resulted from the use of old-fashioned hand-pulled pumps. They insisted the fire department modernize. The Cincinnati Fire Department was organized in 1853 and is the nation's oldest fully paid professional fire department. Furthermore, in 1853, the first steam-powered fire engine in America came into service. Designed by Cincinnati engineers Able Shank and Alexander Latter, it was affectionately named "Uncle Joe Ross." "Uncle Joe" was able to shoot water 225 feet to a fire and required only three men to operate. The advantages of the new-fangled technology could no longer be ignored. By the end of the century, steam fire engines could be found in numerous municipal fire stations, although horses continued to be used well into the 20th century. Chicago retired its last team on February 6, 1923.

Once the popularity of the new engines was established, inventors and firefighters started thinking of ways to further ease the difficulty of their mission. The aerial ladder wagon made an appearance in the fire service by 1870, followed by the hose elevator in 1871.

An Englishman, Captain George Manby, had invented the first portable fire extinguisher in 1816. It consisted of an air-pressurized copper vessel containing three gallons of pearl ash (potassium carbonate). The late 19th century saw the introduction of the soda-acid extinguisher, which consisted of a cylinder containing a solution of sodium bicarbonate in one to two gallons of water.[7] Inside the cylinder was a vial of concentrated sulfuric acid. The device was activated by striking a plunger to break the vial or inverting the extinguisher, which released a lead bung from the vial. When the sulfuric acid met the sodium bicarbonate, it generated carbon dioxide gas, which forced the water out through a nozzle.[8]

FIGURE 1-8 Early 1913 American LaFrance gasoline fire apparatus. *Courtesy of Andrea Franks/ Franks Photography and Weatherford Fire Department.*

Internal combustion fire engines, built in the United States, first saw service in 1907. At first they were used either as pumping engines or as tractors to pull pieces of equipment. By 1910, both functions were combined, and one gasoline-powered engine was used to both propel the truck and drive the pump. On the same day as the city retired its last team of fire horses (February 6, 1923), the Chicago Fire Department, serving 500,000 residents, implemented the use of their motorized fire department (see Figures 1-8 and 1-9).[9] Nevertheless, in 1911 Savannah, Georgia, was the first department to have a fully mechanized fire department in the United States.

Firefighting vehicles today have evolved to serve highly specialized functions and to enable firefighters to respond quickly to a variety of emergencies. Usually diesel powered, vehicles include ladder trucks with aerial platform apparatus that can access high-rise buildings, sending heavy steam applications to heights up to 130 feet. Rescue trucks, brush trucks, mobile command vehicles, smoke ejectors, high-pressure spray trucks, and foam trucks add versatility to the firefighter's arsenal. Industry-specific fire stations and equipment make it possible to handle fires of highly specialized origin. For example, airports under the direction of the Transportation Security Administration (TSA) have introduced fire trucks especially designed to handle fires created by airplane crashes. Refineries have also adapted to their own specialized needs by developing various chemical applicators not used by the typical fire service.

The diesel pump on a modern fire engine is capable of delivering up to 2,000 gallons of water per minute through a lightweight hose line reinforced with artificial fibers and

FIGURE 1-9 1924 gasoline fire apparatus. *Courtesy of Andrea Franks/ Franks Photography and Weatherford Fire Department.*

measuring up to three inches in diameter. A fireboat, not limited to hydrant supply, can easily deliver as much as 10,000 gallons per minute.

Even the fire extinguisher has evolved into a state-of-the-art fire suppression tool. Fire extinguishers now come in various sizes and designs and employ a variety of extinguishing agents, such as water, dry chemicals, dry powder, carbon dioxide, foam, and special halocarbon agents.

These tools have immeasurably increased the effectiveness and ease of firefighting, but the innovations most appreciated by working firefighters are the improvements to their personal protective gear.

Respiratory Protective Equipment or Self-Contained Breathing Apparatus

Early firefighters faced not only fire but also the effects of smoke with little or no personal safety equipment. Try to imagine responding to an out-of-control structure fire without water- and heat-resistant bunker gear, a state-of-the-art fire helmet, and self-contained breathing apparatus (SCBA).

Early photography shows Victorian firemen using their own long beards as their only defense against smoke inhalation. Before entering a burning structure, they would wet their beards with water, then clench them in their teeth to act as air filters.[10] Firefighting was never an occupation for the faint-hearted!

apparatus Aldini ■ One of the earliest recorded attempts to improve breathing difficulties in toxic environments was tested in 1825 in France. This was not the most desirable breathing apparatus for firefighters to wear, but scientific testing was conducted under actual fire conditions.

One of the earliest attempts to reduce breathing difficulties in toxic environments was the French **apparatus Aldini**, which was tested under actual fire conditions in 1825. The Aldini cannot have been comfortable to wear; it consisted of an asbestos mask covered by a second mask made of iron wire, but it may have provided some measure of protection. The space between the two masks trapped a small amount of breathing air. In the 1870s, fire departments began buying and using "Neally's Smoke Excluding Mask." The mask was a simple invention, consisting of a snug-fitting facemask with mica or glass eyepieces and a water bag suspended from rubber tubes that led to a mouthpiece. Two sponges kept wet by the water bag filtered clean air to the wearer. It was marketed as "a most perfect apparatus" and sold for an amazing $15.[11] The first successful American SCBA, known as the Gibbs, came into use in the first decade of the 20th century. The air pack, introduced in late 1945, was the product of research aimed at helping World War II pilots breathe at high altitudes. This unit consisted of a pressurized oxygen cylinder made of steel and a pressure-regulating device that allowed firefighters to breathe fresh air in smoke-filled environments.

The modern SCBA bears little resemblance to these predecessors. The air pack's heavy steel cylinder has been replaced by a lightweight cylinder developed by the National Aeronautics and Space Administration (NASA) and the space industry.[12] Pressure regulators provide positive pressure to the wearer and allow the firefighter to monitor the amount of air remaining in the cylinder through a "heads-up" display. The amount of air in the cylinder has greatly increased from a 15-minute supply to one hour or more.

Firefighting has certainly come a long way since the days of breathing through wet whiskers![13]

Fire Facilities

Just as changes in personal equipment have kept pace with modern technology, so has the design of the firehouse. In the early volunteer days, firefighters needed only a place to store their fire pumps. A small shed was perfectly suitable, and stables for the horses were added later. The advent of the professional full-time firefighter created a need for sleeping quarters and for personal space where each individual could store equipment and fire apparatus. To meet this need, many communities built a firehouse to serve the needs of firefighters serving 24-hour shifts. A firehouse usually consists of a kitchen, a bedroom, and an apparatus bay where fire equipment is stored. As technology and engineering design improve, so does the state of the firehouse. (The design of the modern fire and emergency services facility is discussed in more depth in Chapter 5.)

Fire Service Insignias and Traditions

Firefighting and tradition seem to go hand-in-hand. Since time immemorial, certain emblems have been virtually synonymous with firefighting. The best known of these are the red fire engine, fire badges, **Maltese cross**, and Dalmatian fire dog.

RED FIRE ENGINE

The tradition of painting fire engines red dates back to at least the early 1920s. When Henry Ford made motor cars affordable for the average American family, he let it be known that his vehicles could be bought in any possible color—just as long as it was black. Soon the nation's highways and byways were crowded with black vehicles, and quite logically the fire service began painting its vehicles red in an effort to stand out. Today's consumers can buy vehicles in just about any imaginable color, and so can the fire service. In addition to the traditional red fire engine, it is not uncommon to see white, yellow, blue, orange, green, or even black fire trucks. Many fire departments, however, have stuck loyally to the time-honored red engine.

BADGES

Throughout fire service history, the firefighter has been identified by the badge in much the same way as medieval knights were identified by the insignia on their shields. Insignias allowed knights to be recognized on the battlefield despite the heavy armor that concealed their person. Firefighters' badges serve a similar identifying function and provide a means to distinguish rank and departmental affiliation. Today, badges are used more for Class A uniforms and special occasions; however, they still hold the symbolism of rank and honor and the traditions of the fire service. Badges are a symbol of courage and a commitment to protect the public. (See Figure 1-10.)

MALTESE CROSS

The Maltese cross has become the symbol of protection. It means the person who wears this cross is willing to lay down his or her life for you. The Maltese cross was originally the insignia worn by the crusading Knights of St. John. During the battle for Jerusalem, the Saracens atop the city's walls poured oil down on the attackers and dropped lighted torches to ignite the oil. The knights, witnessing this horror, dropped their weapons and ran into the flames to rescue their comrades. On their way home from the Crusades, the knights established a hospital on the island of Malta to care for the injured, and they flew their insignia on a flag to let all know that they would render aid to the needy. Their

Maltese cross ◼ Symbol representing the traditions and ideals of the fire service. The arms and the tree of the Maltese cross are equal in length. The arms and the tree widen as they extend from a central point. In most cases, the edges are flat or curved slightly outward.

FIGURE 1-10 Various badges. *Courtesy of Jason Loyd.*

FIGURE 1-11 Maltese cross. *Courtesy of Jason Loyd.*

descendants continued this charitable work for the next 400 years, and the cross of St. John became a symbol of generosity to friend and foe alike and a mark of protection for the weak. Not knowing the history of the insignia, many referred to it as the Maltese cross and the name stuck, becoming an international symbol of aid. The fire service adopted this symbol because it proudly upholds the very same standards and traditions as the Knights of St. John. The Maltese cross is a firefighter's badge, signifying that he or she works in a profession of courage, pride, honor, and danger. Over the years the Maltese cross has taken on different forms for various agencies that uphold the same moral standings such as hospitals, ambulances, hospice organizations, police, and other emergency personnel. However, the spirit of the crusade that initiated the Maltese cross still lives strong today. (See Figure 1-11)

DALMATIAN DOG

For years, the loyal Dalmatian has been the trusted companion of firefighters. Few realize that this breed was originally chosen because of the strong bonds that the dogs formed with fire horses, protecting them and keeping them company at the station. The dogs were also expected to rouse the horses at the sound of the alarm bell, then run out and bark a warning at anyone who might be obstructing the firehouse exit. The dogs would then chase the fire apparatus all the way to the scene, sometimes barking the whole way. They served the same function, essentially, as the emergency traffic signals located outside many fire stations today and the sirens on fire trucks.

When horses were replaced by steam- or gasoline-driven fire engines, many departments opted to keep their beloved mascots. It is not unusual even today to see a proud Dalmatian riding on a fire engine as it races to the scene of an emergency.

The Fire Dilemma in the United States

The United States has a severe fire problem that if not addressed, will continue to worsen drastically. Fire statistics show that our nation, one of the richest and most technologically sophisticated countries in the world, lags behind its peer nations in fire security. Nationally, there are millions of fires, thousands of deaths, tens of thousands of injuries, and billions of dollars lost each year—figures that far exceed comparable statistics for other industrialized countries.[14] In 2001, the direct value of property destroyed in fires was $11 billion ($44 billion if the World Trade Center loss is included).[15] More recently in 2010, U.S. fire departments responded to an estimated 1,331,500 fires—482,000 structure fires and 849,500 nonstructure fires—which caused an estimated loss of $11.6 billion. Many of these fires were small or resulted in little or no reported property damage. However, 17 of them resulted in losses of $10 million or more each, for a total of roughly $652 million in direct property losses.[16]

Americans tend to remember the great fires in our history—the Chicago fire of 1871 and the fire set off by the San Francisco earthquake of 1906—but the thousands of smaller tragedies that occur yearly have failed to generate the reforms our nation so badly needs. We get an eye-opening look at how U.S. citizens view fire danger in a 1973 report issued by the U.S. Fire Administration, *America Burning*. As examples of views and attitudes that the average American holds about fires, the report cites the following:

- "It can't happen to me."
- "Odds are that it won't happen to me."
- "The insurance company will take care of me."

- "It is not a disgrace to have a fire."
- "I can set fires for revenge."
- "They're just children playing; they didn't know any better."[17]

Such casual attitudes contrast starkly with the more responsible approach taken by European and some Asian nations, where advanced training is required of fire professionals, and building codes tend to be more stringent than in the United States. (For a more in-depth discussion of building construction, see Chapter 7.)

In 1987, *America Burning—Revisited* was issued and because of this together with the 1973 report, perceptions and attitudes of the fire service began to change. One was that fire prevention was now advocated by firefighters. They don't just sit at the firehouse waiting for the alarm to ring. They now go to schools and teach kids about fire safety and they go to community activities to inform the public about dangers and prevention awareness. Cultural influences have changed the way firefighters take their message to the masses. Education should always be paramount because people need facts to act and facts build a rationale to change. Multilingual pamphlets are distributed, personnel with bilingual speaking abilities are used, and cultural diversity classes are taken prior to field activities.

The terrorist attacks of September 11, 2001, have greatly affected the fire and emergency services in the United States. The passing of the Patriot Act and the inception of **Homeland Security** have changed the way national fire policy is written and implemented. New laws and newly formed government entities affect fire service administration, and the new threats posed by terrorism have required greater vigilance from all emergency personnel as well as training in new levels of hazardous material handling and transport, new security measures, and new strategies for emergency response.

Homeland Security ■ The department created after the events of September 11, 2001, that is responsible for assessing the nation's vulnerabilities.

STOP AND THINK 1-1

List two ways we can change the fire problem in the United States today.

The Scope of Fire and Emergency Services Today

In U.S. communities, the fire and emergency services department is the only entity trained and equipped to deal with multiple kinds of disasters. Fire departments respond to all manner of emergencies and confront all manner of risks and hazards.

The fire service safeguards not only our homes and businesses but also our economy and its critical infrastructure—the electrical grid, interstate highways, railroads, pipelines, and petroleum and chemical facilities—are all protected by the fire service. In fact, the fire service is itself considered a vital part of the nation's infrastructure. The service also responds to emergencies in federal buildings, provides aid to military bases, and protects interstate commerce. Every passenger airliner that takes off from a runway is protected by a fire department. Every hazardous materials spill results in a call to the fire service to protect lives and property and deal with the dangerous clean-up process.[18]

Today some municipalities, faced with tight budgets and limited personnel, are shifting to a triple-certification requirement for their emergency responders. It is obviously more cost-effective to hire one person to act in three roles—firefighter, police officer, and paramedic—at an emergency scene. Like most things that sound too good to be true, however, there is a downside to triple certification. It has resulted in a higher than usual turnover rate, with firefighters tending to burn out and leave the department either for a higher paying position or because they do not want to be police officers.

With or without triple certification, however, there is no doubt that the traditional job description of firefighters has changed radically in recent decades. For instance, they must be prepared to provide emergency medical services such as advanced life support. Many states and organizations now require, as a minimum standard, that all job applicants be certified as Emergency Medical Technicians (**EMT**) or National Registry Emergency Medical Technicians (**NREMT**). Previously, firefighters just needed to be volunteers with no formal medical training to apply for a job. Changes in educational requirements are also affecting the promotional hierarchy. Firefighter seniority was once key to moving up the ranks. Today, higher educational achievement is an equally important critical criterion, often providing a boost up the ranks ahead of candidates who have had less formal schooling. Advanced education can also help firefighters in their public relations role and in educating the public on fire prevention. Today, when we are seeing an increase in the retirement of the veteran firefighters of the baby boomer generation and an influx of new recruits, training and higher education are more important than ever. (This topic will be discussed in more detail in Chapter 4.)

Since September 11, 2001, fire and emergency services have been given another tool to use to help all emergency responders at all levels communicate and work together. On February 28, 2003, President George Bush issued Homeland Security Presidential Directive 5, which created a National Incident Management System (**NIMS**). NIMS provides a consistent nationwide template to enable all government, private sector, and nongovernmental organizations to work together during incidents that threaten domestic security. This allows for the same terminology to be used whether you are a volunteer, paid, or from the North or the South. As a result, fire professionals must now be trained in emergency strategies designed to mitigate disasters at the national as well as the local level. (Chapter 10 will provide more information about the NIMS and the firefighter's role in Homeland Security.)

Not only have firefighting management and education changed, but also have basic firefighting techniques, gear, and equipment. In the past, the firefighter would leave the firehouse in a knee-length bunker coat and rubber boots. In some calls, modern firefighters are more likely to wear Emergency Medical Service (EMS) gloves than bunker gear. The typical call is still an emergency to which firefighters respond with sirens wailing, but when they get to the scene, they may not go in with axes and fire hoses and spray down the building. Instead, they may take out stethoscopes and blood pressure cuffs and begin questioning victims about injuries. It may be time to change the title of firefighter to fire and emergency services technician or emergency preparedness specialist. Some already refer to firefighters as all hazard personnel because they respond to just about every type of emergency call you can imagine. In just the last 10 years, the firefighting profession has evolved into the nation's largest prehospital EMS provider as well as a frontline defense in homeland preparedness.

Some of the new standards and regulations that today's fire and emergency services are embracing were established during the annual Fire and Emergency Services and Higher Education (**FESHE**) Conference. The role of FESHE is to provide a new strategic approach to professional development. The organization will help move fire and emergency services from a technical occupation to the status of a full-fledged profession similar to that of physicians, nurses, lawyers, and architects, who, unlike fire service personnel, have standardized course requirements for their respective degree programs. (A thorough discussion of FESHE will be presented in Chapter 4.)

EMT ■ Emergency Medical Technician. A specified level of medical training that usually consists of around 100 hours of classroom and on-site training.

NREMT ■ National Registry of Emergency Medical Technicians.

NIMS ■ National Incident Management System.

FESHE ■ Fire and Emergency Services Higher Education. Working with coordinators of two- and four-year academic fire and emergency medical services (EMS) degree programs, the U.S. Fire Administration's National Fire Academy (NFA) has established the FESHE network of emergency services–related education and training providers. The FESHE mission is to establish an organization of postsecondary institutions to promote higher education and to enhance the recognition of the fire and emergency services as a profession to reduce loss of life and property from fire and other hazards.

ON SCENE

Tradition and history are deeply entrenched in the fire service, but in today's society, the fire service has also evolved into a new service. Fire and emergency services in the United States deals not only with fire suppression as in the past, but it also deals with emergency medical services, hazardous materials, code enforcement, wildland fires, wildland/urban interface, fire prevention and education, technical rescue, urban search and rescue, aircraft firefighting, arson investigation, explosive response, industrial fire safety, and much more. Some of the same challenges from the past still plague the fire service today. These challenges include staffing, funding, service demands, facilities, health and safety, and training.

1. As a firefighter in the 21st century, are you ready and equipped to carry out the duties and responsibilities of the new role a firefighter plays while still carrying on the traditions that the fire service holds dear? What organizations can help the firefighter hold on to the traditions yet move forward and learn from the past?
2. If problems from the past continue to plague the fire service today, where can you find information on what have we learned to help facilitate a resolution to these long-standing problems?

Summary

Fire has played a major role in the development of civilization as we know it today. As humans race, we have learned to use fire to our benefit, but we need to further increase our respect for its power and to improve the ways in which we deal with its potential dangers.

The nature of firefighting requires an individual to be available to handle any crisis at a moment's notice and to handle it with professionalism. Never before in the history of our country have the actions of firefighters been held to the level of scrutiny that is applied today. Although recent events across the nation have brought national attention to firefighters' jobs and how they perform their jobs, the modern fire and emergency service is being challenged to take on an increasing number of responsibilities every year. Fire and emergency services must handle the training needed to use the ever-changing technology of new equipment. Current trends suggest that the fire service industry of the future will emphasize fire prevention and education as much as fire suppression. This trend—as well as the requirement that firefighters be proficient in emergency medical procedures and be prepared to respond to homeland security crises—have increased the need for education and training.

The traditions and culture of the fire service are changing. The once white male-dominated profession is now open to women and to all minorities. This diversity has broadened the scope of the fire service by introducing new ideas and cultural concepts.

There is a long, colorful, and distinguished history associated with the fire service, but there is plenty of room for improvement. Although no crystal ball can predict the world of tomorrow, the 21st century will surely carry the fire and emergency services field in directions yet imagined.

Review Questions

1. The development and use of the first hand-operated fire pump dates back to _____.
2. The first public volunteer fire department in North America was formed in _____ at _____.
3. What is a "rattle watch"?
4. What was the "apparatus Aldini"?
5. Which insignia was worn by the crusading Knights of St. John?
6. What is the role of FESHE?

Suggested Reading

Coleman, Ronny J., & Granito, John A. (1988). *Managing fire services* (2nd ed.). Washington, DC: International City Management Association.

Cote, Arthur E. (2004). *Fundamentals of fire protection*. Quincy, MA: National Fire Protection Association.

Fire Analysis and Research Division. (2011). *Third needs assessment of the U.S. Fire Service: Conducted in 2010 and including comparisons to the 2001 and 2005 needs assessment surveys*. Quincy, MA: National Fire Protection Association.

Endnotes

1. Fenton, Marc. (2002, November 27). *Great fire of Rome*. Thirteen WNET New York, Retrieved January 26, 2007, from http://www.thirteen.org/pressroom/release.php?get=252.
2. Klinoff, Robert W. (2003). *Introduction to fire protection* (2nd ed.). Clifton Park, NY: Thomson Delmar Learning.
3. Ibid.
4. Ibid.
5. Coleman, Ronny J., & Granito, John A. (1988). *Managing fire services* (2nd ed.). Washington, DC: International City Management Association.

6. Answers Corporation. (2005, February 1). *History of fire brigades.* Retrieved January 27, 2007, from Answers.com website: http://www.answers.com/topic/history-of-fire-brigades.

7. Firehouse International, Inc. (2007). *Fire extinguisher questions and answers index.* Retrieved January 27, 2007, from How Stuff Works Related to the Fire Industry website: http://www.firehouseinternational.com/user-site/how_stuff_works/Fire/Fire_extinguishers/Index/extinguisher_Index.aspx.

8. Ibid.

9. City of Chicago. (n.d.). *History of the Chicago Fire Department.* Retrieved March 29, 2013, from City of Chicago website: http://www.cityofchicago.org/content/dam/city/depts/cfd/general/PDFs/HistoryOfTheChicagoFireDepartment_1.pdf.

10. Hashagen, Paul. (n.d.). *SCBA history: The development of breathing apparatus.* Retrieved on March 29, 2013, from http://lishfd.org/History/scba_history.htm.

11. Ibid.

12. Ibid.

13. Ibid.

14. United States Fire Administration. (2004). *Fire in the United States 1992–2001* (13th ed., p. 29). Arlington, VA: TriData Corporation.

15. Ibid.

16. Badger, Stephen G. (2011, November). *Large-loss fires in the United States—2010.* Quincy, MA: National Fire Protection Association Fire Analysis and Research Division.

17. United States Fire Administration. (2004). *Fire in the United States 1992–2001* (13th ed., p. 29). Arlington, VA: TriData Corporation.

18. Bruegman, Randy R. (2003, April 30). *Meeting the needs of the fire service.* Retrieved January 27, 2007, from the International Association of Fire Chiefs website: http://www.iafc.org.

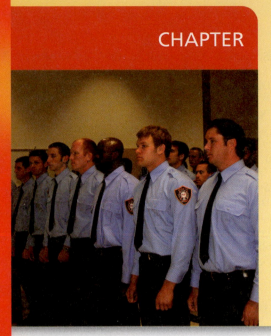

2

Fire and Emergency Services Career Opportunities

Courtesy of Jason Loyd.

KEY TERMS

bioterrorism, *p. 32*

Department of Homeland Security, *p. 32*

ergonomic, *p. 27*

FAA, *p. 22*

fire ecology, *p. 22*

fire lines, *p. 22*

GIS, *p. 31*

helitack, *p. 22*

hotshot, *p. 22*

pictometry, *p. 31*

prescribed fire, *p. 22*

smoke jumpers, *p. 22*

Smokey Bear, *p. 29*

Sparky the Fire Dog, *p. 29*

standard operating procedures (SOP), *p. 20*

OBJECTIVES

After reading this chapter, you should be able to:

- Identify fire protection jobs in the public and private fire service.
- List duties and requirements for the positions of firefighter trainee and firefighter.
- List duties and requirements for the position of firefighter/paramedic.
- Give examples of fire service jobs other than firefighter.
- List duties and requirements of search and rescue teams.
- List duties and requirements of the wildland firefighter.

Introduction

Firefighting jobs are not easy to obtain. They call for high levels of physical fitness and mental stamina, as well as a strong desire to serve others. Increasingly, these jobs also call for advanced education and specialized training. Candidates who meet these rigorous requirements, however, will find employment prospects to be generally favorable (see Figure 2-1). Employment of firefighters is expected to grow by 9%, slower than the average

FIGURE 2-1 Firefighter/paramedic working at a desk. *Courtesy of Jason Loyd.*

for all occupations. Physically fit applicants with high test scores and some postsecondary firefighter education and paramedic training should have the best prospects. Most of this growth will occur in suburban areas where volunteer firefighting positions are rapidly turning into paid positions. Replacements will also be needed for firefighters who retire, stop working for other reasons, or transfer to other occupations.[1]

This chapter is designed to describe traditional firefighting jobs and to introduce some of the new positions that have developed over the last decade. In the fire service, as in any other business or industry today, the range of possible career paths, job duties, and skills required is expanding rapidly. This is the result of varied factors, including technological changes, increased concern for the environment, and radical changes in the nature of the dangers confronted by the fire service. In addition, fire service personnel are being asked to take on duties far beyond those they performed 50 or even 10 years ago. Frequently, firefighters double as emergency medical care specialists. They may also lend their expertise in areas as diverse as public education, forestry, architecture, engineering, insurance, and law enforcement.

However, please note that the names of some positions may differ per locale. The fire service offers a great variety of opportunities and challenges, but all professionals in the field share a common goal—bringing order to chaos and creating a safe environment for the public.

Firefighting Career Opportunities

FIRE CADET/RECRUIT

The fire cadet, an entry-level training position, may be either a private individual interested in pursuing a career in the fire service or a recruit sponsored by a fire department or another organization. In the latter case, the recruit's fire academy education may be financed by the sponsoring group, which also typically guarantees his or her employment after graduation. Students without sponsors will have to finance their own education and conduct their own job search.

FIGURE 2-2 Cadets attending fire academy. *Courtesy of Jason Loyd.*

Fire academies may be part of a fire department, or they may be college based. In either case, the training is intellectually and physically arduous and demands commitment on the cadet's part. Typically, fire academy training is 12 to 16 weeks but may vary from state to state and even within the same state. A high school diploma or General Educational Development (GED) certificate of completion is generally the minimum educational requirement to apply for cadet training; however, this can be different from state to state. If the academy is sponsored by a fire department, the department's training officers supervise fire recruits until they have mastered basic firefighting skills and completed their examinations. Trainees are then assigned to a duty station where their progress is supervised by a fire company officer. Often, fire recruits remain on probationary status for their first year of service while their performance, both as an individual and as a team member, is assessed. The academy could even be administered through a career-technical training program through vocational schools. If the academy is based in a college or university, the graduating cadets begin the long road toward finding an entry-level job and establishing their own career paths with a fire department. Regardless of how cadets find employment, the tasks they face in their new job are the same. In addition to actual firefighting, cadets may be assigned rescue work, first aid, ventilation, forcible entry, salvage, overhaul, site inspections, emergency medical care, public fire education, hazardous materials work, dispatching—and the list goes on (see Figure 2-2).

FIREFIGHTER

Hiring requirements for firefighters are comprehensive and the testing competitive, reflecting the critical nature of the job and the need to work under stressful and physically demanding conditions. Applicants typically must pass a written test, an oral interview, a physical ability test, a medical evaluation, and a background investigation. Those who achieve the highest scores are ranked at the top of the candidate list and have the best chances of being hired. New firefighters work under close supervision learning the department's policies and procedures, which are usually referred to as **standard operating procedures (SOP)** or standard operating guidelines (SOG). As with many other jobs, there is usually a probationary period that can last from six months to a year, depending on department policy. If the probationary period is successfully completed according to the probationary agreement, the probation status is lifted and the firefighter becomes a

standard operating procedures (SOP) ■ Rules by which an organization or fire department operates for its day-to-day functioning. Usually these procedures are documented in a handbook or related source.

full-duty fire professional. However, this can vary from town to town or even state to state. There are different levels of basic firefighting, which include National Fire Protection Association (NFPA) levels such as Firefighter I (FF I) and Firefighter II (FF II). FF I and FF II differ on their levels of supervision and ability to operate independently. FF1 follows *NFPA 1001: Standard for Fire Fighter Professional Qualifications* and covers Chapter 5 of that standard. FF2 covers the same skills but goes into more depth of knowledge as covered in Chapter 6 of *NFPA 1001*. It should be noted that a new edition of *NFPA 1001* Fire Fighter Standard is to be issued in 2013,[2] so new rules and protocols might need to be reviewed at that time.

Firefighters respond to emergency calls of all types. Although their primary responsibility is to fight fires, they are more often called on to assist with other kinds of emergencies, such as medical calls and vehicle accidents. As a public service, firefighters often visit schools, day care centers, and senior centers to teach fire safety. Firefighters' schedules vary, but the typical shift is 24 hours on duty, followed by 48 hours off. Of course, firefighters cannot just leave the scene of an emergency and go home when their appointed shift ends. Long and irregular hours are not unusual, and applicants should consider this before entering the profession. Moreover, a firefighter's duty and responsibilities fluctuate, and written job descriptions can vary considerably. Candidates need to gather as much information about a department's requirements before applying for a job. The Internet and job boards are good resources, and it is also helpful to visit agencies in the area.

FEDERAL FIREFIGHTER

The federal government employs both civilian and military firefighters at some of their installations, especially at military bases. These jobs require the firefighter to deal with specialized on-base emergencies, such as aircraft crashes, in addition to performing the same tasks and responsibilities of a municipal firefighter. Goodfellow Air Force Base in Texas is the home of the Louis F. Garland Department of Defense Fire Academy. The academy serves all branches of the Department of Defense (DoD), including the Army, Marine Corps, Navy, Air Force, and Civil Service. It provides training for about 2,000 students a year, including allied forces, and can be a path to a variety of fire service careers in the DoD branches. DoD firefighting opportunities are readily available on military bases for both civilians and military personnel.

FIREFIGHTER/EMT OR HIGHER

Fire departments in the United States commonly respond to medical emergencies. In fact, about half of them provide ambulance service. This has led to the dual position of firefighter/EMT or EMT–P (paramedic), in which the same person serves in two capacities in an emergency situation. The position comes with increased responsibility and requires advanced medical training that can take from at least six months to a year to complete. The actual amount of hours required for training can vary from state to state, so please check your local requirements. There may also be an intermediate level offered in your area. In addition to performing basic fire-suppressing duties, firefighter/medics must be prepared to assess injuries and acute illnesses, supervise ambulance transports, and take full responsibility for patient care until relieved by proper medical personnel. They provide emergency treatment working under standing orders or under the directions of a supervising physician. Personnel with this type of training are in great demand, with many fire departments offering pay incentives to qualified applicants for the position. Specific training and certifications requirements are dependent on your state or jurisdiction.

WILDLAND FIREFIGHTER

Fires set by lightning or careless human behavior consume thousands of acres of our precious wildlands annually. For example, in October 2007, more than 500,000 acres in

California burned, and this is a small percentage of the total acres burned in 2007. In national forests and parks, nonfirefighters are the first line of defense against these fires. Forest rangers patrol assigned areas to ensure that travelers and campers comply with fire regulations. Forest fire inspectors and prevention specialists spot fires from watchtowers and report their findings to headquarters by telephone or radio. When a fire is reported, wildland firefighters are brought in to suppress the blaze. In addition to basic firefighting training, these specialists study **fire ecology**, the science of how fire behaves in natural environments and how it affects both living and nonliving things in the environment.

The work is physically demanding and can be emotionally taxing. Wildland firefighters are frequently required to work for days at a time—sometimes up to 14 days straight, without a day off or home leave. A typical shift is 16 hours or longer, and shifts over 24 hours are not unknown. In addition to the long hours, the crews must brave hazardous environments, enduring extremes of heat as well as smoky, dirty, and dusty conditions. It is important to note that there may be age requirements for this specific role of firefighting—please check your local state or district.

A wildland firefighter may not always work directly in the front lines of an advancing fire. Other assignments might include creating **fire lines**—cutting down trees and digging out grass and other combustible vegetation in the path of the fire—to deprive it of fuel. This is one of the most efficient means of battling a blaze. Wildland firefighters may also work at setting and monitoring **prescribed fires**. These are fires purposely ignited by fire personnel or agencies under controlled conditions for specific management objectives. For example, setting a prescribed fire is a good, natural way to help put nitrates back into the soil. Furthermore, a prescribed burn can help decrease wildfires because it removes dead and dying trees to prevent heavy fuel loads during dry seasons.

There are a number of challenging specialist careers within the wildland firefighting field. Helicopter attack teams, or **helitack**, dump enormous buckets or belly tanks of water onto the fire. Some helitack crews rappel out of helicopters, gaining access the only way possible to fires in remote areas or in rugged terrain. **Smoke jumpers** parachute out of planes to reach areas not otherwise accessible, and **hotshot** crews hike into the area where the fire is burning to begin extinguishment. Employers for individuals interested in a wildland firefighting career include the U.S. Department of Agriculture Forest Service, California Department of Forestry and Fire Protection, and contract companies. In 2012, the U.S Forest Service predicted a busier fire season than ever because of the drier than normal winter. Furthermore, having a reduction in its large air tanker fleet has not helped its resources. In Texas, the Texas A&M Forest Service has implemented the Texas Intrastate Fire Mutual Aid System (TIFMAS), which allows training to be conducted to make statewide use of local resources and to make sure everyone has had the same comprehensive training. Other states have implemted such programs such as CAL FIRE in California.

AIRCRAFT RESCUE FIREFIGHTER

The aircraft rescue firefighter (ARFF) goes through a highly specialized course of training to earn certification and then be assigned to a civilian airport. The Federal Aviation Administration (**FAA**), an agency of the U.S. Department of Transportation, sets minimum ARFF requirements for each airport, based on an index determined by the overall length of the aircraft it handles and the average daily departures of those aircraft.[3] ARFF training guidelines are outlined in the FAA's Federal Aviation Regulations, Part 139 (FAR Part 139). This requires that trainees receive airport and aircraft familiarization instruction, training in emergency aircraft evacuation assistance, and airport emergency communications systems. They also must be experienced in using the specialized equipment needed to fight aircraft fires and to minimize aircraft cargo hazards. In addition to this basic training, aircraft rescue firefighters must typically participate in an annual 24-hour ARFF refresher course as well as annual live flammable-liquids fire-suppression training

fire ecology ■ The study of the interrelationship of wildland fires and living and nonliving things in the environment.

fire lines ■ Boundaries around a fire area to prevent access except for emergency vehicles and relevant professionals.

prescribed fire ■ A fire that is purposely ignited by fire personnel or agencies under controlled conditions for specific management objectives.

helitack ■ Used during wildfires when the location of the fire(s) is inaccessible to firefighting crews.

smoke jumpers ■ Firefighters who parachute from airplanes to suppress forest fires in remote locations.

hotshot ■ Highly trained firefighters used primarily in handline construction.

FAA ■ An agency in the U.S. Department of Transportation that oversees all aspects of civil aviation.

FIGURE 2-3 ARFF apparatus. *Courtesy of TEEX-Texas A&M System.*

to maintain their certification. In addition, they usually take part in annual multiple-casualty exercises at the airport (see Figure 2-3).

FIRE APPARATUS ENGINEER

The fire apparatus engineer is a promotion from the firefighter position and requires extensive training and experience. Apparatus engineers work in all kinds of firefighting situations (structural, wildlands, and so on) and have sole responsibility for driving and operating fire apparatus such as engines, water tenders, rescue squad vehicles, and aerial trucks. Emergency driving skills, and knowledge of traffic laws, departmental SOPs, and hydraulics are examples of additional training that a person must have to be successful in this position. It is important to note that this specific role in firefighting requires an enormous understanding of pumper hydraulics and the person must also be able to figure and understand pump calculations quickly in their head. Special driving certifications/licenses are required, and good driving records are compulsory.

SEARCH AND RESCUE

Search and rescue (SAR) teams consist primarily of volunteers, usually drawn from other emergency services. They are responsible for conducting missions in a wide range of emergency situations and almost any imaginable setting, including aircraft accidents or incidents, swift-water and open-water rescue, dive rescue and recovery, confined space rescue, mine rescue, high- and low-angle rescue (mountain and structure), and avalanche rescue (see Figure 2-4). The broad scope of their activities calls for a high degree of proficiency in a wide range of rescue techniques. Frequent and intensive training is required (see Figure 2-5). Each department or agency may have its own local SAR team. In addition, state-supported or regional SAR teams are deployed to cope with disasters of a larger scale.

One such state-supported team is Texas Task Force 1. Task Force 1, sponsored by the Texas A&M Engineering Extension Service (TEEX) a member of the Texas A&M University System, serves as the state's urban search and rescue team under the Governor's Division of Emergency Management and is one of 28 such teams in Federal Emergency Management Agency's (FEMA's) national urban search and rescue system. This task force, with more than 200 members drawn from 60 organizations across Texas, responded to one of the largest natural disasters in U.S. history. When Hurricane Katrina struck the

FIGURE 2-4 Water rescue training. *Courtesy of TEEX-Texas A&M System.*

Gulf Coast of Louisiana and Mississippi on August 29, 2005, Texas Task Force 1 swung into action, performing a variety of functions as the emergency developed:

- An 80-member urban search and rescue team was deployed by the FEMA on August 27 in advance of the hurricane's landfall.
- A 41-member water rescue team was deployed by the Texas Governor's Division of Emergency Management on Monday, August 29.
- Another 75-member urban search and rescue team was deployed by FEMA to New Orleans on Tuesday, September 13.

The teams assisted with more than 13,000 rescues in downtown New Orleans, where thousands were stranded after the hurricane's storm surge broke levees and left much of the city under water.[4] More recently, a major deployment was on September 6, 2011. Task Force 1 was activated to deploy to the Bastrop, Texas, wildland fires. They searched over 2600 damaged, destroyed, or remaining structures over a six-day period. Deployment involved 91 personnel[5] (see Figure 2-6).

FIGURE 2-5 Collapsed structures training. *Courtesy of TEEX-Texas A&M System.*

INDUSTRIAL FIREFIGHTER

An industrial firefighter is specialized in meeting the needs of the manufacturing, refining, and petrochemical industries. The oil industry began in America in the late 1800s, and with it came fierce and uncontrollable petroleum fires. As recently as 20 years ago, the technology and specialized training needed to extinguish such fires was not available, and they often burned until the fuel ran out. The cost to the environment, business, and human health was immense. The training and certifications required for this type of role varies from states and privately owned industries so it is important to note that a potential industrial firefighter would need to research his or her specific place of employment. The need for industrial firefighting was obvious, but there were no facilities capable of handling the large-scale training exercises that would be required. This need is now met nationally and internationally through training facilities such as TEEX at the Brayton Fire Training Field, a part of the Texas A&M University System (see Figure 2-7).

HAZARDOUS MATERIALS TECHNICIAN

Hazardous materials technicians are concerned with the discovery, regulation, and recovery of hazardous materials. They must meet the standards for the hazardous materials

FIGURE 2-7 Simulated chemical complex fire training structure. *Courtesy of TEEX-Texas A&M System.*

technician level, as set by the National Fire Protection Association. States have different requirements, so please verify with your jurisdiction. In Texas, the minimum requirement to achieve full status on a hazardous materials response team (HazMat) is approximately 80 hours of training. A 40-hour annual recertification class is also required. In addition to hazardous materials training, personnel must possess basic first aid and rescue skills. As a rule, these technicians are also required to have specialized training in dealing with weapons of mass destruction as well as biological warfare agents and other hazardous materials that might be used in terrorist incidents (see Figure 2-8).

FIRE INVESTIGATION UNITS/ARSON UNITS

Fire Investigation Units (FIU) or Arson units are charged with combating arson and other fire-related criminal offenses. A large city might have its own arson team, but in smaller towns and in rural districts, these units tend to be multijurisdictional. The units fulfill their mission by bringing together and utilizing the talents and resources of experienced and dedicated fire, police, and prosecutorial personnel in a manner that maximizes results while minimizing taxpayer expense. Certification in this area often includes peace officer training, which gives arson team members the power to arrest.

Nonfirefighting Careers in Fire Protection

Other career paths in the fire service are available that do not require active, on-site firefighting. The physical, educational, and certification requirements for these positions differ. All require an understanding of fire and fire safety, and each serves a critical function in protecting the public.

DISPATCHER/COMMUNICATIONS OPERATOR

One of the most challenging civilian jobs in emergency services is that of fire dispatcher. The dispatcher's job is to support fire management operations, primarily by relaying information from those seeking emergency services to the appropriate responders. In most cases, dispatchers are not solely dedicated to fire emergencies but must handle police and medical emergency calls as well. It is their job to receive each incoming 9-1-1 telephone call, obtain essential incident information from the caller, evaluate what type

FIGURE 2-9 Dispatch communications center. *Courtesy of Jason Loyd.*

of emergency service is required, determine the priority of the call, and then dispatch the appropriate unit. They must also relay all essential information about the emergency to the responding unit via voice radio or intercom while at the same time monitoring all other emergency frequencies. Dispatchers work at a fast pace and are under constant pressure. They must be expert communicators and proficient multitaskers and be able to handle high-stress situations calmly (see Figure 2-9). Training includes many different topics, including philosophy of law enforcement and firefighting, law, technical topics (radio, telephone, computers), management of stress, handling difficult callers, emergency medical dispatching instructions, and listening and speaking skills. Although firefighting skills are not required for the position, many firefighters—whether paid or volunteer—are sometimes given special preference for this job because of their background.

OCCUPATIONAL HEALTH AND SAFETY SPECIALIST

Occupational health and safety specialists are employed in both the public and private sector. Their expertise is analyzing potential hazards in work environments and developing programs and procedures to eliminate or minimize these hazards. They may also serve in an enforcement capacity, conducting inspections and monitoring compliance with health and safety regulations and laws. The safety specialist is able to identify any condition—physical, chemical, biological, or **ergonomic**—that may cause disease or injury. The experience and education required may be different depending on the needs of the public or private sector area. Candidates should always find out what their potential employers may require.

ergonomic ■ The scientific principle that bases design around human needs, and the profession that applies science and data to this concept.

FIRE EQUIPMENT SALESPERSON

Because firefighters and other emergency personnel can have flexible work schedules, they may be able to work at dual occupations. A firefighter might elect to continue work in the fire service field by working as a part-time fire equipment salesperson. Companies such as Pierce Manufacturing and Tyco Fire & Security, which design and service fire protection equipment, are always looking for experienced personnel to sell their products. Firefighter certification is not required for this position.

FIRE EXTINGUISHER SERVICE TECHNICIAN

Another occupation not directly related to firefighting duties is that of the fire extinguisher service technician who inspects and services firefighting equipment, including all types of hand-held fire extinguishers and fire hoses. Fire extinguisher service technicians must have the technical knowledge to disassemble extinguishers and examine them for defects, replace defective parts, clean the fire extinguishers, refill them with fire extinguishing agents, and then test them, using special testing equipment, to ensure that they meet specifications. This is another occupation that does not require firefighter certification.

FIRE MARSHAL

The fire marshal typically heads the fire prevention division. However, there are no concrete sets of rules on fire marshal positions. In addition, this type of position could be at a fire department, government entity, or both. Numerous different titles exist, but the basic definitions are the same. Fire marshals conduct inspections of the premises within their jurisdiction to identify unsafe conditions and ensure compliance with fire codes. Additional duties may include fire and arson investigation. Fire marshals also work with developers and city planners to check and approve plans for new buildings. In conjunction with the goals of the fire prevention division, fire marshals often speak before public assemblies and civic organizations to promote fire prevention. This position may require formal education such as a bachelor's degree as well as experience in fire prevention, plan review, inspections, and supervision (see Figure 2-10).

STOP AND THINK 2-1

Several different avenues are available for a career in fire and emergency services.

Select two different choices in the fire and emergency services field that an individual may pursue besides civil service employment. Explain your answer in detail.

FIGURE 2-10 FESHE professional model for fire marshal.

PUBLIC INFORMATION OFFICER

The job of a fire department's public information officer (PIO) is to communicate accurate information to the community and to members of the fire service in a timely, efficient manner. This information may relate to ongoing emergencies or to other aspects of the department's operations. A public information officer is often interviewed on television or on the radio, responding to questions from reporters. A public information officer must be available at any time of the day or night to cover significant events and serve as liaison between the incident commander and media outlets. The position requires proficient verbal skills and media knowledge. Applicants for the job typically have a bachelor's degree and meet the *NFPA's Standard 1035: Standard for Professional Qualifications for Public Fire and Life Safety Educator*. Although firefighting skills are not required for this position, many PIOs are hired out of the firefighter arena.

FIRE PREVENTION SPECIALIST/PUBLIC FIRE EDUCATION SPECIALIST

A fire prevention specialist is a unique kind of fire service professional who fights fire by promoting fire prevention training and education. The position involves visiting schools and community organizations to talk about fire safety, and it may also require performing routine checks of fire systems, such as sprinklers and hydrants. Fire prevention specialists use campaigns such as **Smokey Bear**, **Sparky the Fire Dog**®, and Fire Prevention Week to build awareness and educate the public. The Smokey Bear campaign, which was created in 1944, is the most well-known public service prevention program geared toward wildland and forestry fires. In August 2012, a week-long celebration across the United States was held for Smokey Bear's 68th birthday bash. The birthday celebration allowed a renewed opportunity to reiterate that most wildfires are preventable. Examples of job opportunities as a fire prevention specialist are federal or state forestry services, city or county fire departments, municipalities (such as inspection bureaus), private companies, or through partnership with several of these.

Smokey Bear ■ Created in 1944. The longest running public service campaign in U.S. history for fire prevention.

Sparky the Fire Dog® ■ Created in 1951. Used for NFPA's Risk Watch® and Learn Not to Burn® programs and has a special appeal to children's programs.

FIRE DEPARTMENT TRAINING SPECIALIST

A fire department's training specialist or training officer develops and coordinates programs to maintain and enhance skilled performance by the department's firefighters. The position also involves developing standard operating procedures for fire suppression and assuring that these procedures are in compliance with state and federal fire and safety regulations. This is typically a midmanagement position that reports directly to the fire chief. The specific functions of the training officer are described in more detail in Chapter 4, "Training and Higher Education."

INSURANCE ADJUSTER

Insurance companies frequently hire people with backgrounds in the fire service to work in their loss prevention departments. They especially seek those with knowledge in the fields of fire protection engineering, fire risk management, fire prevention, and arson investigation. Such expertise is invaluable when it comes to inspecting fire-damaged properties to assess whether arson may have been involved. Adjusters also estimate the amount of loss, negotiate with policyholders, and when necessary, gather evidence to support contested claims. They may be called to testify in court on behalf of their employer.

FIRE INSPECTOR

The major goals of fire inspectors are to increase public awareness so that potential fire hazards can be avoided. They also have to keep sound records of their inspections and physically verify that fire protections systems are working and maintained in the building he/she is inspecting. To achieve these goals, fire inspectors examine the interior as well as the exterior of buildings and structures for fire hazards. Code enforcement plays a major role in their work. Fire inspectors may also collect fees for building licenses and permits. A fire inspector is normally employed by a fire department and may be a member of a team of inspectors. Overall, the main objective of this position is to be proactive and prevent fire from occurring.

FIRE INVESTIGATOR

Fire investigators are essentially detectives whose job it is to determine how a fire originated and, in particular, to determine whether it was accidental or the result of negligence or arson. These types of jobs are both private and public. Fire investigators visit fire scenes, photograph and sift through evidence, and interview property owners, building occupants, and witnesses. Once they arrive at a plausible theory, they have the expertise to reconstruct the fire scenario in order to test their theory. They also receive training in interrogating suspects and interviewing witnesses. In some states, fire and arson investigators have the authority to arrest suspects or to subpoena witnesses.[6] In states where they do not have arresting powers, they work closely with law enforcement officials.

Private Sector Fire Protection Careers

STATISTICIAN

The National Fire Incident Reporting System (NFIRS) provides resources used by U.S. fire departments to maintain records in a uniform manner. All 50 states and the District of Columbia utilize this database. This career is unique that it can be private, volunteer, or paid. This position gives the ability of the fire service to track and trend its data.

FIRE PROTECTION ENGINEER

Fire protection engineers (FPE) often enter the profession after working in some other sector of the fire service. Their work, however, is very different from firefighting. Their job is to design structures to optimize their fire resistance. The tools they use are mathematical and scientific knowledge. Fire protection engineers may be employed by a large fire department—usually in their fire prevention or code enforcement departments—or they may work as employees or consultants for engineering firms, fire equipment and systems design companies, forensics investigators, hospitals and health care facilities, manufacturing firms, research and testing laboratories, or in the insurance industry.

CONTRACT FIREFIGHTER OPPORTUNITIES (OVERSEAS)

The market for U.S. firefighters overseas is booming. In most cases, they are employees of U.S. companies with operations or contracts abroad. For example, Kellogg Brown & Root (KBR) is under contract to provide fire service for various DoD facilities. The aerospace technology company Lockheed Martin employs firefighters stateside and abroad, as does Boots & Coots International Well Control, Inc., both based in Texas. The Boots & Coots company is known for its role in controlling oil well fires in Kuwait after the Gulf War. These companies provide firefighters exciting career opportunities abroad, including airport rescue firefighting (ARFF), fighting industrial fires, and serving as HazMat technicians, safety specialists, and inspectors. Some companies even seek specialists at the crew chief, fire chief, or fire captain level.

STOP AND THINK 2-2

As with many specialized occupations, outsourcing is on the rise in firefighting. Why, do you think, is there an increasing need for overseas firefighters?

INVENTOR

As technology advances, firefighting becomes more complex, and the need for innovative equipment and techniques increases. Inventors working to fill this need have found firefighters to be a rich source of ideas. After all, firefighters know best what would make the job safer and more efficient. One new innovation is the use of the Global Positioning System (GPS) in fire apparatus. Several large cities, including Chicago, have invested in state-of-the-art emergency response systems built with GPS-based Automatic Vehicle Location (AVL). A city the size of Chicago may have to respond to over 14,000 emergency calls a day, so a system that speeds response to those calls is well worth the investment; the use of GPS pays off in both public safety and the efficient use of firefighting equipment and labor. A related innovation is geographic information systems (**GIS**), which provide access not only to street maps but also to geographic data such as water sources, hazardous materials storage sites, residences with disabled persons, and municipal hydrants. Another new tool is **pictometry**, an enhancement of GIS that generates a three-dimensional (3-D) perspective of a given area. Understandably, pictometry is an enormous help to firefighters, allowing them to assess a fire scene even before arriving at the site. The size of the building, access to the building on all sides, and the characteristics of the surrounding area are revealed. These inventions all require a person with vision to understand the need and create the product that answers that need.

GIS ■ Provide access not only to street maps but also to geographical data such as water sources, hazardous materials storage sites, residences with disabled persons, and municipal hydrants.

pictometry ■ An enhancement of GIS that generates a 3-D perspective of a given area.

Volunteer Opportunities

The Community Emergency Response Team (CERT) is a FEMA outreach program designed to train citizens in elementary emergency response procedures. CERT team members are volunteers who receive education in disaster preparedness for the kinds of hazards that may affect their area. They learn basic disaster response skills, such as fire safety, light search and rescue, team organization, and disaster medical operations. Using this training, which is taught both in the classroom and during exercises, CERT members can assist others when an emergency arises in their neighborhood or workplace and professional responders are not immediately available to help.

Fire Corps is another avenue for volunteerism. Also under the direction of FEMA, Fire Corps creates valuable resources and foster programs that will bring community members into local fire and emergency service departments to assist them with nonemergency tasks. CERT groups and Fire Corps often work together to form a comprehensive citizen program.

The Future of Fire and Emergency Careers

The future of careers in the fire service is bright. As baby boomers retire, new job opportunities and opportunities for promotions will become available. The fire service will undoubtedly continue to take on new tasks and revise old ones, and this evolution will offer further opportunities as the need for technical specialists grows. The increase in educational requirements will lead to career opportunities in the educational field as new degree programs and specialty training courses are created at fire academies, colleges, and universities.

**Department of Home-
land Security** ■ Cre-
ated in 2002 primarily
from a conglomeration
of existing federal
agencies in response to
the terrorist attacks of
September 11, 2001. Its
purpose is to protect
the nation against
threats to the home-
land. Its efforts affect
firefighters and emer-
gency personnel.

bioterrorism ■ Relating
to or involving the use
of toxic biological or
biochemical substances
as weapons of war.

After the disaster of September 11, 2001, the **Department of Homeland Security** has come to the forefront of national issues and is the center of a growing source of new career opportunities. New jobs have already been created, not only in the fire and emergency fields but also in military and law enforcement, scientific research, cybertechnology, biomedical technology, disaster assistance and relief, air marshal operations, and intelligence operations, to name but a few. For example, as recently as 10 years ago, the threat of **bioterrorism** was not of much concern to our nation's firefighters. Today, all emergency personnel must be familiar with its deadly potential and trained in ways to contend with that danger—as well as with the other threats posed by hostile governments and extremist groups. These dangers, already grave, are likely to escalate. Fire and emergency service professionals will need to be prepared to meet these new challenges. A reality check shows that firefighters are first in, last to know, and required to do the most with the least, but it is their responsibility to anticipate the worst and plan the best possible response. This topic will be discussed more in depth in a later chapter.

ON SCENE

This chapter covers the professional opportunities available in the fire and emergency service. Most firefighters started out as volunteers in their local community—and many remain volunteers. For those who do decide to take up firefighting as a paid profession, laws and policies are in place to protect their jobs in the event that they are called to help in national disasters. At present, however, there is no similar protection against termination or demotion for the volunteer firefighter or EMS worker.

On March 9, 2006, Senate Bill S. 2399, the Volunteer Firefighter and EMS Personnel Job Protection Act, was introduced to protect the jobs of volunteer emergency services personnel responding to a presidentially declared national disaster for up to 14 days per calendar year. This bill would protect volunteers from having to choose between their careers and serving their country. This bill was referred to committee; however, it died in that stage of its life. On May 12, 2009, it was reintroduced as S. 1025, the Volunteer Firefighter and EMS Personnel Job Protection Act. As of January 2013, it still has not made it passed the Committee on Health, Education, Labor, and Pensions.

1. Give concrete examples of how volunteer firefighters can be deployed away from their homes to respond to national disasters and why these volunteer firefighters should not have their livelihood jeopardized.

LIFE AND CAREER OF EDDIE BURNS WHO ROSE THROUGH THE RANKS TO BECOME THE FIRST BLACK DALLAS FIRE CHIEF

Eddie W. Burns, Sr., became chief of the Dallas Fire-Rescue Department on April 19, 2006. He is 46 years old with 27 years of firefighting experience with the Fort Worth Fire Department. He started his career in the fire service in July 1979. After serving nine years in the Operations Division, where he rose to the rank of lieutenant, he accepted the demanding position as a member of the department's Fire Investigation/Bomb Squad. The assignment required him to become a certified peace officer, and he obtained certification as an explosive ordnance technician through training from the Federal Bureau of Investigation (FBI) and U.S. Army. On his promotion to captain in 1994, he returned to the Operations Division and worked at one of the busiest stations in the city. In 1997, he obtained the rank of battalion chief and was assigned as the shift commander for the city's westside battalion. He was selected in 1999 to assume the challenging role of chief training officer. In August 2000, he was promoted to deputy chief of Educational Services. He was reassigned to the Executive Services Division in September 2003, where he served as the executive deputy fire chief until coming to Dallas. He has been active in the fire service as an adjunct instructor with Weatherford and Tarrant County Colleges. He obtained an associate degree in fire administration from Weatherford College in August 2004. He graduated cum laude from Dallas Baptist University and is a graduate of the Executive Fire Officer Program from the National Fire Academy in Emmitsburg, Maryland, and John F. Kennedy School of Government from Harvard University.

He is married to Jackie Burns, and they have three grown children: Talia, Eddie, Jr. (firefighter in the Houston fire department), and Cory. He volunteers with the Sickle Cell Foundation as a camp counselor and serves on the advisory committee for the Boy Scouts. Chief Burns has his own foundation called "All That I Am I Owe." This foundation was formed to provide training and certification to underprivileged young men and women.

Summary

Although fire and emergency service jobs may be hard to obtain, the profession will continue to be a stable source of livelihood. Firefighting is forecasted to remain a very competitive field for the next several years and to provide a considerable amount of job security while also providing an honorable sense of public service. The nation's economic status plays a role in the fire and emergency service profession. A healthy economy results in the growth and development of communities, and this growth increases the need for additional fire services and individuals to fill those new positions. Over the years, the firefighter has taken on new roles and responsibilities and has begun the evolution from tradesperson to professional. Titles in this profession and the nature of the job can vary greatly depending on circumstances. Personnel may be employed in either the public or the private sector, but all fire and emergency professionals protect communities against the loss of life, injury, and destruction of property by fire. These professionals have the vital goal of bringing order to chaos. The challenges are real—and so are the rewards.

Review Questions

1. What is the difference between the firefighter recruit and firefighter positions?
2. In the fire and emergency professions, employees can work with various entities or departments on a single incident. Name a few examples that firefighters and emergency personnel might encounter or be governed by on an emergency call.
3. What additional requirements are placed on a firefighter paramedic over that of a regular firefighter?
4. Because of recent terrorist attacks in the United States and abroad, new career opportunities have been formed. Can you name a few?
5. What are the differences between a fire prevention specialist and a fire department training specialist?
6. What is the longest running public service campaign in history used for teaching forest fire prevention?
7. What jobs are available in the fire service that do not require actual firefighting?
8. What are some of the private sector jobs that would help you to prepare for the position of firefighter?

Suggested Reading

Lasky, Rick. (2006). *Pride & ownership: A firefighter's love of the job.* Tulsa, OK: PennWell Corporation.
Lepore, Paul S. (2012). *The aspiring firefighter's 2 year plan* (3rd ed.). San Clemente, CA: Freespool Publications.

Singerman, Philip. (1990). *An American hero: The Red Adair story: An authorized biography.* New York, NY: Little Brown & Co.

Endnotes

1. Bureau of Labor Statistics, U.S. Department of Labor. (2012). Firefighters. In *Occupational outlook handbook, 2012–13 edition. Retrieved* January 12, 2013, from Bureau of Labor Statistics website: http://www.bls.gov/ooh/protective-service/firefighters.htm.

2. National Fire Protection Association. (2008*). NFPA 1001 Standard for Fire Fighter Professional Qualifications.* Quincy, MA: Author.
3. Nilo, James R. (2000, February 1). Have trainer, will travel. *Fire Chief Magazine.* Retrieved January 27,

2007, from http://firechief.com/mag/firefighting_trainer_travel/ index.html.

4. Texas Engineering Extension Service. (2005). *Hurricane Katrina*. Retrieved January 29, 2007, from Texas Engineering Extension Service (TEEX) website: http://www.teex.com/teex.cfm?pageid=Agency&area=teex&templateid=1547.

5. Texas Task Force 1 Urban Search & Rescue Response System. (2011). *Tx-TF1 conducts wide area search operations in Bastrop County*. Retrieved August 11, 2012, from Texas Task Force 1 Urban Search & Rescue Response System website: http://usar.tamu.edu/about-us/Documents/2011%20Deployments/2011_Bastrop_Wildland_Fires.pdf.

6. Minnesota Internet System for Education and Employment Knowledge. (2005). *Career: Fire investigators*. Retrieved January 29, 2007, from Minnesota Internet System for Education and Employment Knowledge website: http://www.iseek.org/sv/13000.jsp?id=100320.

3

The Selection Process

Courtesy of TEEX-Texas A&M System.

KEY TERMS

ability/agility test, *p. 38*
background check, *p. 41*
ceiling breach and pull, *p. 38*

criminal conviction record check, *p. 41*
diversity, *p. 43*
NFPA 1582, *p. 41*

probationary period, *p. 42*
stair climb, *p. 38*

OBJECTIVES

After reading this chapter, you should be able to:

- List the standard phases of the hiring process.
- Describe the importance of preparation in the selection process.
- Give examples of body gestures that can be interpreted in an interview.
- Explain the purpose and importance of the probationary period.

Introduction

Firefighting is a demanding career that requires very special qualities. Therefore, fire departments devote great care in choosing their entry-level employees. They look for candidates who like to work with people and with machinery, enjoy teamwork, are physically fit, and have an interest in public safety. Potential firefighters should also be adept at following instructions and making quick decisions and should be willing to take risks while remaining calm in stressful situations. The hiring process is extremely competitive. It is not unheard of for hundreds, if not thousands, of applicants to vie for a small handful of jobs. Success depends on much hard work and thorough preparation.

In this chapter, we will consider the various stages of the selection process. It is important to remember that the hiring process, rigorous though it is, is only the first step toward a firefighting career. On the average, becoming a full-time paid firefighter takes anywhere from three to seven years.[1] Nonetheless, getting hired is the first step. We will explore the do's and don'ts of each phase of the process and suggest strategies to optimize success. We will recommend how to build a résumé filled with the right kinds of education and experience and suggest tips for succeeding at test-taking and in face-to-face interviews.

Preparation for a Firefighter Position

PREREQUISITES

The job search begins with finding where job openings are posted in your area. Most fire departments and fire service organizations postemployment opportunities on their city's website and in local newspapers. Some even hold job fairs at local colleges or community events. For a broader search, it is extremely useful to subscribe to job opportunity mailing lists or firefighter examination notification services, such as www.firecareers.com or www.firerecruit.com. Both of these services provide websites with notifications of when departments nationwide are accepting applications, and both are worth every penny of their subscription price.[2] They will save you the time and effort of calling each fire department and asking when they are next hiring.

Once you have identified a job for which you want to apply, it is wise to do a considerable amount of research on the position and the organization. The more you know, the better prepared you will be for the selection process. Talk to people who know the organization, look up its website if it has one, and familiarize yourself with its requirements. Entry-level firefighting jobs usually require that you are at least 18 years of age, possess a valid driver's license, and have, at the minimum, a high school diploma or GED. Many departments may have their own academy, which will suffice all requirements for obtaining a job; however, before applying for a job it is imperative that you check all prerequisites needed such as firefighter certification, and level of EMT training if necessary. *Also be sure to check local and state requirements, as they do vary.*

Once you have found a job opening for which you want to apply, obtain the necessary paperwork and take note of the application deadline. Most applications require high school transcripts to be attached. However, a candidate must read carefully the necessary paperwork because it may also require references or a résumé. Supplying the proper paperwork is the first impression that an employer has of you. Be certain that you get your application to the correct address by the deadline. The application should be typed or neatly handwritten in ink. Make additional copies of the application to use as trial drafts while you are gathering your information. That way, any mistakes or changes that you make will not appear on the final copy. Remember that all the information you provide will be considered fact, and facts should always be accurate and truthful. Once you have completed the application, make a copy and keep it in your files. This will help you fill out future applications accurately and more quickly.

If you meet the minimum qualifications and have submitted a complete and timely application packet, you will be invited to participate in the next phase of the selection process, a written test.

WRITTEN TEST

Two basic kinds of written tests may be given: (1) tests of general knowledge and (2) skills and tests of material related to firefighting. Most fire service organizations use standardized tests prepared by professional testing services. These tests are backed by extensive research and are inexpensive and easy to administer. They may be scored by the department or by a testing company. These tests can be state civil service tests or standardized tests that are bought from testing companies. Again, it is the candidate's responsibility to research the type of test that will be given for that particular position.

The typical exam consists of approximately 150 to 200 multiple-choice questions and is timed. Most fire departments require that you score at least 70% to continue in the selection process. (Keep in mind that preference points may be given to military veterans, so it is theoretically possible for them to score over 100%.) Scoring well on a written exam does not necessarily mean that you will be an excellent firefighter, of course, but it does mean that you are eligible to go on to the next phase of the weeding-out process. Moreover, some departments use the written exam score as a part of an overall candidate ranking score, so a high score gets you off to a good start.

Some areas that may be assessed on the written exam are as follows:

- Reading comprehension
- Mechanical reasoning and spatial scanning
- Reasoning and problem solving
- Mathematical calculations and comparisons
- Basic emergency medical care (for those departments requiring emergency medical training only)
- Interpersonal skills/teamwork

Reading comprehension items usually relate to fire service terminology and problems commonly encountered by firefighters; they evaluate your ability to understand fire department data and training manuals. Mechanical reasoning items test your knowledge of the principles of mechanics and physics commonly encountered by firefighters. Reasoning and problem-solving skills test your ability to react and respond in emergency situations. The mathematical computation items determine your ability to perform the kinds of calculations that measure friction loss, engine pressure, nozzle pressure, and so on. Interpersonal skill items test your ability to understand orders and instructions issued on the fire ground and the ability to communicate with other firefighters. If the department handles medical emergencies, a test of basic medical knowledge may be included as well.

Candidates for a job opening will be ranked according to their score on the exam, from the highest to the lowest grade, so a good score is vital to going on to the next phase of the selection process (see Figure 3-1). As with any test, the key to success is preparation.

FIGURE 3-1 Potential candidates taking a written exam. *Courtesy of Jason Loyd.*

It is the rare candidate who can walk in and get a high score without first doing some serious studying. Bookstores carry a wide selection of test preparation materials, including study materials and sample tests. It is also possible to borrow these from a public library.

PHYSICAL ABILITY/AGILITY

Fire departments usually administer an **ability/agility test** to determine the candidate's physical agility and general ability to meet the physical requirements of the job. A widespread test is the Candidate Physical Ability Test (CPAT). This standardized test is accepted by most employers and is being used increasingly in the selection process. The CPAT measures along eight job-specific areas. Other tests can be used rather than the CPAT and typically this involves completing an obstacle course consisting of tasks similar to those a firefighter may encounter in the line of duty. These can include the dry hose drag, charged hose drag, halyard raise, roof walk, attic crawl, a ventilation exercise, a victim removal exercise, ladder removal and carry, a **stair climb** with hose, a **ceiling breach and pull**, a crawling search, and a stair climb with air bottles and hose hoist. Whatever the details of the course, this is a grueling test of strength and stamina. It is normally done wearing full firefighting gear, although some departments have changed to using a weighted vest to simulate the weight of the bunker gear. Scoring methods vary, as the test can be pass/fail (CPAT requirements), based on the candidate's ability to complete all the tasks within a specified time limit, or candidates may be assigned a score based on how long it took them to complete the course, with higher points given for faster times.

To do well, you must get in shape and stay in shape. You must develop a rigorous workout routine that emphasizes muscle groups used in lifting, pushing, pulling, carrying, and other actions related to firefighting work. Your program may include distance running, sprints, rowing, stair climbing with a weight pack, and weightlifting to develop both the upper and lower body. You may be allowed to visit the testing site to familiarize yourself with the testing layout and even to practice with the equipment that will be used in the actual test. Practice is bound to result in a better time.

On the day of the test, arrive early, well nourished, and well rested. Take appropriate clothing such as sweats, good athletic shoes, and gloves, if required. If the testing process will be long, make sure you have water and high-energy snacks on hand. If the day is hot, make sure you drink plenty of water before you start the test. Warm up and stretch just as you would before any strenuous workout (see Figure 3-2).

FIGURE 3-2 Physical ability/agility test. *Courtesy of TEEX-Texas A&M System.*

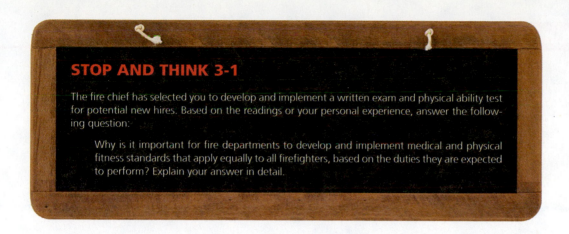

Interviewing Skills

Many individuals who want to go into the fire and emergency services discover that they do not have the skills necessary to go through a job interview with confidence. Although curriculum requirements vary from state to state, rarely is time devoted to interviewing techniques, so it is hardly surprising that many cadets find themselves uncomfortable when facing the interview situation for the first time. Some become so frustrated that they actually give up on a fire/emergency career path completely. Homework and practice can make all the difference between success and failure in the interview arena.

FIRST INTERVIEW OR CHIEF'S INTERVIEW

Prior to getting that first interview, a résumé is often asked for in the applicant process. If you are just entering the fire service and don't have much of a fire background, be sure to write down all your other accomplishments—no matter how small they may seem to you. There are many websites or programs that can help you properly develop a great résumé. It is important to note that you should continually be updating your résumé with current information throughout your career. Never underestimate the importance of first impressions. The candidate's initial interview, which can be held with the chief alone or with a committee of department representatives, is crucial. Typically, if the first interview is held with a panel of committee members, they will narrow the selection to a select few. Then, a second interview may occur, depending on the organization or fire department, with the candidate and the fire chief alone. This interview can actually alter an applicant's position in the rankings. A candidate should learn as much as possible about the department and the chief before the interview. For example, what are some of the core values of the organization? What are some of the key issues or projects with which the department is involved? Although you will be answering many questions, don't be afraid to ask questions as well. Make it clear that you know something about the department and are eager to learn more. The fire service is looking for good communicators. Also, don't be overly modest. Talk about yourself in terms that demonstrate your ability to work as part of a team and your knowledge of firefighting terminology and firehouse etiquette. Show that you care about honesty and integrity.

Another way you can prepare for the interview is to have experienced firefighters give you a mock oral interview. The mock interview should be with people who have been there and who know what to expect. Their questions can give you insights about issues that may be raised during the interview and prepare you to give thoughtful answers. Sometimes writing out a practice answer or rehearsing your delivery in front of a mirror or using an audio or video recorder can be helpful. The more practice you have and the more feedback you get from others, the more confidence you will have during the actual interview.

1. Describe your work history and experience.
2. Describe credentials you hold that qualify you for this job (education, certifications, and so on).
3. What are your major strengths?
4. What do you consider development areas?
5. What characteristics do you feel mark a good supervisor?
6. Describe a conflict you experienced with a previous supervisor and how it was resolved.
7. What do you think is a satisfactory attendance record?
8. How do you define "pressure"? How well do you perform in that type of situation?
9. In what type of work environment do you excel?
10. Describe when you went above and beyond the call of duty to get a job done.

The interview is typically casual in tone, but be sure you dress appropriately—for example, wear a suit and tie (see Figure 3-3). Don't overlook the importance of body language. Sometimes candidates can be so concerned with conveying their knowledge and skills that they forget the nonverbal message their appearance is sending to the interviewer. Seeming at ease and confident can be just as important as giving the right answers. Interviewing tips are listed below:

Posture.   Sit in a comfortable posture, relaxed but alert. Avoid stiffness or nervous, fidgeting movements such as foot tapping. Breathing should be calm and even. These are signals that you are ready to communicate.

Arms.   Sit with your arms uncrossed and your hands relaxed and open. This suggests that you have nothing to hide.

Eyes.   Look into the other person's eyes, particularly when they are speaking. Glancing away indicates that you are nervous or not interested. And don't wrinkle your brow!

Nodding.   Using nods to punctuate key things the other person has said signals agreement, interest, and understanding.

Smile.   A smile signals that you're warm and personable.

Leaning closer.   Reducing the distance between yourself and the interviewer, particularly when the interviewer is speaking, indicates that your interest is up and

FIGURE 3-3 Potential candidate dressed in a suit for a first interview. *Courtesy of TEEX-Texas A&M System.*

your barriers to communication are down. Leaning back or slouching in your chair can signal boredom or a lack of respect.

Gestures. Talking with your hands, particularly with your palms open, indicates involvement in the conversation and openness to the other person.

Don't interrupt. Be sure the interviewer is finished speaking before you answer or make a comment.[3]

Personal History

MEDICAL EXAM

A preemployment medical exam is required for fire and emergency service employment. The National Fire Protection Association (NFPA) has established standards for the exam to ensure that it is an accurate test of a candidate's physical ability to perform firefighting tasks. **NFPA 1582**, *Standard on Comprehensive Occupational Medical Program for Fire Departments*. This standard covers minimum medical requirements for firefighters, including full-time or part-time employees, and paid or unpaid volunteers. This physical is similar to any other comprehensive medical examination. It typically begins with a patient history that includes, but is not necessarily limited to, the following:

- Medical history
- Surgical history
- Hospitalization history
- Social activities including smoking, alcohol use, exercise, stress, and diet
- Drug screening
- Employment history
- Health conditions including head/nose/mouth/throat, ears/hearing, eyes/vision, respiratory system, cardiovascular system, gastrointestinal system, liver/spleen/gallbladder, reproductive system, neurological system, psychological/mood, musculoskeletal system, endocrine/metabolic system, blood/lymph system, cancer, skin, allergies, and infectious/childhood diseases

You will be required to give blood and urine samples. The exam may also include a body fat composition analysis, hearing and vision tests, a treadmill stress test, a physical condition assessment, and any other medical tests that the department may request. The purpose of the exam is to assure that candidates are healthy and to establish a health baseline against which future test results can be compared. It is important to note that medical exams cost the potential department so if an applicant gets to this point it surely means they are moving in the right direction. Typically this is the last phase of the hiring process. Under guidelines of the American Disabilities Act (ADA), the job offer must be made first and the offer is made contingent on passing the medical physical.

NFPA 1582 ■ *Standard on Comprehensive Occupational Medical Program for Fire Departments.* Standard covers minimum medical requirements for firefighters, including full-time or part-time employees, and paid or unpaid volunteers.

BACKGROUND HISTORY

A **background check** is used to verify identity and to examine records for information about an applicant's character and behavioral history. References will be checked, and the information provided on the candidate's application will be confirmed: education, past employment history, and any certifications or licenses. An address history and social security number trace may be run, and a **criminal conviction record check** will be made at the county, state, and federal levels, including the sexual offender registry, county and federal civil records, drivers' history, credit reports, workers' compensation records, and results of previous drug testing.[4] Some departments or organizations require that potential employees submit to a polygraph exam. Be sure to check local and state requirements, however, as they do vary. It is important to note the accuracy of information that an applicant puts down. In many cases, investigators talk to a considerable amount of people about every

background check ■ Verification that a candidate's experience or credentials are what they represent.

criminal conviction record check ■ A search of an individual's history of any convictions under the law for felony, misdemeanor, and motor vehicle convictions.

aspect of your life. Sometimes if inaccurate information is given, it can drop you from the selection process entirely. Checking social media outlets such as Facebook, Twitter, and other outlets sometimes occur as well.

Hiring Conditions

ELIGIBILITY LIST

After the final candidate list is compiled, it will be certified for a specified time frame, such as one year. If a position becomes available during that period, the top candidates including any with tied scores, will be called for final interviews. As more positions become available, the department will move down the list. A candidate should call periodically to check the status of his or her name on the list. The eligibility list is typically valid for six months to one year; however; a candidate should verify the terms of the list. If time expires for the eligibility list, the department will normally schedule testing procedures again, and the whole process starts over.

PROBATIONARY PERIOD

New hires are on probationary status for a period of time that can run between three months and one year. Probation gives supervisors and fellow firefighters a chance to assess the candidate's abilities and character in real-life situations. This trial period should be considered a part of the selection process, just like the interview or the medical exam. Probation is also an opportunity for the new employee to become familiar with standard operating procedures and department guidelines and to take part in ongoing training programs. During the **probationary period**, an individual may be terminated for any reason. The probationary employee has the burden of proof to show that he or she can do the job. Supervisors, however, should keep track of all probationary employees and document their progress.[5]

probationary period ■ Introductory period of employment that allows the employee and agency to determine if the employee is suited for the job.

Marketing Yourself for the Firefighting Job

There is more than one path to a career in the fire and emergency services. Some firefighters start out in a civilian job, such as dispatch, administration, or fire prevention, before making the transition to firefighting. Working as a paramedic, nurse, medical assistant, or in any direct service capacity in an emergency environment are other possible routes to firefighting. A candidate who comes to the fire service already experienced in these fields has a definite edge over a raw recruit. Other avenues of valuable training and experience are listed below.

EMT STATUS

Many fire departments now require a minimum of EMT Basic status as a prerequisite for application. There are varying levels of EMT certification and obviously the higher certification that you are the better marketing abilities you have for yourself. EMT levels of certification will be discussed in a later chapter. Depending on a department's needs, paramedic status or even police officer certification can further improve your chances of getting a job.

FIREFIGHTER I, OTHER CERTIFICATIONS, AND HIGHER EDUCATION

Some departments that do not conduct their own cadet training may require that the candidate be certified to the Firefighter I level before they will even accept his or her application. Other departments may require Firefighter II and/or EMT or paramedic, so be sure to check local and state requirements, as they do vary.

Other certifications that improve the chances of employment include hazardous materials technician, investigator, and prevention specialist. Training and higher education will be discussed in depth in Chapter 4 but it is important to mention that in the selection process it is paramount to stress all higher education endeavors whether it is certifications or associate's degrees, bachelor's degrees, or higher.

VOLUNTEER EXPERIENCE

Volunteering is a great way to expand your horizons. In addition to the experience you gain, you will be demonstrating community spirit and a readiness to help others. Joining a volunteer fire department is an obvious move in the right direction, but you may also contact your local fire department and ask if you could volunteer in other areas, such as in administration or in a fire prevention or HazMat program. In fact, almost any kind of volunteer experience will enhance your résumé, showing that you are well rounded and service oriented. Using volunteer experience—especially if it is fire related—is a useful stepping-stone to future employment with career departments. As noted in Chapter 1, about 87% of today's fire departments are totally or mostly volunteer, so it fitting for the applicant to stress fire-related experiences in the interview process. But whether community related or fire related, these experiences can be great marketing tools.

BILINGUAL ABILITY

In our increasingly diverse society, language skills are becoming valuable in many occupations, and the fire service is no exception. Some departments give bilingual candidates preference in the hiring process.

CLEAN DRIVING RECORD

A clean driving record means no accidents and no moving violations. This is important for safety reasons, because at some point you may be driving fire apparatus. Also, if you have too many accidents and moving violations on your record, you may not be eligible for coverage by the department's insurance. This can eliminate you from the hiring process. An applicant should be mindful of any other behavioral issues that might be deemed undesirable or need explanation such as past criminal activities, which include minor offenses in possession, driving under the influence, and recreational drug use. Additional offenses that could be harmful to an applicant are

- Any alcohol-related driving offense
- Hit and Run, Failure to Stop, Failure to Render Aid, or Failure to Disclose I.D. at the scene of an accident
- Negligent Manslaughter involving a motor vehicle accident
- Three or more moving violations
- Two moving violations and one or more chargeable vehicle accidents

EXCELLENT PHYSICAL FITNESS

If you don't have a physical fitness routine now, you need to get one! Pair up with a buddy or two and develop a regular training schedule. Obtain guidance from a professional on what to do and how to do it to obtain the maximum results. Weightlifting alone won't be enough. You need a combination of aerobic, cardiovascular, and strength-building activities. Also, a well-balanced diet will enhance and speed up the results of your fitness training. Physical training is critical preparation for the academy.[6]

DIVERSITY

Many fire departments are increasing the **diversity** of their applicant pool through targeted recruitment of minorities and women. To be sensitive to a community's needs, the department needs a workforce reflective of that community. This means that the ideal fire

diversity ■ Ethnic, gender, racial, and socio-economic variety in a situation, institution, or group.

department should include both women and men from the various racial, ethnic, and socioeconomic groups that the department serves.[7] Women and minority applicants should be proud of what they have to offer and should not hesitate to point out that their cultural or ethnic background will be an asset to the department, breaking down barriers that would otherwise isolate the department from sectors of the community.

SOCIAL MEDIA

In today's fast-paced technology world of Facebook, LinkedIn, and Twitter, social media has added another avenue to the hiring process. Using social media for hiring purposes is controversial. If a potential employer looks at your Facebook or other social media account, his or her hiring decision can only be based on job-specific information—information about race, age, gender, religion, or disability cannot be used, as that could be considered a form of discrimination. On the other hand, an applicant should be aware that potential employers could peruse these types of accounts and should be mindful of status updates, foul language, inappropiate pictures, excessive drinking, and other acts that could be construed as undesirable in an employee. It is to the applicants' benefit to have their social media accounts maximize their attractiveness and get rid of anything that could put them in a bad spotlight.

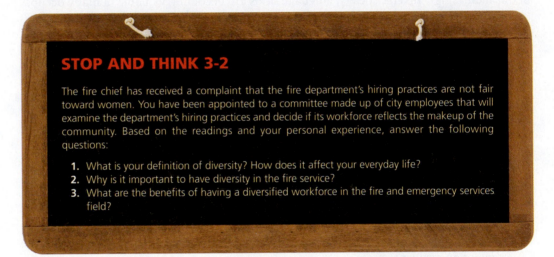

STOP AND THINK 3-2

The fire chief has received a complaint that the fire department's hiring practices are not fair toward women. You have been appointed to a committee made up of city employees that will examine the department's hiring practices and decide if its workforce reflects the makeup of the community. Based on the readings and your personal experience, answer the following questions:

1. What is your definition of diversity? How does it affect your everyday life?
2. Why is it important to have diversity in the fire service?
3. What are the benefits of having a diversified workforce in the fire and emergency services field?

ON SCENE

Although there will be an increase in hiring of personnel in fire and emergency services over the next decade, competition will be keen. The fire service attracts all types of applicants from candidates with a GED to individuals in the white-collar professions such as accountants and attorneys with graduate degrees.

1. Why, do you think, does the fire service attract so many different types of individuals?
2. Because the fire service attracts various individuals, what can you do to make yourself more marketable?

Summary

The selection process for fire and emergency personnel is not easy and takes time. Given the critical nature of the job, the responsibility to the public, the dangers and rigors involved, the need for specialized knowledge, the willingness to be ready to face any emergency, to work irregular and long hours as part of a team, and to be an articulate communicator, it is not surprising that the process is so demanding. The department needs to be certain that each new hire has the necessary skills and qualifications to do the job and, very importantly, that he or she has the character traits that make for a good community role model.

Although the process is challenging, the various stages of the process are fairly predictable. This chapter explained what a candidate should expect from the application process and suggested ways to improve performance at each stage. In addition, various ways in which the prospective firefighter can strengthen his or her résumé were listed, particularly through related training, certifications, and volunteer experience.[8]

According to the *U.S Department of Labor Statistics Handbook for 2012–13*, employment of firefighters is expected to grow by 9%, slightly slower than the average occupation.[9] Firefighting is a highly respected, rewarding profession with a secure future. One of the keys to success, however, is to understand the challenges and to prepare for them. The road to that full-time paid job may take you months or even years. It will involve hard work—research, study, and physical exercise—and personal sacrifice.

Review Questions

1. What two pieces of documentation do most departments require of applicants before they may apply for the firefighter exam?
2. What is generally the first step in the selection process?
3. List two ways to find out when application periods are open for firefighter exams.
4. List a resource for study material used to prepare for the written firefighter exam.
5. List two ways to prepare for the physical agility and ability test.
6. List two ways to prepare for the oral exam.
7. What is the purpose of the probationary period?

Suggested Reading

Barr, Paul. (2010). *How to become a firefighter or get hired as a firefighter: The application, requirements, and equipment training for your exams, interview, and physical test*. United Kingdom: Psylon Press.

Bruegman, Randy R. (2009). *Fire administration I*. Upper Saddle River, NJ: Pearson Education.

Chikerotis, Steve. (2006). *Firefighters from the heart: True stories and lessons learned*. Clifton Park, NY: Thomson Delmar Learning.

Lasky, Rick. (2006). *Pride and ownership: A firefighter's love of the job*. Tulsa, OK: PennWell Corporation.

Mahoney, Gene. (2007). *Fire department interview tactics*. Clifton Park, NY: Thomson Delmar Learning.

Pearn, Kory. (2012). *The complete guide to becoming a firefighter*. Markham, Ontario, Canada: Fitzhenry & Whiteside Publishing.

Endnotes

1. Prziborowski, Steve. (2006, August 15). *So, you want to become a firefighter? 5 guidelines to assist you in becoming a firefighter*. Hayward, CA: Chabot College.
2. Ibid.
3. Alyn, Kim. (2009, November 18). *Tips for successful firefighting panel interviews*. Retrieved March 30, 2013, from Firerecruit.com website: http://www.firerecruit.com/articles/604979-Tips-for-Successful-Firefighting-Panel-Interviews.

4. Net Detective. (2006, May 16). *Types of background check: You can find out the truth about anyone!* Retrieved January 29, 2007, from the Net Detective website: http://www.softsite.org/background-check-online.

5. Graham, Gordon J. (2004, January 15). *Why things go right—why things go wrong: Risk management considerations in the fire service operations.* Texas Association of Fire Educators Conference, TX.

6. Alyn, op cit.

7. Crawford, Brian. (2004). *Patchwork force.* Retrieved May 12, 2008, from http://firechief.com/staffinghuman-relations/patchwork-force.

8. Bureau of Labor Statistics, U.S. Department of Labor. (2012). Firefighters. In *Occupational outlook handbook, 2012–13 edition.* Retrieved January 12, 2013, from Bureau of Labor Statistics website: http://www.bls.gov/ooh/protective-service/firefighters.htm.

9. Ibid.

4
Training and Higher Education

KEY TERMS

andragogy, *p. 56*

Associate of Applied Science, *p. 55*

baccalaureate programs, *p. 56*

burn building, *p. 49*

certification programs, *p. 55*

drafting pit, *p. 49*

drill tower, *p. 49*

higher education, *p. 48*

International Fire Service Accreditation Congress (IFSAC), *p. 56*

manipulative training, *p. 49*

National Fire Academy (NFA), *p. 50*

NFPA 1041, *p. 54*

NFPA 1403, *p. 49*

Pro Board, *p. 60*

tactical training, *p. 49*

training, *p. 48*

training bureau, *p. 54*

OBJECTIVES

After reading this chapter, you should be able to:

- Discuss National Fire Protection Association standards geared toward training.
- Explain the difference between training and higher education.
- Describe the various types of certification and degree programs available.
- Explain FESHE's professional development model.

Introduction

In this chapter, we will explore firefighter education, including both training and higher education. We will additionally explore the training options that are available to prospective firefighters, as well as the opportunities for higher education that are available to fire professionals. We will also describe certification programs, two-year associate's degree programs, and higher degree programs. We will describe the roles of the National Fire

Academy, International Fire Service Accreditation Congress, and Pro Board in improving and standardizing fire service education in the United States.

It is critical to understand that a firefighter's education is open-ended, a lifelong, or at least career-long, process. Skills once learned must be refreshed, and then with the advent of new technology, new skills need to be acquired. The knowledge that was sufficient to a firefighter 10 years ago is not sufficient for today's challenges.

training ■ Obtaining knowledge for the maintenance of skills required for actual tactical tasks performed during an emergency response.

Firefighting is physically and mentally challenging work and, of course, highly dangerous. By **training**, we refer to the knowledge and skills that enable the firefighter to do this vital work with efficiency and safety. Some of the knowledge modern-day firefighters need may surprise you. For example, an understanding of building construction is essential because how a building is designed and the materials used in its construction influence how a fire will behave. The more firefighters know about these factors, the better they will be able to anticipate a fire's location and its movement.

As a firefighter's career advances, he or she will want to take advantage of opportunities to earn specialty certifications in areas such as hazardous materials and fire investigation. We will look at the requirements of some of these certifications programs.

Higher Education

higher education ■ Obtaining knowledge for certifications, professional development, or college credit. The act of accumulating new facts to augment your personal and mental database.

By **higher education**, we refer to advanced learning at the college, university, or equivalent level. The terrorist attacks of September 11, 2001, taught us how important it is to have a well-educated fire service workforce—one that is always prepared. This horrific event took away many mentors and teachers of the fire service. The talent that was lost that day has encouraged members of the fire service and communities to support education and preparedness for the worst-case scenario occurrences. This is true at every level of firefighting, but it is particularly critical within the ranks of management. We will consider the growing value of associate's, bachelor's, and even advanced degrees for firefighters seeking promotion to higher or more specialized positions within their organizations.

Despite the importance of training and education, agencies and fire departments vary in their support of education and training for their members. Often the amount and kind of training they encourage are based on the department's mission and financial capabilities. Some departments protect only airports, whereas others do structural firefighting and provide emergency medical services. Understandably, training and education generally must conform to the goals and mission of the organization.

As a firefighter, it will be your personal responsibility to identify your career goals and to research opportunities for advanced training and education. These opportunities may be available from your employer, or they may involve on-site or online learning offered by an educational institution or by the National Fire Academy. The more you learn, the better prepared you will be to succeed and further your career in the fire science field. One program, like many others, is the National Fire Academy sponsored Volunteer Incentive Program (VIP). This program offers compressed course work in six days for volunteers. The program is held in Maryland but is free and has college accreditation available. There are many other courses offered, many of which are free or online, so a student interested should research all avenues.

Training

Firefighter training is the single most important factor in determining if firefighters are prepared to meet the challenges of the environments in which they work. Training means the difference between success and failure. In fact, in the fire service and emergency service professions, training can be synonymous with safety.

FIGURE 4-1 There are a variety of burn buildings such as this 3-story burn facility. *Courtesy of Jason Loyd.*

We can distinguish two basic types of training: tactical and manipulative. **Tactical training** is another word for know-how. The goal of tactical training is to teach what works and what doesn't in responding to different emergencies. This is the mental part of firefighting. **Manipulative training** is hands-on practice. In manipulative training, skills are learned by actually operating firefighting equipment and tools and carrying out rescue procedures. The more often firefighters practice a procedure, the more effective and confident they will be in a real emergency situation. Both types of training are designed to teach firefighters to know the capabilities and limitations of their equipment. Just as important, training teaches them to recognize their own capabilities and limitations.

Formal training takes place either in a traditional, instructor-led classroom or in field exercises. Most fire departments have classrooms or at least a designated space within the department for training. Field exercises require a specially designed setting with equipment and props that simulate fire conditions. These could include a **drill tower** for ladder training or multistory operations, a **burn building** for actual live fire training (see Figure 4-1), and a **drafting pit** for maintenance of pump-operation and pumper-testing skills. A drafting pit is usually a large square concrete hole in the ground that may hold 50,000 to 70,000 gallons of water from which a fire apparatus can draft water. These structures can be expensive to build and maintain, so a small fire department may partner with larger departments to gain access to their facilities. It is paramount that fire instructors know, understand, and not deviate from the **NFPA 1403**, *Standard on Live Fire Training Evolutions*. This standard provides the framework to improve safety during live fire evolutions and exercises. NFPA 1403 covers conditions such as safety, site preparation, water supply, training plan, fuel, and ventilation. For further information referring to this standard, please refer to the National Fire Codes under NFPA 1403.

As in any field of study, informal training also has a role to play in preparing firefighters for their jobs. Mentoring and peer networking are invaluable. Beginning firefighters should take every opportunity to discuss their career goals and concerns with senior professionals. Training may even be obtained before entering the fire service. Preservice learning can begin by joining the Fire Department Explorers, a scouting program designed for teens interested in firefighting, or by signing up for reserve or cadet programs open to young adults. Joining a volunteer fire department can also be the first step to a professional career.

tactical training ■ Comprehensive training, hands-on learning, and analysis for emergency incidents.

manipulative training ■ Hands-on learning of operations, equipment, and tools. In the fire service, this type of learning exhibits safety and proper techniques.

drill tower ■ A tower and training facility for firefighters.

burn building ■ Training tower used so that new recruits can practice their skills.

drafting pit ■ A poured-in-place concrete pit used to simulate drafting operations.

NFPA 1403 ■ A standard that provides the framework to improve safety during live fire evolutions and covers conditions such as safety, site preparation, water supply, training plan, fuel, and ventilation.

For most new firefighters, however, the initiation to bona fide training is the fire academy or rookie school. This is only the first stage, however. After a firefighter completes academy training and enters the fire service, the need for continuing training becomes obvious. New technology generates new hazards, such as chemicals and other hazardous materials. Technology also provides new and improved methods and equipment for dealing with these hazards. Firefighters must constantly update their knowledge and skills. They will spend a good proportion of their time attending in-service programs run on-site at their place of employment or at state or national training facilities. Training is not just an accessory to the firefighter's job; it is an integral part of that job.

FIRE DEPARTMENT TRAINING OR SPECIAL OPERATIONS BUREAUS

Most larger fire departments have training divisions or special operations bureaus whose mission is to provide the highest quality training to their firefighters. These divisions are in charge of developing and teaching the department's educational initiatives, utilizing both on-site classroom training and any needed field exercises. The bureaus also provide training that is periodically mandated by the state.

A secondary but very important responsibility of the training or special operations bureau is to archive the department's training records. These provide proof of the education given to each firefighter. Training records are essential for compliance with the law in most states and must be as complete as possible. Complete records are important when the department is being evaluated, audited by an outside agency, or investigated as a result of a workplace accident, harassment issue, or patient care situation. Recording training records have changed throughout time—the days of hand writing training records is gone. With iPads and computers, training records have become computer based. This allows the training officer to use specific programs built around its needs. In today's litigious culture, training records become a vital safeguard for the firefighter and the department (see Figure 4-2).

THE U.S. FIRE ADMINISTRATION AND THE NATIONAL FIRE ACADEMY

In the early 1970s, the federal government recognized that the nation's fire problem was changing and appointed a commission to study the situation. The commission published a document titled *America Burning: The Report of the National Commission on Fire Prevention and Control* (1973), which recommended the formation of a federal fire agency that would provide support, including training and educational support, to state and local governments and to private fire organizations. A year later, Congress passed Public Law 93-498, The Federal Fire Prevention and Control Act of 1974. This law created the U.S. Fire Administration (USFA), which today operates under the direction of the Federal Emergency Management Agency (FEMA). The branch of the USFA that is concerned with education—designing fire control courses and programs and making them available to firefighters across the nation—is the **National Fire Academy (NFA)**.

National Fire Academy (NFA) ■ Concerned with education—designing fire control courses and programs and making them available to firefighters across the nation.

The NFA maintains a campus in Emmitsburg, Maryland, where it offers education and training courses aimed mainly at management-level fire personnel and instructors. It also provides classroom courses for volunteer and career firefighters that are held at locations across the country, usually in conjunction with state or local fire services. Online courses and a variety of audio and video training materials are available as well. It is estimated that the NFA has helped to train over 1,400,000 students since 1975. This organization has benefited not only the students but also the American public. Countless lives have been saved and property losses prevented as a direct result of this training and education.[1]

One important program to note is the U.S Fire Administration Executive Fire Officer Program (EFOP). This program is designed to teach senior authority figures in the fire service leadership techniques on how to deal with difficult adaptive problems within their jurisdiction. The EFOP spans a four-year period. It consists of four core courses and

FIGURE 4-2 Training forms. *Courtesy of Jason Loyd.*

College Station Fire Department Training Roster

For Office Use Only

Date Entered: _____
Entered By: _____
Paperwork Filed: Y / N
Class #: _____
Approval #: _____

Class Name: _____

Class Location: _____

Start Date: _____   Start Time: _____   Number of Days: _____

End Date: _____   End Time: _____   CE Hours: _____

In House: Yes / No   Shift: A / B / C   Officer-in-Charge: _____

Class Host: _____   Phone: _____   Cost: _____

Lead Instructor: _____   Instruction Hours: _____   # of Students: _____

2nd Instructor: _____   Instruction Hours: _____   Contact Hrs: _____

3rd Instructor: _____   Instruction Hours: _____

Student Name: (Print)	Initials	Employee ID #	CE Hrs	Test Score
1.				P / F / NA
2.				P / F / NA
3.				P / F / NA
4.				P / F / NA
5.				P / F / NA
6.				P / F / NA
7.				P / F / NA
8.				P / F / NA
9.				P / F / NA
10.				P / F / NA
11.				P / F / NA
12.				P / F / NA
13.				P / F / NA
14.				P / F / NA
15.				P / F / NA

Instructor Signature: _____

Note: All training rosters require a signature of an approved instructor or an attached copy of a training certificate

White / Original to Training   Yellow / Page 2 to Instructor   Pink / Page 3 to Company Officer

requires an applied research project for each. There are rank requirements and higher education requirements for selection into this program.

Advancement and Promotion

Advancement in most fire agencies depends on successfully completing competitive examinations or by earning advanced degrees. A firefighter seeking a promotion is often required to participate in continuing education programs that lead to **certification** or in higher education programs that lead to college degrees.

CERTIFICATION

Fire personnel may obtain specific certifications by successfully passing a written and practical state certification test in the specialty discipline. Certification tests for competencies are provided by the bureau or by a professional association or institution that acts as the host entity. Candidates must meet minimum scoring requirements on the written exam, and practical skills are typically graded on a pass/fail basis.

TABLE 4-1	Specialty Certifications
LEVEL OF CERTIFICATION	**BASIC REQUIREMENTS**
Firefighter I (NFPA 1001)	Obtain a general knowledge of fire department organization, operations, tools and equipment, the role of the firefighter, hazardous materials awareness and operations, and the mission of the fire service.
Firefighter II (NFPA 1001)	Meet the requirements of Firefighter I. Obtain a more advanced knowledge of fire ground operations and fire suppression, and rescue techniques.
Fire Instructor I (NFPA 1041)	Learn basic techniques and methods of teaching, use of teaching aids, and proper presentation skills.
Fire Instructor II (NFPA 1041)	Meet all requirements for Fire Instructor I. Emphasis is on curriculum development and design.
Fire Investigator (NFPA 1033)	Learn basic techniques and skills to determine origin, cause, and development of a fire.
Public Fire and Life Safety Educator I (NFPA 1035)	Obtain basic knowledge and skills in a realm of different areas such as fire prevention education, babysitter training, and CPR.
Marine Firefighting for Land-Based Firefighters—Awareness Level (NFPA 1005)	Meet the requirements of Firefighter I and II. Obtain knowledge and skills that structural firefighters may encounter during marine fire suppression in a maritime environment.
Fire Officer I (NFPA 1021)	Obtain basic knowledge and skills that demonstrate the ability to perform tasks at the supervisory level.
Fire Officer II (NFPA 1021)	Meet all requirements for Fire Officer I. Obtain advanced knowledge and skills that demonstrate the ability to perform tasks at the supervisory/managerial level.
Inspector I (NFPA 1031)	Obtain basic knowledge and skills in fire prevention, fire safety, plans review, hazardous materials storage and handling, fire extinguishing systems, building construction, building codes, and code compliance.
Hazardous Materials Technician (NFPA 472)	Meet all requirements for Hazardous Materials Operations Level. Become trained in stopping the release or potential release of a hazardous material.
Airport Firefighter (NFPA 1003)	Meet the requirements of Firefighter I and II. Obtain a more advanced knowledge of fire ground operations and fire suppression, hazardous materials operations, and rescue techniques. In addition, meet the basic knowledge and skills of airport firefighter.
Fire Department Incident Safety Officer (NFPA 1521)	Obtain basic knowledge and skills that demonstrate the ability to recognize and handle safety responsibilities and duties.
Driver/Operator-Pumper (NFPA 1002)	Meet the requirements of Firefighter I. Learn general principles of pump operations and obtain the knowledge and skills needed to be a safe fire apparatus driver/operator-pumper.

Although it is true that a college degree does not put out a fire, the increase in skills and training required to meet today's operational needs in a fire department successfully has made formal education necessary. Entry-level firefighters who are equipped with a higher education degree may receive faster promotions to management positions and to specialized roles such as fire inspector, fire investigator, or even fire chief.

Whatever their educational background, firefighters must be articulate, educated, skilled, and credentialed to earn the credibility required by the public safety and public service communities. These requirements are essential for administrators but are also important for shift commanders, company officers, and firefighters. Individuals holding these positions are the most visible in the fire service because they develop relationships with other agencies and with the public.[2]

SPECIALTY CERTIFICATIONS

Tables 4-1 and 4-2 list NFPA certifications and the requirements for each certification. The certifications in Table 4-1 reflect levels of general professional preparedness and preparedness for special job designations, such as instructor or investigator. The certifications in Table 4-2 refer to high-level technical specializations.

Note: For the most current and up-to-date information on the standards for each of these certifications, please refer to the NFPA. It is a good idea to subscribe to NFPA (website: www.nfpa.org) to have the most current information.

TABLE 4-2	Other Firefighting Levels
LEVEL OF CERTIFICATION	**BASIC REQUIREMENTS**
Rope Rescue Technician (NFPA 1006)	Meet the requirements of fire department medical first responder. Learn general knowledge and skills pertaining to technical rescue and rope rescue.
Surface Water Rescue Technician (NFPA 1006)	Meet the requirements of rope rescue technician. Learn general knowledge and skills pertaining to swiftwater/flood rescue.
Vehicle and Machinery Rescue Technician (NFPA 1006)	Meet the requirements of basic technical rescue. Learn general knowledge and skills pertaining to vehicle rescue.
Confined Space Rescue Technician (NFPA 1006)	Meet the requirements of rope rescue technician and hazardous materials operations. Learn general knowledge and skills pertaining to confined space rescue.
Structural Collapse Rescue Technician (NFPA 1006)	Meet the requirements of basic technical rescue, rope rescue technician, confined space rescue technician, and trench rescue technician. Learn general knowledge and skills pertaining to structural collapse rescue.
Trench Rescue Technician (NFPA 1006)	Meet the requirements of basic rope rescue and trench rescue operations. Learn general knowledge and skills pertaining to trench rescue.
Dive Rescue Technician (NFPA 1006)	Meet the requirement of open-water certification and basic technical rescue. Learn general knowledge and skills pertaining to rapid deployment search and rescue/recovery.
Wilderness Rescue Technician (NFPA 1006)	Learn general knowledge and skills pertaining to wilderness rescue.

TRAINING INSTRUCTOR

NFPA 1041 ▪ A standard that requires instructors to be cognizant of the safety of their students to ensure that classroom and practical evolutions are conducted in a safe, controlled manner. Prerequisites and requirements are mandated.

The NFPA has also established codes or standards that outline the responsibilities of the training officer or instructor. **NFPA 1041**, *Standard for Fire Service Instructor Professional Qualifications*, identifies three distinct levels of responsibility for instructors:

▪ Instructor Level I. Must assemble necessary course equipment and supplies, review prepared instructional materials, and deliver instructional sessions based on the prepared materials.
▪ Instructor Level II. Expected to be able to develop course materials independently and deliver those materials to students.
▪ Instructor Level III. Applies primarily to administrators and requires personnel to develop agency policy for training programs.

At all levels, NFPA 1041 requires that instructors are cognizant of the safety of their students. Safety is stressed to ensure that classroom and practical evolutions are conducted in a safe, controlled manner (see NFPA 1403).[3]

TRAINING OFFICER

training bureau ▪ A division of an organization or department whose mission is to provide training to firefighters so that they have the knowledge, skills, and abilities to mitigate emergency incidents safely while minimizing the risks to themselves, civilians, and the environment.

The training officer occupies a midmanagement staff position, normally reporting directly to the fire chief. He or she is responsible for developing and coordinating all training for the fire department, including training in fire suppression, fire prevention, hazardous materials, rescue, and other associated areas. This position may require performing various command functions as well as all the regular duties of a firefighter. It is the responsibility of the **training bureau** or training officer to determine adequate levels of training based on several criteria—for example, whether the task is mandated by law, how often the task is performed, the time needed to complete the task, whether any incidents occurred while performing this task, and safety to others. These tasks would be considered job performance requirements (JPRs).

Post-Secondary Education

As stated previously, the world changed on September 11, 2001. The terrorist attacks on that day affected all American citizens, but in no profession was the impact greater than in the U.S. fire service. New responsibilities and expectations were placed on the fire service, and along with these came the requirement for better educated professionals to meet these responsibilities. A common system of communication and sharing of knowledge was an important discovery in the emergency services field.

The increasing importance of advanced tactical and manipulative training for the career firefighter has been emphasized. In the past, vocational training, continuing education courses, and in-service training sessions fulfilled most of these educational needs. Today there is an ever-increasing need for higher education, a need that goes hand in hand with the maturing of the fire service industry from the status of a skilled technical job to a full-fledged profession. The rewards of higher education come in many forms. For the public, the reward is improved safety. For the firefighter, the rewards include career advancement, promotions, pay incentives, and personal satisfaction. For EMS personnel, higher education is a critical part of their career advancement and ever-changing duties as well. A more in-depth discussion pertaining to higher education in the EMS field will be discussed in a later chapter.

Undergraduate college degrees at the associate's or bachelor's level are recommended for entry-level firefighters and for career firefighters interested in promotion. Advanced degree programs, such as master's and even doctoral degrees, will assist fire professionals interested in further advancement.

A recent survey of metropolitan fire departments shows that approximately 75% of fire service professionals currently possess some type of two-year degree. Fewer than 25% of the same group has a four-year degree. This percentage is certain to rise, and it may rise quite rapidly as the need for higher education becomes increasingly clear and as fire departments translate that need into policy. The Charlotte (NC) Fire Department, for instance, stipulated that beginning in 2005, a four-year degree would be mandatory for advancement to a chief officer position.[4] This kind of proactive strategy will ensure that emerging fire service leaders are well equipped to handle the multitude of ever-changing and increasingly challenging problems that the future will bring. As of 2013, this was a common practice for advancement to a chief officer position throughout the United States.

THE FIRE SCIENCE TECHNOLOGY CURRICULUM

A firefighter can pursue numerous avenues when it comes to obtaining a degree related to fire service. Some colleges offer undergraduate **certification programs** in addition to credit-based degree programs. Certification programs generally focus on expanding students' understanding of one area of fire service work, such as firefighting, fire inspection, or fire investigation. Courses universal to these certification programs include fire prevention, fire chemistry, and firefighting strategies and tactics.

certification programs ■ Shorter educational programs that provide a certificate verifying expertise in a specific trade or discipline.

Most colleges allow students to transfer credits earned in a certification program to a degree program from state to state. Students may earn a certificate without completing all of the requirements of the degree program, but if they later decide to work toward an associate's degree, their accrued credits will count toward that degree. For example, a college may offer a Basic Fire Fighter certificate program in which a student completes fire academy and emergency medical technician training. If he or she decides to pursue an associate's degree in fire science, the college credit hours of the certificate program may be transferable toward that degree. This flexible approach encourages firefighters to expand their knowledge base and to obtain the higher education that will help them to advance their careers.

Courses required for a two-year **Associate of Applied Science** degree in fire science focus on providing the student with a theoretical understanding of fire science—how fire behaves in different environments—as well as technical and practical training in such areas as rescue procedures and the basics of fire extinguishment. Hands-on instruction and internships are normally part of the required curriculum.

Associate of Applied Science ■ Degree program usually two years in length.

The following is a list of standard fire science courses that may be offered for an associate's program. This list of courses is not all-inclusive and may be different from state to state.

- Fundamentals of Fire Protection. An introductory course on the philosophy, history, and fundamentals of public and private fire protection.
- Public Education Programs. Covers firefighter and fire officer awareness as well as how to implement fire and public safety programs in an effort to reduce the loss of life.

- **Fire Administration I and II.** Provide an introduction to the organization and management of fire departments and then cover fire service management in greater depth, considering such issues as budgetary requirements, the organization of divisions within the fire service, and relationships among the fire service and outside agencies.
- **Firefighting Strategies and Tactics I and II.** Analyze the nature of fire problems and the strategies and tactics that are effective in dealing with them. Included is an in-depth study of the efficient, effective use of staffing, equipment, and incident command systems to mitigate emergency situations.
- **Legal Aspects of Fire Protection.** Covers the legal rights and responsibilities of firefighters and fire protection agencies and addresses liability concerns.
- **Fire and Arson Investigation I and II.** An in-depth study of basic fire and arson investigation practices. Include instruction on preparing reports, courtroom demeanor, and the role of expert witnesses.
- **Fire Prevention Codes and Inspections.** A study of national, state, and local building and fire prevention codes with an emphasis on fire prevention inspections, practices, and procedures.
- **Firefighter Health and Safety.** Covers occupational safety and health in emergency and nonemergency situations, as well as the role firefighters play in risk management and in preventing accidents.
- **Building Codes and Construction.** Examines national, state, and local building codes and requirements, construction types, and building materials.
- **Fire Chemistry I and II.** Explores the chemical nature and properties of inorganic compounds and carbon compounds, particularly as they affect firefighting. Primary emphasis is on the hydrocarbons and their application in various industrial processes.
- **Hazardous Materials I and II.** Studies the chemical characteristics and behavior of various materials, with in-depth coverage of mitigation practices and techniques to control hazardous materials spills and leaks effectively.
- **Wildland Fire Control.** Covers wildland firefighting strategies and tactics, including the different types of wildland firefighting.
- **Internships.** Real-world working experiences for advanced students in a specialized field. Based on a written agreement between the educational institution and a business or industry. Mentored and supervised by a workplace employee, the student achieves objectives that are defined and documented by the college and that are directly related to specific occupational outcomes.

Several four-year **baccalaureate programs** and a handful of master's programs in the United States offer degrees related to fire and emergency management. Accreditation of these programs by the **International Fire Service Accreditation Congress (IFSAC)** aids in promoting sound curriculum and degrees that are widely respected. One of the best-known, fully accredited institutions is Oklahoma State University, which offers a bachelor of science degree in fire protection and safety technology, a master's degree in political science in fire and emergency services administration, and now a doctoral in fire and emergency management administration. There are many other IFSAC-accredited bachelor programs such as Southern Illinois University—Carbondale, Eastern Kentucky University, and Utah Valley University, to name a few. The NFA Higher Education website has a comprehensive list of institutions offering associate's and higher degrees.[5]

DISTANCE AND ONLINE EDUCATION

The fire service needs to be proactive in utilizing new adult teaching methodology, called **andragogy**. Most adults are interested in study that has immediate relevance to their job or personal life. They prefer courses that are problem centered rather than content oriented. Distance education, particularly, is suited for andragogical methodology.[6]

baccalaureate programs ▪ Degree programs usually four years in length. Students receive a BS or BA degree.

IFSAC ▪ Focuses on ensuring that training and certification within its member jurisdictions meet strict National Fire Protection Association Professional Qualification standards. IFSAC is divided into two assemblies: the Certificate Assembly and the Degree Assembly.

andragogy ▪ The term relevant to lifelong learning for adult education methodology. Distance education is particularly suited for andragogical concepts.

STATE UNIVERSITY OF NEW YORK
EMPIRE STATE COLLEGE

ON THE RECOMMENDATION OF THE FACULTY
AND BY VIRTUE OF THE AUTHORITY VESTED IN THEM
THE TRUSTEES OF THE UNIVERSITY HAVE CONFERRED ON

JASON B. LOYD

THE DEGREE OF

BACHELOR OF SCIENCE

AND HAVE GRANTED THIS DIPLOMA AS EVIDENCE THEREOF
GIVEN IN THE CITY OF SARATOGA SPRINGS IN THE STATE OF NEW YORK
THIS MONTH OF SEPTEMBER TWO THOUSAND AND ONE

Thomas F. Egan
Chairman of the Board of Trustees

George C. Sinnott
Chairman of the College Council

Chancellor
of the State University of New York

Joseph B. Moore
President of the College

An excellent source for higher education in fire and emergency services is the NFA's Degrees at a Distance Program (DDP), which is an independent-study degree program. The NFA has agreements with seven accredited colleges and universities throughout the country to offer bachelor's degrees with concentrations in fire administration/management and fire prevention technology. The DDP provides a way for fire service personnel to earn a bachelor's degree or to pursue college-level learning without having to attend on-campus classes. In recent years, independent study and distance learning have become popular with working adults nationwide, and the DDP is particularly practical for fire department personnel whose work shifts normally make classroom attendance almost impossible. DDP institutions emphasize faculty–student interaction through written, electronic means (e-mail), and telephone contact. Students receive detailed guidance and feedback on the required assignments and take proctored final exams at hometown locations.[7] Distance education courses for college credit are becoming the norm for firefighters who want to advance through the ranks and move up at a faster rate. Online education (Internet) courses are the wave of the future. Their validity was at first disputed, but in today's technological society, they are fast becoming accepted as a practical and effective teaching tool (see Figure 4-3).

Setting Standards for Higher Education in The Fire Service

The NFA's mission is to support professional development for the fire and emergency response community and its allied professionals. In April 1999, the Academy sponsored its first conference on Fire and Emergency Services Higher Education (FESHE).

Throughout the conference, many participants voiced the same concern—they did not have adequate information on the current status of fire service degree programs in the United States. What types and levels of degree programs were being offered? How was distance education being used in these programs? Should a model curriculum be developed?[8] Since then, an annual conference has been held to answer these questions and react to the growing needs of higher education in the fire service.

This annual conference (FESHE) and the International Fire Service Accreditation Congress (IFSAC) have taken on the task of establishing a national professional development model for fire service professionals and of maintaining uniform high standards for their training and education.

STOP AND THINK 4-2

Why is it important for the fire service to have a national professional development model? Based on the readings and your experience, answer the following questions:

1. Which fire science technology courses listed could enhance or benefit your career?
2. How has the ability to obtain online fire education changed the face of education for firefighters?

FESHE GUIDELINES

A result of the 2000 FESHE conference was the development of a model curriculum for an associate's degree in fire science. FESHE attendees identified six core associate-level courses in the model curriculum. In February 2011, the model curriculum was reviewed and revised, including the following:

- Building Construction for Fire Prevention. Explores the components of building construction that relate to fire safety. The focus is on firefighter safety and on learning which elements of construction and design are key factors to consider when inspecting buildings, preplanning fire operations, and operating at emergencies.
- Fire Behavior and Combustion. Teaches the fundamental principles behind fire behavior: how and why fires start and spread, and how they can best be controlled.
- Fire Prevention Codes and Inspections. Teaches the fundamentals of history and philosophy of fire prevention, organization, and operation of a fire prevention bureau, use of fire codes, identification and correction of fire hazards, and the relationships of fire prevention with built-in fire protection systems and fire investigation.
- Principles of Fire and Emergency Services Safety and Survival. Introduces the basic principles and history related to the national firefighter life safety initiatives, focusing on the need for cultural and behavior change throughout the emergency services.
- Fire Protection Systems. Covers the design and operational features of fire alarm systems, water-based fire suppression systems, special hazard fire suppression systems, water supplies for fire protection, and portable fire extinguishers.
- Principles of Emergency Services. Provides an overview of a broad range of topics relating to an emergency service work, including career opportunities in fire protection and related fields; the philosophy and history of the fire protection/service; fire loss analysis; the organization and function of public and private fire protection services; fire departments as a branch of local government; laws and regulations

FIGURE 4-4 The National Professional Development Model pyramid illustrates the progression of education and training in the fire and emergency services as noted by the United States Fire Administration/National Fire Academy for Higher Education. *Courtesy of USFA/NFA, Emmitsburg, Maryland.*

affecting the fire service; fire service nomenclature; specific fire protection functions; basic fire chemistry and physics; an introduction to fire protection systems; and an introduction to fire strategy and tactics.

The major publishers of fire-related textbooks are committed to writing texts for some, or all, of them. Fire science associate's degree programs are encouraged to require them as the theoretical core curriculum on which their major is based. Many schools already offer these courses in their programs, while others are in the process of adopting them. Once adopted, these model courses will answer the need for a high national standard of fire service education. They will also enable problem-free student transfers between schools and promote crosswalks for those who want to apply their academic course work toward meeting the national qualification standards necessary for firefighter certifications and degrees (see Figure 4-4).[9, 10]

This model is not a promotion model addressing credentials, but rather an experience-based model that suggests an efficient path for fire service professional development. This path is supported by collaboration among fire-related training, higher education, and certification providers. The model recommends what these providers' respective roles should be and how they should coordinate their programs.[11] Another important initiative of FESHE is the creation of the High School Pathway Program. This program encourages the incorporation of fire service courses into the high school curriculum. It cultivates an early interest in emergency services and provides an avenue for students to pursue a college education after high school.

International Fire Service Accreditation Congress

The IFSAC focuses on ensuring that training and certification within its member jurisdictions meet strict NFPA Professional Qualification standards. The IFSAC, based at Oklahoma State University, is divided into two assemblies: the Certificate Assembly and the Degree Assembly. It should be noted that IFSAC does not provide training of any sort. It is the responsibility of IFSAC to only accredit certificate programs. As of 2012, the IFSAC Certificate Assembly has a presence in most of the U.S. states, plus all Canadian provinces, U.S. DoD, Canadian National Defence, and several countries outside North America. The Degree Assembly, which provides accreditation to programs that are within the realm of higher education, also has the same diverse background in its membership. Many of these entities are community colleges that conduct accredited courses and programs under the auspices of the larger fire service entities in their jurisdiction.

Pro Board Accreditation

Pro Board ■ National Board on Fire Service Professional Qualifications Accrediting Fire Service Training Organizations.

The **Pro Board** is very similar to IFSAC. Its primary focus is to serve the needs of accreditation of organizations that certify uniform members of public fire departments, both career and volunteer. As with IFSAC, Pro Board does not provide training of any sort. It is the responsibility of Pro Board to only accredit the certificate programs. Other organizations with fire protection interests may also be considered for participation. Pro Board works with the individual states or provinces to issue certifications under that jurisdiction. As with IFSAC, a common goal is to bring professionalism to the fire service. Pro Board encourages reciprocity among certifying entities. Having IFSAC and Pro Board certifications shows the individual is prone to professional growth and advancement. Certification proves that the member is highly motivated and serious about measuring up to national standards. One must keep in mind, when testing Pro Board skills, that they are more team driven and seem to instill the "team" mentality of firefighting.

It should be noted that *not* all higher education institutions have FESHE, IFSAC, or even Pro Board accreditation. Each state is different. When researching which higher education institution is right for you, one should look into the institution's accreditations. For example, in Texas, the Southern Association of Colleges and Schools (SACS) is the granting authority. In the United States, there are six regional accrediting agencies. These six bodies of agencies that can award regional accreditation are recognized by the U.S. Department of Education and the Council for Higher Education (CHEA). It's up to the individual to make sure the higher institution will have quality assurance, compliance, and educational standards in place. All these accreditation associations have one thing in common—they are all a standardization factor for success. In recent years there has been a controversy over regional versus national accreditation. National

 ON SCENE

Fire Chief Martin D. Davila, who has 23 years' experience in military and civilian fire protection. He states, "Training and higher education are the cornerstones of success. Alone they don't make you smarter . . . but with self-motivation they open many doors for you. I believe that you cannot succeed in our business without all three of them." (1) Training teaches you to do the job. (2) Higher education (college) opens you to new ideas and theories, enhancing your past experiences and influencing your future decisions. (3) Motivation (the desire to make it better) is the glue that brings it all together. Training has taught us what to do. Higher education will make us think of the "what-ifs" and help us formulate new approaches. Motivation will push us to implement our new ideas. Training + Higher Education + Motivation = Success. Great leaders aren't born, but through training, experience, and higher education, a great leader is formed.

1. What model does the NFA have in place to promote training and education?
2. Why, do you think, are more fire departments today setting training and education standards for initial hiring?

accreditation is not based on geography. This type of accreditation was designed for specific types of school and colleges. The Distance Education and Training Council (DETC) accredit colleges that offer distance education. The Accrediting Commission of Career Schools and Colleges of Technology (ACCSCT) accredit career schools and technology programs. The main issue when deciding which type of accreditation is right for you is the transferability of credits from one school to another and the acceptability by prospective employers.

Summary

As the fire service moves from a skilled trade to professional status, a college or university education has become essential. Greater administrative demands, advances in technology, and the changing social and political environments affecting our profession necessitate advanced training and higher educational opportunities, especially for those firefighters who will be tomorrow's leaders of the fire service. In any form, whether it is learning in a higher education environment or a certification class for continuing education hours, training equals safety for the firefighter and the community he/she serves.

The fire service professional's access to traditional higher education is limited by geographic, financial, and time constraints. The establishment of national standards, as well as increased acceptance of distance and e-learning, should reduce some of the difficulties. Agencies or departments that work together on a normal basis would best serve their own interests and those of their employees by arranging for joint training and educational opportunities. Such cooperation would give firefighters access to increased, more effective learning experiences, would lead to greater safety, and would minimize the loss of life and property. The role of NFA, IFSAC, and Pro Board all have the same objective of improving the fire service and standardizing education for the discipline of emergency service fields.

As in any profession, a person develops a plan of success and follows it from entry-level position to retirement. Fire and emergency service personnel are no different. They get into it for life, whether it is career or volunteer. A term dubbed in the business world, *proactive futuring*, means planning for tomorrow and that is what our firefighters need to do today. Ultimately, however, it will be the responsibility of individual firefighters—whether new recruits or experienced professionals—to take the steps necessary to get the education they need to meet their own goals and to serve the public to the best of their ability. Education opens the door to promotions, certainly, but its real value is in teaching the leadership skills that the motivated firefighter will need to take on increased responsibility.

Review Questions

1. The fire science technology curriculum is aimed at providing the student with what types of skills?
2. Training in the fire service can lead to certifications. Name a few certifications in the field that can be beneficial to career advancement.
3. What is the basic college degree in fire science technology?
4. Training programs are offered at various levels, such as federal, state, and local. List two others.
5. List two pre-service opportunities for gaining firefighting experience.

Suggested Reading

Ford, Travis M. (2011). *Fire and emergency services safety & survival*. Upper Saddle River, NJ: Prentice Hall.

International Fire Service Training Association. (2012). *Fire and emergency services instructor* (8th ed.). Stillwater, OK: Fire Protection Publications.

Endnotes

1. NFA Higher Education, USFA Training & Education. (2006, April 19). U.S. Fire Administration. Retrieved November 13, 2006, from the Higher Education website: http://www.usfa.dhs.gov/training/nfa/higher_ed/.

2. Lindsey, Jeffery. (2006). *Fire service instructor*. Upper Saddle River, NJ: Pearson Education.

3. Murphy, D. L., & Stanton. L. M. (2004). A hybrid approach to distance education technology: Tailor-made

for the United States fire service. In J. Mathews (Ed.), *Session T1C: Distance learning: Methods and technologies 1* (pp. T1C1–T1C4). Savannah, GA: Frontiers in Education Clearing House.

4. Ibid.
5. NFA Higher Education, USFA Training & Education. (2006, September 20). U.S. Fire Administration. Retrieved November 13, 2006, from the Higher Education website: http://www.usfa.dhs.gov/training/nfa/higher_ed/model/.
6. Murphy, & Stanton, op cit.
7. NFA Higher Education, USFA Training & Education. (2006, January 23). U.S. Fire Administration. Retrieved November 14, 2006, from the Higher Education website: http://www.usfa.dhs.gov/training/nfa/higher_ed/degree_programs/distance/.
8. Sturtevant, Thomas. B. (2001). *A study of undergraduate fire service degree programs in the United States,* pp. 1–152. Retrieved November 14, 2006, from http://www.dissertation.com.
9. NFA Higher Education, USFA Training & Education, September 20, 2006, op cit.
10. Ibid.
11. Ibid.

Courtesy of James D. Richardson.

KEY TERMS

advanced life support units (ALS), *p. 77*

aerial ladder platform, *p. 74*

aerial ladders, *p. 68*

aerial platform apparatus, *p. 74*

articulating boom, *p. 74*

Bambi Bucket, *p. 81*

basic life support units (BLS), *p. 77*

compressed air foam
systems (CAFS), *p. 71*

heavy rescue squad, *p. 76*

hydraulically operated, *p. 73*

initial attack pumper, *p. 70*

Level A ensemble, *p. 87*

Level B ensemble, *p. 88*

Level C ensemble, *p. 88*

Level D ensemble, *p. 88*

Mobile data terminal (MDT), *p. 92*

National Fire Protection Association
(NFPA), *p. 65*

NFPA 1901, *p. 65*

non–fire suppression activities, *p. 66*

OSHA, *p. 80*

OSHA Code of Federal Regulations 29
CFR 1910.120, *p. 80*

personal alert safety system (PASS), *p. 82*

personal protective equipment (PPE),
p. 82

pump-and-roll operation, *p. 71*

regulations, *p. 81*

self-contained breathing apparatus
(SCBA), *p. 82*

standards, *p. 65*

telescoping boom, *p. 74*

thermal imaging camera (TIC), *p. 92*

wildland firefighting apparatus, *p. 70*

OBJECTIVES

After reading this chapter, you should be able to:

- Describe the various types of fire department facilities.
- Identify and describe the uses of pumping apparatus.
- Explain the functions and operations of various aerial apparatus.
- Explain how specialized fire apparatus are utilized.
- Describe the different levels of protection for the work uniform, proximity, structural, wildland, and Level A, B, C, and D ensembles.
- Explain why SCBA and PASS devices are important to firefighter safety.

Introduction

In 2011, the **National Fire Protection Association (NFPA)** reported that there was an estimated 1,100,450 firefighters in the United States. Of this number, 344,050 (31%) were career firefighters and 756,400 (69%) were volunteer firefighters.[1] The modern fire department is a multifaceted organization that is expected to perform an astonishingly wide range of critical tasks. To do its work effectively, the department must be equipped with an equally wide range of resources. In this chapter, we will introduce the student to fire service resources, by which we mean all of the material aids that firefighters use in their work. We will look first at fire service buildings, including their functions and design. Next, we will consider the specialized vehicles that answer the specific needs of firefighting, both standard fire vehicles and those designed for very particular tasks, such as aircraft and marine firefighting. We will look at protective gear and breathing apparatus designed to protect firefighters from the diverse dangers that they encounter in different environments. Finally, we will consider traditional hand-held firefighting equipment, as well as some of the high-tech inventions, such as Geographic Information System (GIS) equipment and thermal imaging devices, that have helped to bring the science of firefighting into the 21st century.

The NFPA establishes manufacturing **standards** for all fire service apparatus in NFPA Standard 1901. The codes that pertain to apparatus discussed in this chapter are listed at the end of each section. Printed or digital copies of the **NFPA 1901** Standard can be requested by contacting the National Fire Protection Association at One Batterymarch Park, Quincy, Massachusetts 02269-9101.

National Fire Protection Association (NFPA) ▪ Organization of insurance companies, manufacturers, and fire service professionals dedicated to firefighter safety.

standards ▪ A recommended course of action that is not mandated by law.

NFPA 1901 ▪ Standard for Automotive Fire Apparatus.

Fire Department Facilities

There are an estimated 30,145 fire departments in the United States.[2] All of these departments have one thing in common—they need facilities to operate. Fire department facilities vary in configuration and design. Smaller fire departments normally consolidate their facilities into one or two buildings, whereas larger departments find it more practical to have separate facilities for each functional area. Regardless of the department's size and finances, its facilities must cover three basic functions: administration, training, and fire suppression.

ADMINISTRATION BUILDINGS

The amount of space required for administration purposes depends on the size of the department's staff and the complexity of its operations. Smaller fire departments may need only office space in a fire station. Larger departments may maintain a separate headquarters building for administrative staff. Each arrangement has its advantages and its disadvantages.

When the headquarters office is located in the fire station, administrative personnel are physically close to the day-to-day line operations of the department, and this may result in better communication. The administrators are better informed about firefighters' concerns, and the firefighters in turn have a better understanding of the reasoning behind administrative decisions. This promotes a greater sense of teamwork and can lead to better compliance with policy and procedural changes. The trade-off is that administrators who work in the fire station may have to put up with distractions. Fire stations tend to be noisy, especially during the day when vehicles and other motorized equipment are being tested and personnel are cleaning the working and living areas. Holding conferences can be difficult under these circumstances. There are disadvantages for the firefighters as well.

FIGURE 5-1 Fire department headquarters offices may be incorporated into a fire station or in their own administration building. *Courtesy of James D. Richardson.*

FIGURE 5-1 Fire department headquarters offices may be incorporated into a fire station or in their own administration building. *Courtesy of James D. Richardson.*

non–fire suppression activities ■ Any fire department activity that is not directly involved in fire suppression.

Department headquarters are often visited by local politicians and news media, whose presence may make the firefighters feel that they are constantly on show and under scrutiny. Some departments believe it is better to keep administration separate from the line positions, if possible.

Other fire service departments that may be located in a separate building include a telecommunications center (dispatch), fire and arson investigation squad, inspection and codes departments, and training facilities. The purpose of having a separate location may also be to consolidate these **non–fire suppression activities** (see Figure 5-1).[3]

TRAINING FACILITIES

The size of a fire department's training facility varies according to the size of the department or the community being served. Each state in the United States has its own requirements for fire service training. In most large cities, the fire department is provided with a training facility. These facilities are designed to accommodate training for several divisions of the department. This would include fire and arson investigation, fire inspections, heavy rescue, fire suppression, and hazardous materials. Because of the high cost of Environmental Protection Agency (EPA)–approved permanent live burn training facilities, smaller communities may choose to invest in portable live burn training simulators. This allows firefighters to receive training in fire ground tactics and strategies. These communities often rely on state-certified regional fire training organizations to provide them with certified training. In many states, the regional fire academies are located at local community colleges.

There is no universal template for the design or construction of the training center, but it must provide a learning environment and equipment adequate to train firefighters to the NFPA's standards. A classroom with adequate lighting, acoustics, and audiovisual equipment should be available, as well as a building or props to use in training nonburn skills. Nonburn skills are those that do not expose trainees to actual fire scenarios, such as ladder carries and raises, hose lays and fire streams, and fire protection systems. At a minimum, the training center should feature a training tower and burn building that permits firefighters to practice extinguishment of Class A structural fires. The Insurance Service Organization (ISO) also gives credit for fire facilities that provide a burn pit for Class B liquid fuel

FIGURE 5-2 The size and complexity of a fire training facility will vary, depending on the requirements of the state certifying agency. *Courtesy of James D. Richardson.*

fires. The facility should also have an adequate water supply, from either a pressurized or static source, capable of providing a minimum of 500 gallons per minute (GPM) for 30 minutes (see Figure 5-2).[4]

NFPA 1402, *Guide to Building Fire Service Training Centers;* NFPA 1403, *Standard on Live Fire Training Evolutions*

FIRE STATIONS

Fire stations are the most complex facilities in a fire department. Because of the different functions they serve, they may include elements of a residential house, an office building, a heavy-duty garage, and a maintenance shop. In rural areas protected by volunteer fire departments, the fire station may not house personnel on a 24-hour schedule. The station may only have apparatus bays, a small office or office area, restrooms, and a training room. Firefighters report to the fire station one or two days a month for training. During an emergency, designated firefighters will respond from their homes or work locations to the fire station to operate apparatus and deliver equipment to the emergency scene while other firefighters respond directly to the emergency scene. In this situation, the fire station does not contain a residential environment. There are many variations in layout and design, but all stations must meet the same safety standards:

- The building must be structurally sound to house heavy vehicles and machinery.
- It must provide protection for the occupants and equipment from changing weather conditions.
- It must provide safe working conditions for the occupants.
- It must have good security.

In addition to these safety requirements, the NFPA recommends the following space accommodations for a typical fire station.[5]

Apparatus Bay

An apparatus bay is a garage area inside the fire station that is used to house fire apparatus and other emergency vehicles. The bay floor must be constructed to safely support apparatus that may weigh 20 tons or more when fully outfitted. It should be large enough to allow for minor apparatus maintenance and include a ventilation system to remove exhaust gases when apparatus are operated indoors.

The bay doors must be oversized (14 feet × 14 feet is recommended) to allow clearance for wide apparatus, and the ceiling of the bay must be tall enough to allow for apparatus that has high head clearance requirements, such as **aerial ladders** or platforms. The area should be well lit and have a heating system that keeps the bay above freezing during winter (see Figure 5-3).

aerial ladders ■ A rotating, power-operated ladder mounted on a self-propelled automotive fire apparatus.

Sleeping Accommodations, Bathroom Facilities, Kitchen, and Living Areas

In most career fire departments, firefighters work 24-hour shifts and then are off duty for a designated period. It is estimated that during their careers, they spend approximately one-third of their lives at the fire station. Therefore, the station must have adequate facilities to support many of the activities of ordinary daily living.

Sleeping accommodations should be separate for male and female personnel and allow firefighters to rest between emergency calls during their long shifts. Some stations are designed with private sleeping quarters, but separate cubicles in dormitories are more common. To prevent injuries that can occur when descending stairs or sliding a pole, many newer fire stations position the dormitory or sleeping area on the same level as the apparatus bay. The sleeping area should be segregated from the apparatus bay to prevent possible exposure to toxic gases. Separate bathing and toilet facilities also must be provided for male and female personnel, and these should be constructed of materials that are easily maintained and have a long service life (see Figure 5-4).

The kitchen area of a fire station is often where firefighters congregate during breaks or for fraternizing after "normal" station work is completed. Many fire departments encourage shift personnel to have common meals, as this helps create a sense of camaraderie. Some companies prepare two or three meals every work shift. The kitchen area, like

FIGURE 5-4 Dormitories in newer fire stations provide more privacy by separating individual sleeping areas into rooms or cubicles. *Courtesy of James D. Richardson.*

the sleeping area, should be segregated from the apparatus bay to prevent possible contamination from vapors, fuels, and oils.

The living area is similar to a residential family room. Lounge furniture allows personnel to relax between responses. A TV and video player are often found in this area and can be used for entertainment or training purposes.

Offices

The station officers should have a room in which to conduct station business. Ideally, this room should be located in a private area of the station to allow for confidential counseling of personnel. All personnel files should be located in this office and secured in a manner that prevents unauthorized access.

Other Areas

There are other areas and equipment that the ideal modern fire stations might provide to enhance the safety and comfort of its personnel:

- Personnel decontamination areas for biological and chemical agents.
- Medical supply storage.
- A specially designed laundry room for cleaning turn-out gear and work uniforms. During firefighting operations, turn-out gear and the work uniform worn underneath absorb contaminants from the smoke and runoff water. These contaminants may be toxic or carcinogenic and pose a risk to those exposed. Therefore, the laundry room must be equipped with a water retrieval system that prevents the water used for cleaning from entering the public wastewater system. The laundry water must be properly stored and disposed of according to EPA guidelines.
- A training room for continuing education.
- An exercise area where personnel can work out to maintain physical fitness.

Fire Apparatus

Fire apparatus come in many makes and models, each designed for a certain type of firefighting operation. We will look first at pumpers, and then we will consider vehicles designed to transport water, to allow access to the upper stories of structures, and to serve in other specific situations.

This standard establishes the basic requirements that must be met by all motorized fire apparatus.

NFPA 1901, *Standard for Automotive Fire Apparatus*[6]

PUMPING APPARATUS

The fire department's Type I pumper is its most basic fire apparatus. It is usually staffed by four personnel: an officer, an operator, and two firefighters. Pumpers are designed to fight structural fires or to support any operation that requires sustained pumping activity. The standard pumper may be equipped with pump capacities ranging from 750 to 2,000 GPM.

Some fire departments expand the versatility of their pumping apparatus by adding an elevating water tower. The towers range in height from 50 to 75 feet and can supply up to 1,000 GPM through a master stream appliance, a nozzle designed to flow 1,000 GPM or more, mounted on the top of the tower (see Figure 5-5).[7]

Other types of pumping apparatus are the Type II **initial attack pumper**, also referred to as a first or fast attack pumper, and Type III **wildland firefighting apparatus**. The Type II initial attack pumper consists of a custom pumping module and a water storage tank mounted on a pickup truck chassis. It is primarily used to initiate a rapid attack on a structure fire to try to hold the fire in check until the Type I fire apparatus arrives. It is not adequate for sustained operations. Initial attack pumpers can also be used in combating

initial attack pumper ■ A fire pumping apparatus that is utilized for fast response and rapid attack on a fire.

wildland firefighting apparatus ■ A fire pumping apparatus with a high ground clearance and off-road capabilities.

TYPE I REQUIREMENTS

- Minimum pump capacity of 750 GPM
- Minimum water supply tank of 500 gallons
- 1,200 feet of supply hose larger than $2\frac{1}{2}$ inches in diameter
- 400 feet of $1\frac{1}{2}$-, $1\frac{3}{4}$-, or 2-inch diameter hose
- Assorted tools and equipment (ladders, forcible entry tools, hose adapters and appliances, self-contained breathing apparatus, pike poles, portable fire extinguishers, salvage covers, and so on)

small grass or vehicle fires. The NFPA requires that the tank have a capacity of at least 200 gallons and that the pump has a minimum capability of delivering at least 250 GPM.[8]

NFPA 1902, *Standard for Initial Attack Fire Apparatus*

WILDLAND FIREFIGHTING APPARATUS (BRUSH TRUCKS)

The National Wildfire Coordinating Group (NWCG) *Fireline Handbook*, NWCG Publication 410-1 lists five classifications (types) for brush trucks:

Type 3: Minimum 500 gallon tank capacity and a minimum 150 GPM pump flow
Type 4: Minimum 750 gallon tank capacity and a minimum 50 GPM pump flow
Type 5: Minimum 400 gallon tank capacity and a minimum of 50 GPM pump flow
Type 6: Minimum 150 gallon tank capacity and a minimum of 50 GPM pump flow
Type 7: Minimum 50 gallon tank capacity and a minimum of 10 GPM pump flow[9]

These apparatus may have off-road (all-wheel-drive) capability and must be able to support **pump-and-roll operations**—that is, it must be able to provide pump operations while moving. In addition, many wildland fire apparatus have a Class A foam system. This extends the water supply and increases the water's ability to penetrate the fuel. Approximately 10% to 20% of new apparatus are equipped with a **compressed air foam system (CAFS)**. Compressed air foam systems provide more extinguishing capability by lowering friction loss, increasing the reach of hose streams, and allowing the use of a lighter hose. In other respects, the designs of wildfire firefighting apparatus vary considerably because they are configured to meet the special needs of the areas they will serve. For example, fire departments in areas that have level terrain, with little variation in topography, do not need an apparatus with a side-slope stability system (a hydraulic system that allows the chassis of the truck to stay level on sloped terrain). However, in areas where the topography varies dramatically, the fire apparatus may be required to have wheels that are powered individually to allow for greater ground clearance (see Figure 5-6A).[10] New

pump-and-roll operation ■ The capability of an apparatus to pump water while it is in motion.

compressed air foam system (CAFS) ■ A high-energy foam-generation system consisting of an air compressor (or other air source), a water pump, and foam solution that injects air into the foam solution before it enters a hoseline.

FIGURE 5-6A Wildland fire apparatus are designed for off-road operations. They are lighter and more maneuverable than structural fire apparatus. *Courtesy of James D. Richardson.*

safety standards are being considered for wildland firefighting apparatus. The standards would require each wildland apparatus to have a roll cage and safety harnesses for firefighters riding outside of the apparatus cab (see Figure 5-6B).

NFPA 1906, *Standard for Wildland Fire Apparatus*
   Notes: NFPA standards 1901, 1902, and 1906 overlap when wildland fire apparatus are equipped with pump capacities of 250 to 500 GPM.
   How the apparatus is used will determine the standard that will be followed.

NFPA 1906 Standard for Wildland Fire Apparatus	Requirements for apparatus ground clearance, angle of approach and departure, side-slope stability, and pump-and-roll capabilities
The National Wildfire Coordinating Group (NWCG) *Fireline Handbook*	Lists five classifications for brush trucks
NFPA 1901 Standard for Automotive Fire Apparatus	More stringent requirements in regard to hose, ladders, equipment, and water tank capacity
NFPA 1902 Standard for Initial Attack Fire Apparatus	Pump capacity minimum of 250 GPM and a minimum tank size of 200 gallons. Hose and equipment requirements are lower than for Type I pumper.

MOBILE WATER SUPPLY APPARATUS

Mobile water supply apparatus, also referred to as tenders, are designed to transport water to an incident scene for use by pumping apparatus. NFPA standards require that a mobile water supply apparatus have a minimum tank capacity of 1,000 gallons. Straight-chassis apparatus with tank capacities of 1,500 to 4,000 gallons must be of tandem rear-axle construction, whereas a tractor-trailer design is recommended for larger tank capacities. If the apparatus is equipped with full-size fire pumps as well as other standard tools and equipment of a fire pumper, it is classified as a tanker-pumper and can operate as either a mobile water supply apparatus or a pumper (see Figure 5-7).[11]

FIGURE 5-7 Tenders vary in carrying capacity from several hundred to several thousand gallons and are the primary water supply for many rural fire ground operations. *Courtesy of James D. Richardson.*

NFPA 1903, *Standard for Mobile Water Supply Fire Apparatus*

AERIAL APPARATUS

Aerial apparatus are vehicles designed to provide firefighters access to the upper levels of a structure that cannot be reached with ground ladders. These apparatus are primarily used for rescue and ventilation and for providing elevated master streams. The NFPA requires that aerial devices be power operated and self-supporting with an elevation capability of at least 50 feet. (Working ladder height is determined by raising the ladder to its maximum elevation and extension and measuring from the ground to the highest rung on the ladder.) The International Fire Service Training Association classifies aerial apparatus into four categories:

- Aerial ladder apparatus
- Aerial ladder platform apparatus
- Telescoping aerial platform apparatus
- Articulating aerial platform apparatus.[12]

Aerial ladders range in length from 50 to over 135 feet and must be able to support 250 pounds of weight at the tip, when fully extended horizontally and at a 45° angle. The ladder is **hydraulically operated**, but is required to have electrically and manually operated backup systems that can be used to lower the ladder if the hydraulic system fails. The apparatus must carry a large complement of ground ladders, forcible entry tools, rescue tools, and ventilation equipment. The support chassis for the aerial ladder may range from a two- or three-axle single chassis to a tractor-trailer apparatus, known as a tiller truck. The tractor-trailer model is equipped with a rear-wheel steering compartment. The steering mechanism in the compartment is operated by a firefighter, who is referred to as "the tillerman." The rear-wheel steering allows the trailer to be more

hydraulically operated ■ Functioning by utilizing pressure created by compressing oil in a cylinder.

FIGURE 5-8 The single frame ladder truck may have a single- or dual-rear axle, and the aerial ladder may be midship or rear mounted. *Courtesy of James D. Richardson.*

maneuverable on narrow streets and in traffic than straight-chassis aerial apparatus (see Figure 5-8).[13]

A variation of the aerial ladder is the **aerial ladder platform**. This apparatus consists of a structurally reinforced aerial ladder, ranging in height from 85 to 110 feet, with a platform attached to the tip that is usually mounted over the rear axles of a single-frame three-axle chassis.

The platform, or basket, is required to support 750 pounds of weight in any position in which it is placed.[14] It is constructed of steel or aluminum alloy and is usually equipped with a heat shield and shower nozzle to protect personnel. Electric, air, and hydraulic outlets are supplied. Many platforms are equipped with $1\frac{1}{2}$- and $2\frac{1}{2}$-inch outlets, which allow handlines to be supplied from the platform to upper levels of a structure when standpipe or sprinkler systems are not available or are inoperable. Normally, these also have a permanently mounted turret nozzle and a water supply system capable of supplying flow rates of 1,000 GPM or more. A hydraulically operated backup system and two primary operating systems, one at ground level and one in the basket, are required (see Figure 5-9).

The **aerial platform apparatus** consists of a platform with a permanently mounted turret nozzle and water supply system mounted on a **telescoping boom** or **articulating boom**. The boom may be of box-beam or tubular truss-beam construction. The only difference between the telescoping and articulating versions boom is how they operate. The telescoping boom is sectioned in a way that allows the sections to telescope into each other in a straight line. The articulating boom is constructed of several sections that are connected by a hinge that operates much like a human elbow. The functions of the aerial platform apparatus are the same as those of the aerial ladder platform (access to upper levels, rescue, ventilation, and master stream operations), but aerial platforms do not have aerial ladder operations capabilities. The boom of the telescoping platform does have a small ladder mounted on it, but its only function is to provide an emergency escape route for firefighters who are trapped on the platform.[15] Some fire apparatus manufacturers have combined the strength of the telescoping beam with the agility of the articulating arm. For example, Bronto Skylift of Finland developed an aerial platform that combines a

aerial ladder platform ▪ A power-operated aerial device that combines an aerial ladder with a personnel-carrying platform supported at the end of the ladder.

aerial platform apparatus ▪ A fire apparatus that carries a hydraulically operated elevating platform that can be utilized for rescue, fire, or observation operations.

telescoping boom ▪ Aerial device raised and extended through sections that slide within each other.

articulating boom ▪ A boom constructed of several sections that are connected by a hinged joint that allows the boom to function much like the human elbow.

FIGURE 5-9 The aerial ladder platform allows the apparatus to be used as an observation post, for heavy stream operations, or for aerial ladder operations. *Courtesy of James D. Richardson.*

multisection telescoping boom with an articulating arm. The apparatus ranges in height from 265 feet to 367 feet (see Figure 5-10).

Each fire department determines the apparatus type and method of operation that will meet its needs. Articulating platforms are more maneuverable in confined areas, such as narrow streets or alleys. However, these types of aerial apparatus are more unstable than telescoping-type apparatus if they are extended past their designed operating limits (which may occur during fire rescue operations). Telescoping platforms and ladder

FIGURE 5-10 Aerial platform apparatus can utilize a combination of telescoping and articulating arm technologies to reach up to 33 stories. *Courtesy of Bronto Skylift.*

platforms are more stable and provide a means of escape for firefighters if they must abandon the platform. However, telescoping aerial platforms or ladder platforms are less maneuverable in confined areas. Aerial apparatus also need a large, clear area to set up for operation. There must be enough room for the apparatus to extend the stabilizing outriggers. The outriggers are mounted on each side of the apparatus, near the front behind the personnel cab and/or pump panel and near the rear between the rear axle and tailboard. When they are extended the "footprint" of the apparatus is greatly expanded.

NFPA 1904, *Standard for Aerial Ladder and Elevating Platform Fire Apparatus*

Specialized Apparatus

Prior to 1970, the main responsibility of a community's fire department was to extinguish fires, teach fire prevention, and administer some basic first aid. A few departments provided medical services, but in most communities, medical response and transport were provided by private agencies or funeral homes. Response times were long, and in some instances, transportation was actually denied because the victim could not pay for the service. In response to the public outcry, community leaders and emergency response professionals began looking for a practical way to provide faster, less expensive emergency medical service. The solution they found was based on the medical response and transport techniques that were already in place in a few urban centers such as Los Angeles County and Jacksonville, Florida, as well as on lessons learned during the Vietnam War. Many returning veterans were now working in the emergency services, and their wartime experience in efficient methods for treating and transporting seriously injured patients proved invaluable. One example was the use of pressurized garments (MAST trousers) to maintain blood pressure in vital organs of patients who had low blood volume or were in severe shock. Other examples were the development of air ambulance helicopters and the use of advanced intravenous and drug techniques by paramedics. As a result of the input gathered from these groups and the vision and creativity of those who expanded on it, the United States today has one of the most sophisticated public emergency medical services in the world. Although some communities have placed the responsibility for this service on other emergency response agencies, such as the police department or a separate emergency medical services department, most communities have turned to their fire departments to implement this service in their 9-1-1 system.

Another consequence of the growing sophistication of the emergency medical services is that fire departments are now able to provide a **heavy rescue squad**.

Many fire departments had light rescue squads capable of performing tasks involving extrication and medical treatment. However, these units were often also called for nonrescue tasks, such as providing extra help during fire suppression operations. In the 1970s, fire departments began to see the value of a specially trained team whose sole responsibility would be to provide rescue services in a variety of situations. This team would need extensive training in industrial and farm equipment extrication, trench and tunnel rescue, building collapse, high- and low-angle rope rescue, confined space and swift and rapid rising water rescue. The team would also need a designated vehicle to carry their equipment.

Between the years of 1977 and 1979, the fire service assumed another special operations function: hazardous materials response and mitigation. In 1977, after a series of devastating railroad tank car explosions in their community, the Jacksonville, Florida, Fire Department organized the first fire service–sponsored hazardous materials team. By

heavy rescue squad ■ A team of firefighters who are highly skilled in various types of rescue operations that require the use of highly specialized equipment and training.

1979, the fire departments in Houston, Texas, and Nashville, Tennessee, followed Jacksonville's lead and developed hazardous materials teams in their cities. These three departments set the standard for others to follow. Although the first teams were ill-equipped and learned as they faced situations, today's hazardous materials teams possess some of the most sophisticated equipment and highly trained personnel in the world. As a result of the events of September 11, 2001, hazardous materials teams are required to respond not only to chemical incidents but also to incidents that involve biological, radiological, and nuclear agents.[16]

AMBULANCE APPARATUS

An ambulance is a vehicle specially designed for delivering emergency medical personnel to an incident scene and transporting the sick and injured to a medical facility. The Type I ambulance is a modular unit that is mounted on a light- or medium-duty truck chassis. The medium-duty chassis is used in many communities because it has a larger module, allowing for more headroom and larger work area. Both models have a window between the module and the cab portion of the apparatus to allow communications between the driver and caregivers. Occupants must exit the vehicle to move from the module to the cab. The Type II ambulance is a standard full-size van body with a raised roof to allow for more interior head clearance. The interior space is one complete unit, so occupants are able to move from the cab to the treatment area without exiting the vehicle. The Type III ambulance is a modular unit on a van chassis. It may have walk-through capability to allow occupants access to all areas of the vehicle. Ambulances must provide enough floor space to accommodate a stretcher for the patient, a work area for medical personnel, and adequate storage for all medical supplies (see Figure 5-11).

Fire department ambulances may be operated as **basic life support units (BLS)** or **advanced life support units (ALS)**. Most BLS units provide first aid and nonemergency transportation to the hospital. Personnel are usually trained only to the Emergency Medical Technician Basic Level. ALS units provide advanced medical services, including cardiovascular monitoring, and under certain circumstances are allowed to administer drugs intravenously. ALS personnel are trained to the Emergency Medical Technician Paramedic level.

basic life support units (BLS) ■ Medical units that provide medical and transportation needs for patients who do not have life-threatening injuries or illness.

advanced life support units (ALS) ■ Medical units that provide advanced medical treatment and transport to the severely injured or ill.

In addition to the standards set forth in NFPA 1901, ambulances must meet the federal requirements set forth by the General Services Administration Standard KKK-A-1822, *Federal Specifications for Ambulances*. The vehicle portion of the apparatus must also meet requirements set forth by the *Federal Motor Vehicle Safety Standards*.

RESCUE APPARATUS

Rescue apparatus are used to move specially trained rescue teams and equipment to the scene of an incident, where the team's job is to remove victims to a place of safety. These vehicles are not intended for patient transport. The number of assigned personnel and the type of equipment that is carried on the rescue vehicle varies, based on the needs of the community. Rescue apparatus are classified as light, medium, and heavy duty. Light- and medium-duty rescue apparatus are used to perform basic rescue operations that require a limited range of tools and equipment, such as hydraulic spreaders and cutters, jacks, saws, ropes and rigging, oxyacetylene cutting equipment, or air bag lifting equipment. These units may also carry firefighting tools and equipment and may be utilized as

FIGURE 5-12 Heavy rescue apparatus carries a variety of tools and equipment for specialized rescue operations. *Courtesy of James D. Richardson.*

dual-purpose apparatus. Heavy rescue apparatus are single-purpose units, carrying equipment and personnel whose only purpose is to extricate victims from dangerous situations. These include a wide range of possible operations, including vehicle and industrial accident extrication, as well as high-angle, confined space, swift and rapid rising water, trench, and tunnel rescues. These apparatus consist of modular units that are mounted on a heavy-duty truck chassis with two or three axles because of the massive amounts of equipment they must transport (see Figure 5-12).[17] There is no NFPA standard specifically for rescue vehicles.

STOP AND THINK 5-1

If the type of fire apparatus purchased by a fire department is determined by the priority of threats and a needs analysis of the community that it serves, what kind of specialized apparatus would a rural community need?

HAZARDOUS MATERIALS APPARATUS

A hazardous materials (HazMat) apparatus may be an enclosed trailer or a module on a three-axle heavy-duty truck chassis. These units carry specialized equipment for detecting hazardous materials and mitigating the damage they cause. This includes patching kits for various types of containers and agents; special protective clothing and breathing apparatus; air, water, and soil testing and monitoring devices; chemical, biological, radiological, nuclear, and explosive (CBRNE) detection equipment; special communication devices; and weather monitoring equipment (see Figure 5-13). Fire departments and fire personnel must meet the requirements of the General Duty Clause, CFR 1910.120 sec 5 (a) (1). This

FIGURE 5-13 Hazardous materials teams' equipment trailers carry a variety of specialized equipment and materials for mitigating chemical and other HazMat incidents. *Courtesy of James D. Richardson.*

clause requires the employer to provide a workplace that is free from recognized hazards that cause or may cause death or serious injury. It also requires that the employee comply with the Occupational Safety and Health Administration standards, rules, regulations, and orders required by the General Duty Clause CFR 1910.120 sec 5 (b) (1).

OSHA ■ Occupational Safety and Health Administration.

OSHA Code of Federal Regulations 29 CFR 1910.120 ■ The federal code that mandates the level of training for anyone working with hazardous materials.

The Occupational Safety and Health Administration (**OSHA**) establishes the levels of training for all public and private agencies that respond to hazardous materials incidents. The code governing these is **OSHA Code of Federal Regulations 29 CFR 1910.120** (also known as HAZWOPER, Hazardous Waste Operations and Emergency Response).

In EPA states where responders are not covered by the OSHA-approved State Occupational Health and Safety Plan, CFR 1910.120 (q) is referenced in EPA regulation Title 40 CFR 311, Protection of Environment, Worker Protection.

AIRCRAFT RESCUE AND FIREFIGHTING APPARATUS

These apparatus are specialized pieces of equipment that are used only for fighting aircraft fires. The units are very large with water tank capacities ranging from 1,500 to 6,000 gallons and foam storage tanks that carry up to 1,000 gallons of foam concentrate. Pump capacities can be as high as 2,000 GPM.[18] The NFPA defines three categories of Aircraft Rescue and Firefighting Apparatus (ARFF) apparatus: major firefighting vehicles, rapid intervention vehicles (RIVs), and combined-agent vehicles. RIVs are designed to respond quickly to the emergency scene and initiate rescue and initial firefighting measures until the major firefighting vehicles arrive. Major firefighting vehicles are large apparatus that carry in excess of 3,000 gallons of firefighting foam. Combined-agent apparatus are similar to the major firefighting vehicles but are capable of simultaneously delivering firefighting foam and dry-powder extinguishing agents on a fire. All are required to have pump-and-roll and off-road capability. It is recommended that they have positive drive to each wheel in order to achieve maximum chassis clearance (refer to Figure 5-6A).[19]

NFPA 414, *Standard for Aircraft Rescue and Fire-Fighting Vehicles*
The size, number, and type of ARFF apparatus needed are determined by Federal Aviation Administration (FAA) **regulations**.

regulations ■ Mandated by law.

FIREBOATS

Most port communities have fireboats to combat fires involving watercraft, wharves, and marinas. These apparatus range in size from high-speed shallow-draft vessels to large oceangoing tugboats. Fireboats can provide high volume pump capacities of over 20,000 GPM through master stream turret nozzles that flow up to 3,000 GPM each.[20] They may also be used for water rescue operations and serve as a drafting source to supply water to land-based fire apparatus for firefighting operations near a shoreline. Fireboats should meet NFPA 1925 Standard on Marine Fire-Fighting Vessels (see Figure 5-14.)

FIREFIGHTING AIRCRAFT

Aircraft, both fixed wing and rotary blade (helicopters), are used extensively in large wildland firefighting operations. Their primary duty is to support ground operations by dropping extinguishing agents or fire retardants on burning vegetation. The fixed-wing aircraft are usually propeller-driven military or commercial aircraft that have been converted to serve firefighting functions. Their carrying capacity varies from 700 to 3,000 gallons of agent (see Figure 5-15). Most firefighting helicopters do not have the carrying capacity of fixed-wing aircraft, but they do have the ability to hover over a fire and more accurately place the retardant or extinguishing agent on the target. The majority of helicopters transport their payload in a bucket that is suspended under the aircraft. This bucket is referred to as a **"Bambi Bucket"** and has carrying capacities

Bambi Bucket ■ A brand of collapsible bucket.

FIGURE 5-14 Fireboats are capable of delivering thousands of gallons per minute. *Courtesy of Tyco/Ansul Corporation.*

FIGURE 5-15 Firefighting aircraft are used extensively in wildland firefighting operations. *Courtesy of Tyco/Ansul Corporation.*

ranging from 300 to 1,000 gallons. Heavy-lift helicopters, using a special "Twin Bambi Bucket" configuration, have demonstrated the ability to deliver up to 5,200 gallons of retardant or water.[21]

Personal Protective Equipment

Firefighters are asked to respond to many different types of emergencies, each presenting its own range of challenges and dangers, and each calling for a different kind of **personal protective equipment (PPE)**. We will look next at specialized outfits, called ensembles, that are designed to respond to these diverse needs. These include the station/work uniform, as well as ensembles suited for wildland firefighting, structural firefighting, and proximity firefighting. Also discussed are ensembles designed to defend against very specific hazards: dangerous vapors, liquid splashes, and chemical or biological agents. We will also describe **self-contained breathing apparatus (SCBA)** and **personal alert safety systems (PASS)**.

Personal Protective Equipment ▪ Safety equipment worn by firefighters.

self-contained breathing apparatus (SCBA) ▪ Wearable respirator supplying a breathable atmosphere that is carried by or generated by the apparatus and is independent of the atmosphere.

personal alert safety system (PASS) ▪ Electronic lack-of-motion sensor that sounds a loud tone when a firefighter becomes motionless.

NFPA 1500, *Standard on Fire Department Occupational Safety and Health Program,* addresses the use, care, and maintenance of most types of fire PPE.

CLOTHING
Station Work Uniform
Station work uniforms are not designed to protect personnel from hazardous atmospheres, but they do offer some degree of protection when worn under other fire

protection ensembles.[22] The station work uniform that offers the most safety to the wearer is made of 100% cotton. Cotton allows the airflow through the material and helps transfer heat from the wearer's body. Cotton uniforms require more maintenance but do not produce static electricity when worn. Station uniforms made of polyester or polyester/cotton blended materials greatly reduce the transfer of body heat and are flammable. These types of materials can build up a static electrical charge. Although cotton uniforms can burn, polyester uniforms will burn and melt, sticking to the wearer's body. Nomex and other fire-resistive materials may also be blended with cotton in station work uniforms. Station work uniforms should comply with NFPA Standard 1975 (see Figure 5-16).

NFPA 1975, *Standard on Station/Work Uniforms for Fire and Emergency Services*

Structural Firefighting Ensemble

Protective clothing used during structural firefighting includes coats, trousers, helmets, gloves, footwear, and hoods. The coat and trousers construction consists of three layers: an outer shell, a vapor barrier, and a thermal barrier. The outer shell is the protective layer of the ensemble. It not only protects the inner layers of the garment, but it also has a fire-resistive quality that resists ignition when in direct flame contact and repels water. The outer shell must be well maintained and properly cleaned (see Figure 5-17A). The vapor barrier is designed to prevent vapor and moisture from reaching the body of the wearer, thus restricting heat flow through the garment. The vapor barrier must be tested annually to ensure its effectiveness. If it fails the annual test, the garment must be taken out of service (see Figure 5-17B). The thermal barrier is a quilted fabric constructed of several layers of material. It is the layer of the garment that is closest to the wearer's body. The thermal barrier absorbs body fluids and should be washed thoroughly after

FIGURE 5-17A The outer shell must be properly cleaned and maintained. *Courtesy of J. D. Richardson.*

FIGURE 5-17B The vapor barrier protects the wearer from moisture. *Courtesy of J. D. Richardson.*

use (see Figure 5-17C). Federal regulations forbid the removal of the protective barriers from the outer shell when being worn during a response. NFPA Standard 1971 recommends two performance tests for structural firefighting turnout coats and pants: a fabric flammability test to establish the materials resistance to ignition under direct flame contact and a thermal protective performance test to measure the ensemble's ability to protect the firefighter from second degree burns in a flash fire.[23] The minimum thermal protective standard of the structural firefighting ensemble coat and trousers requires that the wearer be protected from second-degree burns during the first 35 seconds of heat exposure. The structural firefighting ensemble is not designed to withstand a flashover environment (see Figure 5-17D).[24]

NFPA 1971, *Standard on Protective Ensembles for Structural Fire Fighting and Proximity Fire Fighting*

Wildland Firefighting Ensemble

The wildland firefighting ensemble consists of a lightweight jacket and trousers, high-top boots, a helmet, gloves, web gear (a utility belt for carrying survival equipment and tools), and a fire shelter. The fire shelter is an aluminized sheet of fire-resistant material used during an emergency to protect an entrapped individual if his or her position is overrun by

FIGURE 5-18 The wildland firefighting ensemble is lightweight but offers little protection from heat. *Courtesy of James D. Richardson.*

fire. The individual lies in a prone position on the ground and covers himself or herself with the sheet. The aluminized outer coating will reflect a large amount of the heat and help protect the individual until the fire has passed. The National Wildfire Coordinating Group requires all federal wildland firefighters and anyone involved in a federal wildfire operation to carry a fire shelter and be trained in its deployment and operation. The NFPA recommends that wildland firefighters carry a fire shelter, but cannot mandate it. Local, state, or county fire departments that do not respond to federal wildfire operations may or may not choose to adopt the NFPA standard. The jacket and trousers must be constructed of fire-resistant cotton or fire-resistive material and are designed to be worn over undergarments of 100% cotton or fire-resistive materials. The wildland ensemble should comply with NFPA 1977 (see Figure 5-18).[25]

NFPA 1977, *Standard on Protective Clothing and Equipment for Wildland Fire Fighting*

Proximity Ensemble
The proximity ensemble is designed to be worn in situations where the firefighter may be working in close proximity to high radiated heat, such as that created by petroleum-based fires or during the burning of other liquid fuels. Proximity ensembles are worn by aircraft firefighters and many refinery fire brigades. The coat, trousers, gloves, helmet, and hood have an aluminized coating, which increases heat reflection. The helmet is equipped with a face shield covered with an anodized gold material to protect the wearer from facial burns. Proximity ensembles should comply with NFPA Standard 1976 (see Figure 5-19).[26]

NFPA 1976 *Standard on Protective Ensemble for Proximity Fire Fighting*

FIGURE 5-19 The proximity suit allows the firefighter to advance closer to, but not enter, a high-intensity fire. *Courtesy of Tyco/Ansul Corporation.*

Hazardous Materials Ensembles

Hazardous materials operations require a variety of personal protective ensembles, depending on the degree and nature of the hazard. These ensembles are classified at four different levels, each providing a different degree of protection and each governed by a different NFPA standard.

The **Level A ensemble** offers the highest protection to personnel. It fully encapsulates and separates the wearer from the environment. This ensemble provides vapor protection, as well as splash protection. A self-contained breathing apparatus must be worn with the Level A ensemble see Figure 5-20).

Level A ensemble ■ Highest level of skin, respiratory, and eye protection that can be afforded by personal protective equipment.

NFPA 1991, *Standard on Vapor-Protected Ensembles for Hazardous Materials Emergencies*

FIGURE 5-20 The Level A ensemble protects the wearer against toxic vapors and liquids but offers no fire protection. *Courtesy of Dianne Richardson.*

FIGURE 5-21 The Level B ensemble offers splash protection against toxic liquids. A self-contained breathing apparatus (SCBA) must be worn with this ensemble. *Courtesy of James D. Richardson.*

FIGURE 5-21 The Level B ensemble offers splash protection against toxic liquids. A self-contained breathing apparatus (SCBA) must be worn with this ensemble. *Courtesy of James D. Richardson.*

Level B ensemble ▪ Personal protective equipment that provides the highest level of respiratory protection but a lesser level of skin protection.

The **Level B ensemble** is very similar to the Level A suit except that it does not offer vapor protection. It is designed for splash protection only. It comes in two basic configurations: an overall type and an encapsulated type. A self-contained breathing apparatus must be worn with a Level B ensemble (see Figure 5-21).

NFPA 1992, *Standard on Liquid Splash-Protective Ensembles and Clothing for Hazardous Material Emergencies*

Level C ensemble ▪ Level B personal protective ensemble with filtered respirators for respiratory protection.

The **Level C ensemble**, which comes in a coverall or two-piece style, is worn in areas where there is a danger of splashes, but respiratory hazards are less likely. Respiratory protection can be provided by an air-purifying respirator instead of a self-contained breathing apparatus (see Figure 5-22).

Note: If Level A, B, or C ensembles are used during a possible terrorist incident response, they must also conform to the standard set forth in NFPA 1994, *Standard on Protective Ensembles for Chemical/Biological Terrorism Incidents*.

Level D ensemble ▪ The standard work uniform worn at the fire station.

The **Level D ensemble** is the regular station work uniform and provides no chemical protection. This ensemble is shown in Figure 5-16.

NFPA 1975, *Standard on Station/Work Uniforms for Fire and Emergency Services*

SELF-CONTAINED BREATHING APPARATUS

The SCBA is perhaps the most important piece of safety equipment available to the firefighter. There are two basic classifications for SCBA worn by members of the fire service,

FIGURE 5-22 The Level C ensemble is similar to that for Level B. It offers respiratory protection with a filter mask and cannot be worn in an oxygen-deficient environment. *Courtesy of James D. Richardson.*

the open-circuit SCBA and the closed-circuit SCBA. The open-circuit air supply is provided by a cylinder mounted on a harness that is worn on the back of the firefighter. The firefighter breathes the compressed air from the cylinder. Exhaled air is released to the atmosphere through an exhaust valve on the facepiece. There are two sizes of cylinders that may be used with the SCBA: a 30-minute supply cylinder and a 60-minute supply cylinder. The closed-circuit SCBA utilizes the unused, exhaled oxygen from the wearer. The exhaled air is not released to the atmosphere. The unit contains a canister filled with chemical pellets that removes the carbon dioxide from the exhaled air and recycles the oxygen back into the system. A small oxygen cylinder in the backpack unit provides supplemental oxygen to the firefighter. Closed-circuit SCBA are used for special operations, such as confined spaces, tunnels, and so on, where firefighters will be required to carry out long-term operations without the advantage of exchanging air cylinders every 30 or 60 minutes. Some closed-circuit SCBAs can be worn for up to 5 hours. Every firefighter must be proficient in the care, maintenance, operation, and use of the SCBA. The firefighter should receive extensive training in emergency procedures that must be taken in the event of SCBA malfunction. These procedures should include rescue of a fellow firefighter and self-rescue. SCBA must be worn when the work environment is determined to be immediately dangerous to life and health (IDLH). Common IDLH atmospheric conditions encountered by firefighters include the following[27]:

- Areas that are oxygen deficient. Any atmosphere in which the oxygen concentration is below 19.5%.
- Temperatures above 130°F. If the air temperature is above 130°F, inhaled air will cause pulmonary edema and death by asphyxiation.
- Smoke. Consists of unburned particles and products of combustion, such as tar, carbon, and toxic gases that will cause irritation to the eyes and lungs, or death.
- Toxic atmospheres. Carbon monoxide, hydrogen sulfide, hydrogen cyanide, and carbon dioxide are toxic or oxygen depleting products of the combustion process. Some combustion byproducts are carcinogenic.

The maintenance, testing, and use of SCBA are established by federal regulations and the NFPA. In 1980 OSHA issued regulation 29 CFR 1910.156 regulating fire brigades. However, this only pertained to federal employees and state and local employees who had federally approved state OSHA plans. Only 22 states, Puerto Rico, and the Virgin Islands were affected by this regulation. State and local employees of EPA states were not required to meet the directive. This changed when OSHA issued the hazardous materials emergency response regulation 29 CFR 1910.120. The U.S. Congress directed the EPA to issue a regulation that mirrored the OSHA regulation. The EPA adopted the 29 CFR 1910.120 by reference under EPA regulation 40 CFR 311. It now applies to all state and local government employees and all volunteer firefighters in all 50 states. Both regulations require compliance with 29 CFR 1910.156. The NFPA also recommends that all fire departments comply with NFPA Standards 1500, 1404, and 1981 (see Figure 5-23).[28]

NFPA 1500, *Standard on Fire Department Occupational Safety and Health Program.* NFPA 1404, *Standard for a Fire Department Self-Contained Breathing Apparatus Program.* NFPA 1981, *Standard on Open-Circuit Self-Contained Breathing Apparatus for Fire Service.*

PERSONAL ALERT SAFETY SYSTEM

The PASS is another critical piece of personal protective equipment. A PASS may be a separate device that is attached to the wearer's protective ensemble, or it may be incorporated into the SCBA. The PASS system contains a motion detection device. It is designed to emit a loud audible alarm automatically if the firefighter remains motionless for 30 seconds. The device can also be activated manually by the firefighter if he or she becomes trapped but remains conscious. Some PASS devices also incorporate a bright strobe light that serves as a location beacon. According to the NFPA, PASS devices are not required to be integrated into the SCBA. They can be worn as a separate device on the firefighter's ensemble. PASS devices should comply with NFPA 1982.[29]

NFPA 1982 *Standard on Personal Alert Safety Systems (PASS)*

STOP AND THINK 5-2

What is the major reason why a firefighter should never enter a hazardous area without proper protective clothing and breathing apparatus?

Tools and Other Equipment

Firefighters have access to a variety of tools and equipment adapted for almost any imaginable use. Some are standard tools, such as the hand or hydraulically operated tools used in other work situations for such actions as striking, prying and spreading, cutting, and digging (see Table 5-1 and Figure 5-24). Others are technologically advanced devices that serve specialized firefighting functions.

THERMAL IMAGING CAMERAS

The safety and efficiency of fire ground and emergency response operations have improved immeasurably—and will continue to improve—as a result of technological advances. A notable example is the **thermal imaging camera (TIC)**. Originally designed for locating victims in search and rescue operations on the fire ground, TICs are now used in many other operations, such as locating fires in concealed spaces, observing a fire's path of travel, and locating lost victims in rural areas or at accident scenes. Personnel responsible for operating the unit should be well trained in interpreting the data projected on the camera's imaging screen and should be aware of the limitations as well as the benefits of the device. Training for this device should comply with NFPA 1408. Use, care, and maintenance should meet the standards recommended in NFPA 1801. A TIC provides a one-dimensional view of the area with no depth perception or peripheral view, and should not

thermal imaging camera (TIC) ■ Gives firefighters the ability to see heat generated by an object or person.

TABLE 5-1	Common Tools and Their Uses for Fire and Rescue Operations
TOOL	**USE**
Pick-head ax	Chopping, cutting, puncturing
Flat-head ax	Chopping, cutting, hammering
Halligan tool	Prying
Hydraulic spreaders, cutters, and rams	Prying, pushing, cutting
Pike pole	Pulling and puncturing
Circular saw	Cutting
Chain saw	Cutting
Adze hoe	Cutting and smoothing operations when clearing a fire break
Mattock	Used much as a pickax. It has a broad blade on one end and a pike on the other end
Pulaski	Combines the ax with a mattock

FIGURE 5-24 Firefighters use a variety of tools that are designed for digging, cutting, pushing, pulling, and chopping. *Courtesy of James D. Richardson.*

be used as a navigation tool because the firefighter could become disoriented if the unit should malfunction (see Figure 5-25).

NFPA 1408 *Standard for Thermal Imaging Training.*
NFPA 1801 *Standard on Thermal Imagers for the Fire Service.*

MOBILE DATA TERMINAL

mobile data terminal (MDT) ■ Mobile computer that communicates with other computers on a radio system.

The **mobile data terminal (MDT)** gives emergency dispatchers the ability to remain in direct contact with on-scene personnel. MDTs range in complexity from computerized devices mounted inside the emergency vehicle to handheld units. They consist of a screen that displays data sent to the terminal and a keyboard for entering information to be sent to the dispatcher. Some MDTs use the GPS, which provides the dispatcher with the location of each fire apparatus. GIS may also be integrated into the mobile data terminal, allowing the dispatcher to transfer information such as street maps, addresses, telephone numbers, and building information to firefighters on the scene.

FIGURE 5-25 The thermal imaging camera allows firefighters to see through thick smoke to detect victims. It can also be used to find hidden fires in enclosed areas. *Courtesy of James D. Richardson.*

ON SCENE

Urban flight caused the small rural community of Hill to experience a steady increase in population. With corresponding commercial growth and new residential housing, emergency calls became more than Hill's volunteer fire department could handle. The department had adequate equipment, but response times became longer because personnel were not always available to respond. This prompted the local government to begin the process of converting the volunteer department into a fully paid department. Volunteers grew apprehensive, fearing that their department, which had been serving the community for over 100 years, might be phased out as the paid department grew. A fire chief was hired to implement this transition. On his first day of duty, the new chief began to inventory the community's firefighting resources. He found that the fire station and all of its land, buildings, firefighting apparatus, and equipment were owned by the volunteer department. His first concern was to ensure that the volunteer department would give the paid department access to their resources until the paid department could establish its own resource pool.

Ensuring the availability of adequate resources is a primary function of a fire chief.

1. If you were the chief, how would you deal with this situation?
2. What course of action would you take to provide a smooth transition to a paid department?

Summary

Regardless of its size, every fire department must have access to the basic resources it needs to protect its community. These resources include designated facilities for administrative support, training, and emergency response, as well as the vehicles, tools, and equipment needed to respond safely and efficiently to any predictable emergency. It is humanly impossible to plan for every contingency, but fire departments should be in a constant state of readiness to face incidents that are likely to occur in their jurisdictions.

Review Questions

1. What are the advantages to having the fire department's headquarters office located in the fire station?
2. What is the difference between a Type I pumper and an initial attack pumper?
3. By what methods do aerial platforms operate? How do they differ from aerial ladder platforms?
4. List four types of specialized apparatus.
5. Describe the difference between a Level A and a Level B PPE ensemble.
6. Explain how a PASS device operates.

Suggested Reading

Acomb, David J,. and LeBoeuf, Roger M. (2008). Fire rescue stations and fire service training centers. In *Fire protection handbook* (20th ed., Volume II). Quincy, MA: National Fire Protection Association.

Cavette, Chris. (February 1, 2002).Wet and wild. *Fire Chief.* Retrieved March 10, 2006, from http://firechief.com/mag/firefighting_wet_wild/.

Ditmore, Chris. (June 4, 1997). Twin Bambis deliver, at 40,000 lbs. *SEI Industries.* Retrieved March 19, 2006, from http://www.sonnet.com/usr/wildfire/mi26t.html.

Dodson, David. (2004). *Firefighter's handbook: Essentials of firefighting and emergency response.* (2nd ed.). Clifton Park, NY: Delmar Publishing.

International Fire Service Training Association. (2011). *Fire and Emergency Services Orientation and Terminology* (5th ed.). Stillwater, OK: Author.

Laughlin, Jerry W. (2008). Training programs for fire and emergency service personnel. In *Fire protection handbook.* (20th ed., Volume II). Quincy, MA: National Fire Protection Association.

Police and emergency vehicles. (2006, October 2). *Neumann does communications.* Retrieved October 2, 2006, from Neumann Communications website: http://www.neumanncomm.com/police_and_emergency_vehicles.htm.

Endnotes

1. Darter, Michael J., Jr., and Stein, Gary P. (2012, October). U.S. fire department profile. Retrieved January 19, 2013, from National Fire Protection Association website: http://www.nfpa.org/itemDetail.asp?categoryID=2486&itemID=55953&URL=Research/Statistical%20reports/Fire%20service%20statistics/.

2. Ibid.

3. International Fire Service Training Association. (2011). *Fire service orientation and terminology* (5th ed.). Stillwater, OK: Author.

4. Forsman, Douglas. (2003). Training fire and emergency services. In *Fire protection handbook* (19th ed.). Quincy, MA: National Fire Protection Association.

5. Acomb, David J., and LeBoeuf, Roger M. (2008). Fire rescue stations and fire service training centers. In *Fire protection handbook* (20th ed., Volume II). Quincy, MA: National Fire Protection Association.

6. Tutterow, Robert. (2008). Fire department apparatus and equipment. In *Fire protection handbook* (20th ed., Vol. II). Quincy, MA: National Fire Protection Association.

7. International Fire Service Training Association, op cit.

8. Tutterow, op cit.

9. Avsec, Robert. (2012, December 20). How to spec a brush truck. *Fire Recruit.com*, on-line, http://www.firerecruit.com/articles/1383460-How-to-spec-a-brush-truck.

10. Cavette, Chris. (February 1, 2002). Wet and wild. *Fire Chief*. Retrieved March 10, 2006, from http://firechief.com/mag/firefighting_wet_wild/.

11. International Fire Service Training Association, op cit.

12. Ibid.

13. Ibid.

14. Tutterow, op cit.

15. Ibid.

16. Burke, Robert. (2005, July). HazMat team spotlight: Nashville, TN Fire Department. *Firehouse Magazine*.

17. International Fire Service Training Association, op cit.

18. Ibid.

19. Tutterow, op cit.

20. International Fire Service Training Association, op cit.

21. Ditmore, Chris. (1997, June 4). Twin Bambis deliver, at 40,000 lbs. *SEI Industries*. Retrieved March 19, 2006, from http://www.sonnet.com/usr/wildfire/mi26t.html.

22. Dodson, David. (2004). *Firefighter's handbook: Essentials of firefighting and emergency response* (2nd ed.). Clifton Park, NY: Delmar Publishing.

23. NFPA Standard 1971, National Fire Protection Association, Quincy, MA.

24. Ibid.

25. Rutledge, Marty. (2004). *Firefighter's handbook: Essentials of firefighting and emergency response* (2nd ed.). Clifton Park, NY: Delmar Publishing.

26. Ibid.

27. Teele, Bruce W. (2008). Fire and emergency services protective clothing and protective equipment. In *Fire protection handbook* (20th ed.). Quincy, MA: National Fire Protection Association.

28. Ibid.

29. Ibid.

6

Fire Dynamics

Courtesy of James D. Richardson.

KEY TERMS

absolute temperature, *p. 102*

absolute zero, *p. 102*

atom, *p. 97*

backdraft event, *p. 114*

British thermal unit, *p. 102*

caloric (or calorific) value, *p. 104*

chain reaction of self-sustained combustion, *p. 99*

electron, *p. 97*

endothermic reactions, *p. 106*

energy, *p. 101*

exothermic, *p. 98*

fire flow, *p. 105*

fire load, *p. 104*

fire tetrahedron, *p. 111*

flash point, *p. 99*

flashover (rapid fire progress event), *p. 112*

heat of combustion, *p. 104*

heat sink, *p. 102*

hydrophoric materials, *p. 107*

ignition temperature or fire point, *p. 99*

latent heat, *p. 102*

latent heat of fusion, *p. 102*

latent heat of vaporization, *p. 102*

matter, *p. 97*

molecule, *p. 97*

neutron, *p. 97*

normal atmospheric pressure, *p. 100*

oxidation, *p. 97*

proton, *p. 97*

pyrophoric material, *p. 107*

pyrolysis, *p. 99*

rate of heat release, *p. 105*

resistance, *p. 107*

spalling, *p. 103*

specific gravity, *p. 101*

specific heat, *p. 102*

static electricity heating, *p. 109*

stoichiometric reaction, *p. 104*

Type I combustion, *p. 110*

Type II combustion, *p. 110*

vapor density, *p. 100*

vapor pressure, *p. 99*

OBJECTIVES

After reading this chapter, you should be able to:

- Describe the components of an atom.
- Explain how an atom becomes reactive.
- Describe the process of combustion at the molecular level.
- Describe the three forms of matter.
- Explain how the process of combustion affects the different forms of matter.

- Identify and describe the different types of heat-producing energy.
- Describe the four stages of fire evolution.
- Identify the four classifications of fires and the principles involved in their extinguishment.

Introduction

Fire inspires both fascination and fear. When controlled, it can be the most productive force on earth, but when uncontrolled, it has enormous destructive potential. In this chapter, we will consider the scientific principles behind this awesome force of nature.

Fire is defined as a "rapid oxidation process that is a chemical reaction resulting in the evolution of light and heat in varying intensities."[1] This definition is accurate, brief, and readily accepted by the fire service, but there is much that it does not explain. For example, what is oxidation, and why is the oxidation process self-sustaining? What causes the evolution of heat and light? To answer these questions, the student of fire science must understand the chemistry, thermodynamics, and behavior of fire. An old military adage has it that good generals know and understand the enemy. The same holds true of the good firefighter. To outsmart the enemy, he or she must understand what fire is and how the environment in which it develops determines its behavior.

Atoms and Molecules

The **atom** is the smallest unit of a chemical element. It consists of three components: protons, neutrons, and electrons. The core of the atom, called the nucleus, is usually composed of protons and neutrons. The **protons** have a positive electrical charge, and the **neutrons** have no electrical charge. Orbiting the nucleus at different levels are negatively charged **electrons**, usually equal in number to the positively charged protons. The number of electrons in the outermost orbit of the atom determines if the element will react chemically with other elements. The innermost orbit holds at most two electrons (except in the case of the element hydrogen, which has only one electron). The second orbit can hold a maximum of 8 electrons; the third, a maximum of 18. The oxygen atom has a total of 8 electrons—2 in the inner orbit and the remaining 6 in the outer orbit. Because its outer orbit does not contain the full 8 electrons, oxygen bonds chemically with other elements in a process called **oxidation**, which will be explained more fully because it is crucial to fire chemistry.[2]

A **molecule** is formed when two or more atoms bind together in fixed proportions. The atoms that make up a molecule share pairs of electrons to form a chemical bond. A molecule may consist of atoms of the same chemical element—for example, the oxygen molecule (O_2) is comprised of two oxygen atoms. A molecule may also be composed of different elements—for example, the water molecule consists of two atoms of hydrogen and one atom of oxygen (H_2O). Molecules are in a constant state of motion. The amount of motion is determined by the physical state of the matter (see Figure 6-1). Temperature also has an effect on molecular movement. Molecular movement increases as the temperature increases.

Matter, Energy, and The Combustion Process

Matter can exist in three physical states: solid, liquid, or gas. Because it occupies space and has weight, matter is quantifiable—that is, it can be measured in dimensions, volume, and mass (weight). It has properties that give it a physical description (see Figure 6-2). Matter can be changed chemically or physically. Physical changes are usually reversible. The

atom ■ The smallest particle that constitutes the matter of an element.

proton ■ A particle in the nucleus of an atom that carries a positive electrical charge.

neutron ■ A particle in the nucleus of an atom that has no electrical charge.

electron ■ Negatively charged particles that orbit the nucleus of an atom.

oxidation ■ Chemical process that occurs when a substance combines with oxygen.

molecule ■ Two or more atoms joined together in a fixed proportions.

matter ■ The substance or substances of which all physical things are composed. It occupies space, has weight, and can be measured. It exists as a solid, liquid, or gas.

FIGURE 6-1 The atom is
the smallest particle of
matter. When two or
more atoms are joined, a
molecule is formed.

FIGURE 6-2 Matter can exist in a solid, liquid, or gaseous state. *Courtesy of James D. Richardson.*

physical state of matter is dependent on temperature and pressure. For example, at a normal sea-level atmospheric pressure of 14.7 pounds per square inch (psi) and at temperatures between 33°F and 211°F, water exists in a liquid state. At 32°F, water begins to solidify into ice; at 212°F, water will boil and convert to a gas. The chemical state of the water is not changed. Ice is solidified H_2O, and steam is vaporized H_2O. These processes can be reversed by melting the ice and condensing (cooling) the gas.

Chemical changes in matter are usually irreversible because the physical state of the matter changes as well as its molecular structure. New chemical substances are created. A prime example of this is the process of combustion (burning). We will next consider what happens in combustion at a molecular level.

SOLID MATTER

Solid matter is formed when molecules are so close together that they cannot change places. Free movement is not possible; the molecules can only vibrate in place. This bond gives the matter a definite form and shape. The closer the molecules are packed together, the denser the matter. Form and shape are maintained until a reactive force (such as heat) breaks the molecular bond.

When heat is applied to combustible solid matter, it is absorbed and dissipates throughout the mass. If heat is applied faster than the mass can dissipate it, the molecules begin to move faster and faster, collide with each other, and break apart. Some of the loose molecules bond with the oxygen in the atmosphere (oxidation), while other molecules remain in a free state. The oxidization process is an **exothermic** reaction, which means that heat is released when the bonding occurs. This heat, in turn, oxidizes more free molecules, which release more heat and causes more molecules in the solid to break away and oxidize, and so on.

This process is referred to as the chemical chain reaction, or the **chain reaction of self-sustained combustion**.[3] An example of this process can be seen by observing the combustion process in wood, the most common fuel that feeds the fires that firefighters face.

Wood is composed of cellulose. Derived from a living organism, cellulose contains moisture in its cellular structure. When wood is exposed to an ignition source, its molecular structure begins to change. As the surface temperature increases to 212°F, the moisture in the wood boils and converts to steam. As the temperature increases even more and penetrates deeper into the wood, more moisture is converted to steam. This process effectively dries the wood. When the surface temperature reaches 482°F, the wood begins to pyrolyze. **Pyrolysis** is the chemical decomposition of matter through the action of heat.[4] When pyrolysis is reached, wood releases combustible gases, and a layer of char forms over the heated surface. At this point, if a continuous source of ignition (a pilot flame) is present, the rapidly released combustible gases will ignite. As further heating is imposed on the wood surface, more combustible gases are released, and self-sustained combustion occurs. Many other factors that affect this process will be discussed later in this chapter.

LIQUID MATTER

When matter is in the liquid state, its molecules are in proximity to one another, but they are free to move in all directions, sliding past each other and changing places. This explains why liquids, unlike solids, have no characteristic shape. Instead, a liquid assumes the shape of the container in which it is confined and flows downward to occupy the lowest part of the container. The speed at which the molecules move is directly related to the temperature of the liquid. The higher the temperature, the faster the molecular movement.

The molecular movement in all liquids allows for the escape of molecules through the surface layer in the form of vapor. If the liquid is in an open container, all of the molecules eventually escape, and the liquid evaporates into a gaseous or vapor state. The speed of this process depends on the temperature of the liquid. Increasing the temperature increases the molecular movement, which increases the speed of evaporation.[5]

If the liquid is in a closed container and the temperature is kept constant, some of the molecules that escape from the surface collide with one another or the walls of the container and return to the liquid. An equilibrium is established when the rate of molecules leaving the surface of the liquid balances the rate of molecules returning to the liquid. When this equilibrium point is reached, the pressure exerted on the walls of the container by the molecules remaining in the gaseous state is referred to as the **vapor pressure** of the liquid. Vapor pressure is dependent on the molecular characteristics of the liquid and its temperature. The stronger the bond of the molecules, the lower the vapor pressure of the liquid at any given temperature. As the temperature increases, the molecular movement increases, and vapor pressure increases proportionally. When the vapor pressure of the liquid equals atmospheric pressure, the liquid boils (see Figure 6-3).

It is important for firefighters to understand the concept of temperature and its effect on the vapor pressure and evaporation rates of different liquids. Remember, a liquid does not burn. The gases that the liquid generates when heated to a given temperature create the flammable atmosphere. In this context, "heat" is a relative term, because some liquids generate vapors at very low temperatures. For example, gasoline has a vapor pressure range of 7 to 14.5 psi. It can generate flammable vapors at −40°F, which is its **flash point**.[6] The flash point is the temperature at which a liquid will emit vapors at a rate that will support a temporary flame as long as an ignition source is present. The **ignition temperature or fire point** is the temperature at which the liquid will emit vapors at a rate that will support a sustained flame after the ignition source is removed. The ignition temperature or fire point is slightly above the flash point temperature.

Remember that the atmospheric pressure at sea level is 14.7 psi. If the vapor pressure of a liquid is higher than 14.7 psi, it can exist as a liquid only when it is confined in a vessel

exothermic ■ Chemical reaction between two or more materials that changes the materials and produces heat.

chain reaction of self-sustained combustion ■ When a fuel is heated to the point that it releases flammable vapors, which burn and generate more heat on the fuel, which releases more flammable vapors.

pyrolysis ■ The chemical decomposition of a substance by heat.

vapor pressure ■ The pressure exerted on the walls of a closed container by a liquid when the number of molecules escaping from the liquid reaches an equilibrium with the number of molecules returning to the liquid.

flash point ■ Minimum temperature at which a liquid gives off enough vapors to form an ignitable mixture with air near the liquid's surface.

ignition temperature or fire point ■ The temperature at which the liquid will emit vapors at a rate that will support a sustained flame after the ignition source is removed. The ignition temperature or fire point is slightly above the flash point temperature.

FIGURE 6-3 When the vapor pressure of a liquid is greater than the atmospheric pressure, the liquid will boil until all of the liquid is converted into vapor. *Courtesy of James D. Richardson.*

or when the temperature of the liquid is sufficiently lowered. For example, propane, a flammable gas that is used extensively for heating and cooking, has a very high vapor pressure (24.5 psi at 0°F). It is sold to consumers in a liquid state in a pressurized container. When the valve on the container is opened, the pressure inside the vessel is lowered; the liquid begins to boil and vaporize; and it is converted into a gas. For propane to exist as a liquid at **normal atmospheric pressure**, the temperature would have to be lowered to approximately –20°F. At that temperature, the vapor pressure of propane is 11.5 psi.[7]

GASEOUS MATTER

When a substance is in a gaseous state, the molecules move at high speeds. They collide with one another and the walls of the container (if the gas is confined). When a substance is in a gaseous or vapor state, it occupies much more space than it would in a solid or liquid state. Its **vapor density** is determined by its molecular weight in comparison with the molecular weight of normal air, which is 29. If the vapor has a higher molecular weight than that of air, its vapor density is greater than 1.00 (the vapor density of air), and thus it tends to seek lower levels. If the vapor has a lower molecular weight than that of air, its vapor density is less than 1.00, and thus it tends to rise in the atmosphere (see Figure 6-4).

Vapor density can be determined through a very simple formula:

$$\text{Vapor Density} = \frac{\text{Molecular weight of the gas or vapor}}{29 \text{ (Molecular weight of normal air)}}$$

normal atmospheric pressure ■ The pressure of the atmosphere at sea level is 14.7 pounds per square inch.

vapor density ■ Weight of a given volume of pure vapor gas compared with the weight of an equal volume of dry air at the same temperature and pressure.

FIGURE 6-4 If the vapor density is greater than 1.00, the gas will seek lower levels. *Courtesy of James D. Richardson.*

The vapor density of propane can be determined as follows:

$$\text{Vapor density of propane} = \frac{44.096}{29} = 1.52$$

The vapor density of propane is one and one half times that of normal air. Therefore, propane vapors are heavier than air and will seek lower levels when released from confinement.

However, when we look at the **specific gravity** (its density compared with that of water) of the same substance in a liquid state, we may find that its behavior in the liquid state cannot be predicted by analogy to its behavior as a gas. For example, gasoline in a liquid state has a specific gravity of 0.72 to 0.76 and will "float" on top of water, which has a specific gravity of 1.0. Its vapor density, however, is 3.5. This means that when converted from a liquid to a gas, gasoline vapors will seek the lowest level possible. Its vapors are 3.5 times the weight of air. The liquid is lighter than water, but the vapor is heavier than air (see Figure 6-5).[8]

specific gravity ■ Weight of a substance compared with the weight of an equal volume of water at a given temperature.

STOP AND THINK 6-1

Why should firefighters be concerned with the form of matter with which they are faced on the emergency scene?

Energy is defined as the capacity or ability to do work. Energy itself cannot be observed. We can only perceive and measure the work that it performs. The First Law of Thermodynamics (also known as the Law of the Conservation of Mass and Energy) states, "Energy in all ordinary transformations is neither created nor destroyed, but is merely changed in form."[9] Energy exists as either potential energy (stored and waiting to be released) or kinetic energy (active or in motion). It can take a variety of forms, including heat, light, electrical, mechanical, nuclear, and chemical energy. Chemical, electrical, and mechanical energies are the three most common sources of the fires encountered by firefighters. These energies generate thermal or heat energy, which is why it is crucial for firefighters to understand the measurement and physics of heat. We will consider these four types of energies and their application to fire science.

energy ■ The capacity to do work.

FIGURE 6-5 If the specific gravity of a liquid is less than 1.0, it will float on top of the water's surface. *Courtesy of James D. Richardson.*

Heat Energy

Heat energy is a product of the other types of energy. It transfers from the hottest to the coldest substance and continues to do so until temperature equilibrium occurs. Heat is quantifiable, meaning it can be measured. For this purpose, in the United States, the fire service uses the English Customary system of measurement instead of the internationally recognized Système International (SI), which is based on the metric system. Even though the use of the Customary system is discouraged by the scientific community, it is used in this text because of its wide use and acceptance by the fire service in the United States.

FAHRENHEIT, RANKINE SCALES, AND BRITISH THERMAL UNITS

In the Customary system, temperature is measured using the Fahrenheit and Rankine scales, and heat units are measured in **British thermal units** (BTUs). The Fahrenheit (F) scale establishes a degree as a 1/180th differential between the melting point of ice and the boiling point of water. On this scale, water freezes at 32°F and boils at 212°F. A degree on the Rankine scale is equal to a degree on the Fahrenheit scale, but temperature is measured from absolute zero. **Absolute zero** is defined as the temperature at which all molecular movement ceases. **Absolute temperature** is measured from absolute zero. This has been established as −459.67°F or 0°R. A BTU is defined as the amount of heat needed to raise the temperature of one pound of water at 60°F by 1°F at constant pressure.[10]

SPECIFIC HEAT AND LATENT HEAT

Heat energy can be identified as specific heat or latent heat. **Specific heat** is the amount of heat absorbed by a substance as its temperature increases. **Latent heat** is the amount of heat absorbed or given up by a substance when it changes physical states without a change in temperature. For example, the specific heat of water is 1.0. It takes one BTU of heat to raise the temperature of one pound of water at 60°F by 1°F (specific heat); whereas it takes 143.4 BTUs of heat to convert one pound of 32°F ice to 32°F water. The energy absorbed in this process is known as the **latent heat of fusion**. It takes 180 BTUs of heat to raise one pound of 32°F water to 212°F water (specific heat). However, 970.3 BTUs of heat are needed to convert one pound of 212°F water to 212°F steam. The energy absorbed in this process is referred to as the **latent heat of vaporization**. This ability to absorb large quantities of heat during vaporization makes water a very effective extinguishing agent.

CONDUCTION, CONVECTION, AND RADIATION

The NFPA identifies three methods by which heat may be transferred: conduction, convection, and radiation. There are fire service agencies that identify direct flame contact as a method of heat transfer. This method of heat transfer is not recognized by the NFPA because it is believed that direct flame contact is a result of heat convection and radiation.

Conduction

Conduction is the ability of heat to move through matter. The matter can be solid, liquid, or gas. Although heat conductivity can occur in matter that is in the liquid or gaseous state, it is most recognized as the means of heat transfer through solid matter. The heat conductivity of matter is directly related to its density. When a solid is heated, molecules become agitated and begin to transfer that movement to other molecules throughout the matter in an effort to dissipate the heat. The ability of the matter to dissipate the heat is directly related to its density. The closer the molecules are to one another, the denser the matter and the higher its conductivity. Metals have high heat conductivity because they are the densest substances on earth (see Figure 6-6). Concrete and masonry are also good conductors of heat because they have the ability to accumulate and store heat faster than it can be dissipated and to release the heat when the surrounding environment cools. This phenomenon is referred to as serving as a **heat sink**. For example, when a fire burns inside

Glossary (margin)

British thermal unit ■ The amount of heat required to raise the temperature of one pound of water at 60°F by 1°F at constant pressure.

absolute zero ■ The temperature at which all molecular movement ceases. This is established at −459.67° Fahrenheit, 0° Rankine (based on the Fahrenheit scale), −273° Celsius, and 0 Kelvin (based on the Celsius scale).

absolute temperature ■ Temperature that is measured from absolute zero.

specific heat ■ The amount of heat required to raise the temperature of a specified quantity of material by 1°C.

latent heat ■ The amount of heat energy absorbed or released by a substance when it changes physical states without a change in temperature.

latent heat of fusion ■ The amount of heat absorbed by a substance when converting from a solid to a liquid at the same temperature.

latent heat of vaporization ■ The amount of heat absorbed by a substance when converting from a liquid to a gas or vapor at the same temperature.

Heat sink ■ A material that has the ability to absorb and store heat and release it after the environment cools.

FIGURE 6-6 Conduction is the transfer of heat through a solid object. *Courtesy of James D. Richardson.*

of a concrete building, the concrete readily absorbs the heat and stores it. After the fire is extinguished and the environment begins to cool, the concrete will release the heat through the process of convection. If concrete or masonry absorbs more heat than it can store, the moisture in the material will expand and begin to crack and break away from the structure. This is called **spalling**.

Convection

Heat transfer by convection usually occurs in gases or liquids. When a gas is heated, its molecules become very active and begin to move freely away from one another so that there are fewer molecules in a given volume of the heated gas than in the same volume of cool gas. Because there are fewer molecules, the gas becomes lighter and begins to rise to a higher elevation and to transfer its heat to the surrounding areas in the upper levels. This process is known as convective heat transfer (see Figure 6-7).[11] During the process of combustion, heated gases will follow the path of least resistance. They will continue to travel upward until they are stopped by on obstruction, or cool to the point that they can no longer have buoyancy. Vertical openings, such as elevator shafts, dumb waiters, and utility chases act as chimneys to allow these products of combustion to travel upward.

Spalling ▪ When heat allows moisture in concrete to expand, causing the concrete to crack or break away from the structure.

FIGURE 6-7 Convection is the transfer of heat through circulating air or liquid. *Courtesy of J. D. Richardson.*

Radiation

Heat transfer by radiation does not require direct contact with the heat source or with convective heat currents. Radiant energy is transmitted as an electromagnetic wave that does not need to travel through any medium. The radiated heat is in the form of infrared rays, which are invisible to the human eye but behave in the same manner as visible light rays. Infrared heat rays travel in a straight line until they are absorbed or reflected by an object. They can pass through any object that is transparent. The color, opacity, and exposed surface area of a substance determine the degree of absorption of radiated heat. Infrared rays are readily absorbed by black or dark colors and by solid matter that has a large exposed surface area. White, light-colored, or mirrored surfaces reflect heat or infrared rays. Radiated heat is a major concern to firefighters when defending exposed surfaces. It has been estimated that 30% to 50% of the total amount of heat energy released during large fires of ordinary fuels is in the form of radiant energy.[12]

Some members of the fire service community describe a fourth method of heat transfer, referred to as direct flame contact. There is no firm consensus on this theory and many in the fire service scientific community believe that the transfer of heat is a result of convection currents and radiation.[13]

Conduction, convection, and radiation do not act independently during a fire. All three phenomena act together to increase fire growth and movement.

Chemical Energy

We have noted that chemical energy is a result of oxidation reactions. When a sufficient amount of heat is applied to a substance, it causes the breakdown of the molecular structure of the substance (fuel). The molecules that break free combine with oxygen in a chemical reaction called oxidation. Oxidation is exothermic, meaning that it produces heat, which in turn maintains the fuel at a temperature necessary for further oxidation to continue. This chemical process produces combustion and is the most common source of the fires dealt with by the fire service.

HEAT OF COMBUSTION OR CALORIC VALUE

The amount of heat energy that a substance releases during the combustion process is known as its **heat of combustion** or **caloric (or calorific) value**. This varies with the type of substance and its mass. Calculation of the heat of combustion is based on a **stoichiometric** (complete combustion) **reaction**. This can occur only when the amount of fuel and the amount of oxygen are perfectly apportioned to provide complete combustion, leaving no fuel or oxygen at the end of the process. Although this type of combustion reaction can occur only in laboratory conditions, it is used in the fire service to calculate fire loads.[14]

Fire Load Computation

A **fire load** is defined as the potential heat energy of a fire and is calculated by multiplying the caloric value of a unit mass of the fuel feeding a fire by the amount of fuel available. In SI units, caloric value is expressed in joules per gram; the English Customary system, in BTUs per pound, or per cubic foot if the substance is a gas or vapor.

In the United States, the fire service uses a rule-of-thumb estimate in computing caloric value. If the substance is organic (such as cotton, wood, paper, or plant matter), the estimated caloric value is 8,000 BTUs per pound. If the substance is a plastic or hydrocarbon, the estimated calorific value is 16,000 BTUs per pound.[15] These are theoretical figures because, as we have noted, complete combustion does not occur under real fire conditions. However, they are useful "ball-park" figures to work from when estimating fire loads. For example, 1,000 pounds of cotton and 1,000 pounds of foam plastics produce an estimated total fire load of 24,000,000 BTUs of heat (1,000 × 8,000 + 1,000 × 16,000 = 24,000,000).

heat of combustion ■ Total amount of thermal energy that could be generated by the combustion reaction if a fuel were completely burned.

caloric (or calorific) value ■ The amount of heat necessary to raise the temperature of one gram of water by 1° Celsius by 1°.

stoichiometric reaction ■ A reaction in which all reactants are present in fixed, definite proportions for the reaction to go to completion. The most violent of all combustion processes occurs when the fuel and oxygen proportions provide complete combustion, leaving no fuel or oxygen residue.

fire load ■ The amount of fuel in a compartment expressed in pounds per square foot obtained by dividing the amount of fuel present by the floor area.

Fire Flow Computation

Why is it important to understand the concepts of caloric values and latent heat? We must have these figures in order to estimate the amount of water needed to extinguish a fire. This is known at the fire flow computation. There is a series of simple mathematical steps that are used to estimate fire flow requirements.

The first step is to determine the amount of BTUs one gallon of water will absorb. This can be calculated as follows:

Water in most public water systems is carried in piping that is underground. The temperature of this water usually ranges between 60°F and 70°F. For ease of calculation, we will establish our water temperature at 60°F. To raise one pound of 60°F water to 212°F water takes 152 BTUs of heat: $212 - 60 = 152$. Next we must add the latent heat of vaporization. It takes 970.3 BTUs of heat to convert one pound of 212°F water to 212°F steam. This must be added to our first figure: $152 + 970.3 = 1122.3$ BTUs.

One pound of 60°F water, when converted to 212°F steam, will absorb 1122.3 BTUs of heat. To determine the amount of BTUs one gallon of water will absorb, multiply 1122.3 by 8.33 (the weight of one gallon of water). Therefore, if one pound of 60°F water will absorb 1122.3 BTUs of heat, one gallon of water will absorb $1122.3 \times 8.33 = 9348.759$ BTUs of heat. For simplicity in calculation, this can be rounded off to 9349.

The next step is to calculate our fire load. If a warehouse has 1,000 pounds of cotton and 1,000 pounds of plastics, the fire load would be $1,000 \times 8,000 = 8,000,000$ BTUs and $1,000 \times 16,000 = 16,000,000$ BTUs. The total fire load would be $8,000,000 + 16,000,000 = 24,000,000$ BTUs. To determine the fire flow needed to extinguish this type of fire, divide $24,000,000$ by $9349 = 2567$ gallon per minute (GPM).

Rate of Heat Release

Fire load, the total amount of heat that can be released by a fire, is only part of the equation that must be used when computing **fire flow** (also referred to as critical flow), the amount of water measured in gallons per minute (GPMs) that will be needed to extinguish a fire. The firefighter must also be prepared to estimate the rate of heat release. The **rate of heat release** is the speed at which heat energy is released from the substance. A one-pound block of wood and a one-pound pile of wood chips both have the potential to generate approximately 8,000 BTUs of heat. However, the one-pound block of wood will burn slower and release heat over a longer period of time than the burning pile of wood shavings. It is very important to identify the form of the fuel in order to determine rate of fire flow. Fire flow can be regulated by the use of different diameter hose lines and nozzle tips. For example, one and one-half inch handlines can deliver up to 125 GPM. One and three-quarter inch handlines can deliver up to 200 GPM. Two and one-half inch handlines can deliver up to 250 GPM. Large-diameter hose is used for heavy stream devices that have the capability of delivering up to 2,000 GPM.

Other chemical energy reactions that are of concern to firefighters are spontaneous heating, the heat of decomposition, the heat of solution, and the heat of reaction.

fire flow ■ Quantity of water available for firefighting in a given area.

rate of heat release ■ The speed at which a fuel releases its caloric value when burning.

SPONTANEOUS HEATING

Spontaneous heating takes place when a substance increases in temperature without the assistance of an outside ignition source. When oxygen combines with an organic substance, oxidation occurs, and heat is generated. If this heat cannot escape, the temperature will increase. The increase in temperature increases the rate of the chemical chain reaction process, doubling the rate with every 50°F increase in temperature. When the heat energy is generated faster than it can be dissipated, spontaneous ignition occurs. This process causes the fires that erupt when rags have been soaked in vegetable oils and improperly stored. Similarly, clothing has been known to ignite if it is stored in a

FIGURE 6-8 If heat is not allowed to dissipate, spontaneous heating can occur. *Courtesy of James D. Richardson.*

closed bin immediately after being heated in a clothes dryer (see Figure 6-8). In order for spontaneous ignition to occur, the following series of events must take place:

- The generation of heat through oxidation must be high enough to raise the temperature to the substance's ignition temperature.
- There must be proper ventilation to provide an adequate air supply to support combustion.
- The substance around the fuel must provide insulating properties that prevent heat dissipation.[16]

HEAT OF DECOMPOSITION

endothermic reactions ■ Chemical reactions in which substances absorb heat energy.

The heat of decomposition is generated during the breakdown of unstable compounds that were formed by **endothermic reactions** (reactions that absorb heat instead of giving off heat). These compounds have molecular structures that hold the fuel and the oxidizer in the same molecules. Once decomposition begins, it will generate large amounts of chemical heat energy.[17] In the early years of the movie industry, film was made of cellulose nitrate, an unstable compound. If this film were not stored in a cool, dark, dry environment, it would readily decompose and could spontaneously burst into flames. This fact was realized only after film-locker fires destroyed many classic and historic films. Even films that were not actually burned have been badly damaged by chemical decomposition. Cellulose nitrate film was highly flammable and able to burn in an oxygen-deficient atmosphere. The main hazard was the large amounts of toxic gases that it generated while burning. On May 15, 1929, a fire erupted in the x-ray film storage area of a clinic in Cleveland, Ohio. One hundred twenty-two people died from the toxic fumes that rapidly spread throughout the multi-story building.[18] Production of cellulose nitrate film was discontinued in 1951 and replaced with a slower burning and more stable cellulose acetate compound. Bacterial activity in agricultural products can also generate enough heat to cause ignition. This is common in landscaping facilities that store large quantities of mulch or composing materials.

HEAT OF SOLUTION

The heat of solution is the heat released, or absorbed, when a substance is dissolved in a liquid. The reaction can be endothermic or exothermic. If the reaction is endothermic, the substance will absorb heat and the solution will become cool. If the reaction is exothermic, the temperature of the solution will increase and release heat. For example, if ammonium nitrate or ammonium chloride is mixed with water, the water will absorb heat and become very cold. If sodium thiosulfate is mixed with water, an exothermic reaction will occur and the solution will become heated. Medical technicians use this type of solution reaction, in a controlled state, in the form of "cold" or "hot" packs to treat injuries or medical emergencies involving strains, sprains, heat, cold, or fever. The packs contain two compartments—one contains the chemical and the other contains water. When the pack is squeezed or struck against an object, the layer of film separating the chemical from the water is broken and the contents of the pack form the reactive solution.

HEAT OF REACTION

The heat of reaction is the heat released when a substance comes in contact with a liquid or with air. The intensity of the heat released during this reaction is capable of causing ignition. For example, magnesium is relatively stable when placed in cold water. However, when it is exposed to boiling water or steam, the heated metal reacts with the steam or boiling water, producing hydrogen, which ignites. Other metals, such as lithium or lithium hydride, react with water to form flammable gases. Water-reactive substances are classified as **hydrophoric materials**. **Pyrophoric materials** are substances that can spontaneously ignite when exposed to air. Many pyrophoric materials also react violently with water.

hydrophoric materials ▪ Materials that react with water.

pyrophoric materials ▪ Materials that react with air.

STOP AND THINK 6-2

What is the only way firefighters can protect themselves against an unexpected event on the emergency scene?

Electrical Energy

Electrical energy is the passing of electrons from one atom to another. Some atoms have electrons that are loosely attached. These electrons are in the outermost orbit or shell of the atom, farthest from the nucleus, and therefore have less attraction to the protons than electrons in closer orbits or shells. When an atom loses an electron, it has more protons than electrons and becomes positively charged. All atoms want to be "balanced." The positively charged atom has a strong attraction for free negatively charged electrons. Negatively charged electrons also seek positively charged atoms. The attraction increases in intensity as the number of positively charged atoms and negatively charged electrons increase. This attraction is called the *charge*. When electrons move through matter from one atom to another, one electron is gained, and another is lost. This movement is called a *current*. The matter that the electrons move through is called the *conductor*. Some matter is better suited for electrical conduction than are others.

FORMS OF ELECTRICAL HEAT ENERGY

The six forms of electrical heat energy that are important to a firefighter are resistance, arcing, sparking, static, lightning, and induction.

Resistance Heating

The opposition that a substance offers to the passage of an electric current is called its **resistance**. If the electrons in a substance are tightly held, a current cannot move through it very well. This type of substance is called an *insulator*. Materials such as rubber, glass,

resistance ▪ The opposition to the flow of an electric current in a conductor component.

and dry air are excellent insulators and have very high resistance qualities. Copper, silver, aluminum, and steel—in fact, most metals—are excellent conductors because they hold their electrons loosely, enabling a current to pass readily through the material.[19]

The resistance in a conductor depends on its thickness and length in relation to the force of current flowing through it. As the speed of the electron movement from atom to atom increases (increased current flow), some electrons collide with the conductor's molecules, and the collision causes the molecules to break down. The energy that held the molecules together is released in the form of heat. This process is referred to as *resistance heating*. Resistance heating, when controlled, is very useful. It is used in electric appliances to provide heat for cooking (see Figure 6-9) and for heating structures. The incandescent light bulb is a perfect example of resistance heating. However, when it is not controlled, resistance heating can produce extremely high temperatures, causing failure of the conductor or ignition of materials around it.

Arcing

Arcing heating is caused when an electric current is interrupted. This can occur intentionally when contacts are separated in a switch (see Figure 6-10) or accidentally when electrical contacts are loosely connected. In either case, electrical energy is discharged across an air gap, and the heat of that energy can increase until it is sufficient to ignite insulation, nearby combustible materials, or even a flammable atmosphere.

Sparking

Sparking heating is different from arcing heating in that it is a short-term, one-time event in which there is a high-voltage discharge but the total energy output is low. Under normal environmental conditions, this type of heating is usually not capable of causing ignition because the heat energy is not sustained long enough. However, sparking heating is very dangerous in flammable atmospheres where the vapor/air mixtures are within their flammable range.

Static

Static electricity heating is related to sparking heating. It can occur when two materials are brought together and then separated, when fluids move through a conduit, or when finely divided solid (dust) particles touch as they pass each other. During these occurrences, it is possible for one surface to pick up electrons from the other. The surface that gains electrons becomes negatively charged, while the surface that loses electrons becomes positively charged. If proper bonding or grounding is not provided, sufficient electrical charge may accumulate to produce a spark. This spark is short term, containing very high voltage, but very low current. It is usually not hot enough to ignite ordinary combustible material around it. However, static sparks have been known to ignite flammable vapors, gases, certain explosives, and dust clouds.[20]

static electricity heating ■ The heat generated by static electricity arcing.

Lightning is a form of static electricity that occurs when the action of water and ice crystals within clouds causes an imbalance in the electrical charge of the clouds. As negatively charged clouds pass over the ground, the ground becomes positively charged. Electrons will jump from cloud to cloud or to the ground and back in the form of a large static spark, which is what we know as lightning. The heat that is generated by this phenomenon will readily ignite any combustible material that it encounters. It can even cause masonry to explode. The only way to protect against this powerful type of electrical energy heating is by dissipating the charge to the ground and diverting it away from combustible materials through large metal cables and conductors.

Induction

If a conductor is subjected to an alternating magnetic field, a flow of current is produced. As the frequency of the alternating magnetic field increases, the flow of the current increases and produces resistance heating. Any conductive material within the magnetic field will be heated. This is called *induction heating*. It is most effective on magnetic metal, but can be used with nonmagnetic metal.

STOP AND THINK 6-3

What are the various types of electrical energy that can pose a threat to firefighters on the emergency scene?

Mechanical Energy

Mechanical heat energy, which can be generated by either friction or compression, is responsible for a large number of fires each year. The vast majority of these fires are a result of friction heat.

FIGURE 6-11 Sparks created from friction caused by metal work operations may ignite explosive atmospheres. *Courtesy of Dianne Richardson.*

FRICTION

The National Fire Protection Association's *Fire Protection Handbook* defines friction heat as the mechanical energy used in overcoming the resistance to motion when two solids are rubbed together.[21] Friction is a part of everyday life, but when it is generated faster than it can be dissipated, it becomes a hazard. Friction between two metal objects, for example, can generate sparks ranging in temperature from 2,500°F to 5,400°F, depending on the types of metal involved. Although these temperatures are high, the total heat content of the spark is usually low, and ignition of combustible materials is unlikely unless flammable liquids, gases, or vapors are present (see Figure 6-11).

COMPRESSION

Heat of compression occurs when a gas is compressed suddenly. As the molecules in the gas are forced closer together, they begin to collide and break apart, releasing heat. If the ambient vapor/air mixture is in the flammable range of the gas, combustion will take place. The diesel engine operates on this premise. Firefighters experience this phenomenon when they refill SCBA cylinders. If the cylinder is filled too fast, the air inside the cylinder will heat and expand. When the cylinder cools, the air pressure will decrease and the cylinder gauge will indicate that the cylinder is low on air. If the cylinder is filled according to the manufacturer's guidelines, the air will not be heated and the cylinder will be filled with the proper amount of air and pressure.

COMBUSTION PROCESS

The process of combustion can be separated into two types. **Type I combustion** is a process of direct oxidation, which means that pyrolysis or chemical decomposition is not necessary for the combustible material to burn. It combines directly with oxygen and does not need to convert to a gaseous or vapor state. Sulfur and charcoal are examples of materials that undergo Type I combustion (see Figure 6-12).

    Type II combustion is the process of sequential oxidation, as in wood burning. The combustible solid or liquid material must go through the chemical decomposition process of pyrolysis before combustion occurs; that is, the molecular structure of the substance must be broken down to produce combustible gases, which will react with oxygen in the

Type I combustion ■ Direct oxidation. Pyrolysis or decomposition of the matter is not required to occur in order to sustain combustion.

Type II combustion ■ Sequential oxidation. The molecular structure of the substance must be broken down to produce combustible gases, which will react with oxygen in the air.

FIGURE 6-12 The burning of charcoal is a Type I combustion process. *Courtesy of James D. Richardson.*

air. For example, when a match is ignited, the flame slowly moves down the wooden shaft. If you look closely, you will see that the flame does not touch the shaft. The flame is actually burning the vapors that are being released by the heated wood. Type II combustion occurs in almost all liquids and solids that contain carbon or hydrogen in their molecular structure (see Figure 6-13).

Previously, students of fire science were taught that the combustion process required three elements: fuel, oxygen, and heat. This was called the fire triangle. However, research has proven that another element is needed for combustion to take place: a self-sustained chemical chain reaction. In fact, the NFPA defines the combustion process as a self-sustained chemical reaction producing energy that causes more reactions of the same kind. By adding this fourth component, the fire triangle became known as the **fire tetrahedron**. If one of these components is removed from the process, combustion cannot take place (see Figure 6-14).

Fire tetrahedron ■ Identifies the four basic components required for the combustion process to take place.

BURNING PROCESS

The burning process involves a predicable sequence of four stages: ignition, growth, full development, and decay.

FIGURE 6-13 The burning of wood is a Type II combustion process. *Courtesy of James D. Richardson.*

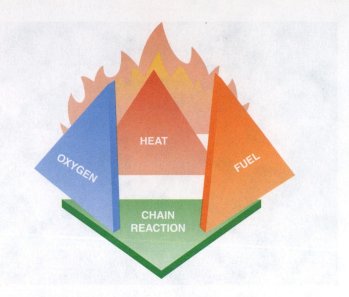

Ignition Stage

The fuel is heated to a point at which flammable vapors are released and ignite (see Figure 6-15).

Growth Stage

At this point, the flame is small but begins to heat other areas of the fuel and combustibles in close proximity. The chain reaction process increases. The speed of the growth stage and size of the fire depend on the amount of oxygen and fuel that is available. Normal air contains approximately 21% oxygen. The combustion process can be maintained at oxygen concentrations as low as 14%. The intensity of a fire inside a structure depends not only on the amount of fuel available but also on the amount of oxygen available to support the combustion process. To understand rapid fire progress, consider a fire burning in a confined area, such as a room. Flammable gases and carbon by-products (including highly flammable carbon monoxide gas) formed by the combustion process collect at the ceiling and begin to travel laterally as the fire increases. If there is adequate oxygen in the room during the growth stage (15% to 21%), the gases are ignited and burn freely as they move along the ceiling. This event is referred to as a rollover and precedes **flashover** by a

flashover (rapid fire progress event) ■ Stage of a fire at which all surfaces and objects within a space have been heated to their ignition temperature and flame breaks out almost at once over the surface of all objects in the space.

FIGURE 6-15 The ignition stage is the first stage of the burning sequence.
Courtesy of J. D. Richardson.

FIGURE 6-16 During the growth stage, the fire is heating surrounding fuel. *Courtesy of J. D. Richardson.*

few seconds. The heat from these burning gases and by-products is radiated down and begins heating the unburned contents of the room. The heated contents begin to pyrolyze and emit flammable vapors. Vapor emission increases with increased pyrolysis until the vapors reach their ignition temperature. This ignition or flashover is all-encompassing. It appears that the entire contents of the room ignite simultaneously. Temperatures can reach as high as 2,000°F. It is at this point that the fire suddenly transitions from the growth stage to the *fully developed stage*. Flashover does not always occur. There must be adequate fuel and oxygen for this event to develop (see Figure 6-16).[22]

Fully Developed Stage

At this point, all available fuels are burning. There is adequate fuel, heat, oxygen, and chemical chain reaction. The fire has ventilated itself and accumulated, unburned gases move along the ceiling and burn as they exit the compartment and ignite. Interior temperatures range from 1292°F to 2192°F[23] (see Figure 6-17).

FIGURE 6-17 During the fully developed stage, all of the fuel is being consumed. *Courtesy of J. D. Richardson.*

FIGURE 6-18 In the decay stage, all fuel is consumed and the fire reduces in size. *courtesy of J. D. Richardson.*

Decay Stage

This is the final stage of the combustion process. Decay occurs as the amount of fuel or oxygen is diminished by the chemical decomposition of combustion. If the same enclosed room is underventilated, the fire will decay because of the lack of oxygen. There is potential for a **backdraft event** to occur. In this case, heat and fuel are in plentiful supply, but there is not sufficient oxygen to support combustion. Superheated gases accumulate and fill the room. Three sides of the fire tetrahedron are present: heat, fuel, and a chemical chain reaction. Only oxygen is missing. The gases are heated well above their ignition temperatures; when oxygen enters the room, the gases will suddenly ignite, perhaps explosively.[24]

When the fuel is exhausted, the fire will extinguish itself (see Figure 6-18).[25]

backdraft event ■ Instantaneous explosion or rapid burning of superheated gases that occurs when oxygen is introduced into an oxygen-depleted space.

Fire Classification

Fires are classified as Class A, B, C, D, and K. Each classification is based on the type of fuel that is expected to burn. The burning characteristics of each fuel class are different and require different methods of extinguishment.

CLASS A: ORDINARY COMBUSTIBLES

The fuels of Class A fires are considered ordinary combustibles and include wood, paper, plastic, and organic solids. The primary means of extinguishment for Class A fires is cooling with water, because water has a high heat absorption capability when it is converted into steam. One gallon of 60°F water can absorb over 9,300 BTUs of heat when it is completely converted into steam.

CLASS B: FLAMMABLE LIQUIDS

Class B fires involve flammable liquids, such as some oils, gasoline, and kerosene. The primary means of extinguishment for these types of fires is through smothering. The oxygen supply must be cut off from the fuel to interrupt the fire tetrahedron. Various types of water-based foam extinguishing agents are used in this process. They function by creating a film over the top of the burning fuel to inhibit the production of flammable vapors. The water in the foam also acts to cool the liquid below its ignition temperature. Dry chemical agents are also used to provide extinguishment by disrupting the chemical chain reaction of combustion.

CLASS C: ENERGIZED ELECTRICAL FIRES

Class C fires are initiated by electricity. In the majority of cases, the electric current is only the original heat source that ignites surrounding material. When the electrical supply is removed, the fire continues to burn but is reclassified as a Class A or B fire.

CLASS D: FLAMMABLE METALS

Combustible metal fires are designated as Class D fires. Because metal fires are caused by chemical reactions, the usual method of extinguishment is disruption of the chemical chain reaction. There is no single extinguishing agent that can be used on all metals, and some agents are designed to be used on only one type. Water can be used only on certain types of metal fires, and only if extremely large volumes are available. Small quantities of water can cause a violent escalation in the fire. The intense heat generated by some burning metals has the capability to break down the molecular structure of water into its constituents of hydrogen and oxygen and burn it away in a high-intensity display of heat and light.[26]

CLASS K: HIGH-TEMPERATURE COOKING OILS

On July 1, 1998, NFPA 10, *Standard for Portable Fire Extinguishers*, mandated that all extinguishing systems designed for the protection of cooking appliances using vegetable oils or animal fats as a cooking medium be labeled as Class K fires.[27] Before this regulation went into effect, cooking oil or grease fires were classified as Class B fires. Dry chemical agents, such as sodium bicarbonate and potassium bicarbonate, were used for extinguishment. However, with the development of high-temperature cooking oils, these dry chemical agents became less efficient. Class K extinguishers are designed to use a wet chemical agent that is a mixture of potassium acetate and potassium citrate. It extinguishes the fire by chemically reacting with the fatty acids in the cooking oil or fat to form a layer of soap or foam over the top of the fuel, separating the oxygen from the fuel. This reaction is called *saponification*. The wet chemical agent also cools the cooking appliance and oil to reduce the possibility of reignition.[28]

 ON SCENE

The Champion Forest Volunteer Fire Department, near Houston, Texas, responds to a call at a large area feed store. On their arrival, the firefighters find the building heavily involved in fire. The older part of the structure was built with heavy timber construction, which provides a very high fire load. A large metal-structure warehouse has been added to the original building. This warehouse is 100 feet wide, 200 feet long, and 24 feet high. So far, the fire is confined to the original part of the structure, which contains a variety of chemicals and insecticides used in ranching and farming. The exposed warehouse is full of bagged chemical fertilizers. Other exposed structures nearby include several bulk fuel storage tanks, area residences, and an elementary school that has scheduled class the next morning.

1. If you were the incident commander, what would be your first concern?
2. What hazards can be expected at this type of fire?
3. Would water be an effective extinguishing agent? If so, what would be the results of using water on this fire?
4. What protective measures would you provide for the firefighters?

Summary

Uncontrolled fire is a living and breathing beast that seeks to consume everything in its path. It destroys matter by altering its chemical composition. Fire can reduce matter to its basic elements or create new compounds through the process of combustion. Knowledge of the chemistry and physics of fire is vital to the firefighter's success because it makes it possible to predict a fire's behavior and to initiate effective extinguishment strategies.

Heat is transferred by three methods: conduction, convection, and radiation. Conduction is the transfer of heat through matter. The heat conductivity of matter is directly related to its density; the more dense the matter, the higher the conductivity. Hence, conduction usually refers to the transfer of heat through solids. Heat transfer by convection occurs by the circulation of gases or liquids. The hotter the temperature of a gas, the fewer molecules there will be in a given volume, making the gas lighter than it would be in a cooler state. During combustion, the lighter, heated gas rises to higher levels and transfers its heat to surrounding areas in the upper levels. Radiated heat is an electromagnetic infrared wave and does not require a transfer medium. About 30% to 50% of the total amount of heat energy released during large fires of ordinary fuels is in the form of radiant energy.

Energy cannot be created or destroyed, only changed in form. Chemical, mechanical, and electrical energy are the most frequent sources of ignition.

Heat encountered by the fire service is commonly generated by chemical energy in the form of an oxidation reaction. In oxidation, oxygen in the air combines with the gases produced by a burning fuel to produce a new gas. For example, the carbon of burning wood is oxidized to produce carbon dioxide (CO_2). The reaction is exothermic, meaning that it produces heat, which in turn heats more fuel, setting off a chain reaction of sustained combustion. Every fuel has a caloric value. If the caloric value is known, an estimate of anticipated heat release can be calculated.

Electrical energy is produced by the passing of electrons from one atom to another. When an atom loses an electron, it becomes positively charged and seeks a negatively charged free electron. The movement of electrons from one atom to another is called current. There are six forms of electrical heat energy: resistance, arcing, sparking, static, lightning, and induction.

There are two types of mechanical energy: friction and compression. Friction is the major cause of most mechanical heat energy fires. Friction sparks can generate temperatures between 2,500°F and 5,400°F. These are extremely hazardous if flammable liquids, gases, or vapors are present. Gases that are compressed suddenly will release heat. If the heat reaches the flammable range of the gas, combustion takes place. The heat produced is known as the heat of compression.

There are two types of combustion. Type I combustion is direct oxidation. Pyrolysis or decomposition of the matter is not required to occur in order to sustain combustion. Type II combustion is sequential oxidation and does require pyrolysis. The molecular structure of the matter must be broken down to produce combustible gases that react with air. Four components are necessary for combustion to occur: fuel, oxygen, heat, and a chemical self-sustained chain reaction. If one component is absent, combustion cannot occur.

Fire develops in four stages: ignition, growth, fully developed, and decay. There are five fire classifications that are defined according to the type of fuel involved: Class A (ordinary combustibles: wood, paper, plastics, and so on), Class B (flammable liquids and gases), Class C (energized electrical equipment), Class D (combustible metals), and Class K (cooking oils and fats).

If a firefighter has a thorough understanding of the chemistry and physics of fire, he or she will be able to predict, with a reasonable amount of accuracy, how the fire will act and react under certain conditions. This knowledge is not only essential to success in fighting the fire but is also a basic ingredient in the recipe for the firefighter's survival. Firefighting is one of the most hazardous professions in the world; the more knowledge firefighters bring to the scene, the more likely they will return home safely.

Review Questions

1. What are the components of an atom? What identifying features do these components have?
2. What occurs at the molecular level during the process of combustion?
3. In what forms can matter be found? How do they differ molecularly?
4. Describe the different types of heat energy.
5. Identify the stages of a fire's evolution. Describe events that can occur in each stage.
6. Identify and describe the four main classifications of fire.

Suggested Reading

Brannigan, Francis. (2007). *Building construction for the fire service* (4th ed.). Quincy, MA: National Fire Protection Association.

California Energy Commission. (2003). Resistance and static electricity. *Energy story*. Retrieved October 30, 2005, from http://www.energyquest.ca.gov/story/chapter03.html.

California Energy Commission. (2003). What is electricity? *Energy story* (Chapter 2). Retrieved October 30, 2005, from http://www.energyquest.ca.gov/story/chapter02.htm.

Fire Equipment Manufacturers' Association. (2006, September 8). Statement on class K extinguishers. *The life safety group*. Retrieved September 8, 2006, from http://www.femalifesafety.org.

Flame Engineering. (n.d.). Propane technical information. *Propane & safety information*. Retrieved October 7, 2005, from Flame Engineering website: http://www.flameengineering.com/Propane_Info.html.

Gorbett, Gregory E., and Pharr, James L. (2011). *Fire dynamics*. Upper Saddle River, NJ: Pearson Education.

Grimwood, Paul. (2003, February). *Flashover—A firefighter's worst nightmare!* Retrieved November 8, 2005, from http://www.firetactics.com.

Hall, Richard, & Adams, Barbara (Eds.). (2008). *Essentials of fire fighting* (5th ed.). Stillwater, OK: International Fire Service Training Association.

National Fire Protection Association. (2008). *Fire protection handbook* (20th ed.). Quincy, MA: Author.

Solids, liquids and gases—Their "molecular" structure. (n.d.). *Science for all*. Retrieved October 16, 2005, from http://www.apqj64.dsl.pipex.com/sfa/slg_structure_m.htm.

Endnotes

1. National Fire Protection Association. (2008). *NFPA Standard 921*. Quincy, MA: Author.
2. Ibid.
3. Miale, Frank. (1999). *Fire fighter's handbook: Essentials of fire fighting and emergency response*. Clifton Park, NY: Delmar Publishing.
4. Scarbrough, Don R. (2008). *Fire protection handbook* (20th ed.). Quincy, MA: National Fire Protection Association.
5. Stowell, Frederick M., and Murnane, Lynne. (2013). *Essentials of fire fighting and fire department operations*. (Chapter 5). Stillwater, OK: International Fire Service Training Association, Oklahoma State University.
6. Equiva Services—MSDS.(n.d.). Material safety data sheet number 401430ET-2. *Product and company identification*. Retrieved October 16, 2005, from Equiva Services—MSDS website: http://www.equivashellmsds.com/getsinglemsds.asp?ID=201791.
7. Flame Engineering. (n.d.). Propane technical information. *Propane & safety information*. Retrieved January 5, 2013, from Flame Engineering website: http://www.flameengineering.com/Propane_Info.html.
8. Equiva Services—MSDS. Material safety data sheet number 401729E. *Product and company identification*. Retrieved October 16, 2005, from http://www.equivashellmsds.com/getsinglemsds.asp?ID=201791.
9. Skrotzki, Bernhardt. (1963). *Basic thermodynamics*. New York, NY: McGraw-Hill.
10. Scarbrough, Don R. (2008). *Fire protection handbook* (20th ed.). Quincy, MA: National Fire Protection Association.
11. Tuve, Richard. (1976). *Principles of fire protection chemistry*. Boston, MA: National Fire Protection Association.
12. Scarbrough, op cit.
13. Klinoff, Robert. (2012). *Introduction to fire protection*. Clifton Park, NY: Delmar Publishing.
14. Scarbrough, op cit.
15. Brannigan, Francis L., and Corbett, Glenn P. (2007). *Building construction for the fire service* (4th ed.). Quincy, MA: National Fire Protection Association.

16. Hall, Richard, and Adams, Barbara (Eds.). (1998). *Essentials of fire fighting* (4th ed.). Stillwater, OK: International Fire Service Training Association.

17. Scarbrough, op cit.

18. National Fire Protection Association. (1929, May 15). The Cleveland Clinic fire. Retrieved from National Fire Protection Association website: http://www.nfpa.org/assets/files/PDF/Research/cleveland.pdf.

19. California Energy Commission. (2003). What is electricity? *Energy story.* Retrieved January 5, 2013, from http://www.energyquest.ca.gov/story/chapter02.htm.

20. Scarbrough, op cit.

21. Ibid.

22. Hartin, Ed. (2008). Fire development and fire behavior indicators. retrieved January 5, 2013, from http://cfbt-us.com/pdfs/FBIandFireDevelopment.pdf.

23. Ibid.

24. Ibid.

25. Classification of fires. (2013). Retrieved January 5, 2013, from http://www.pyrene.ca/classifications.shtml.

26. Grimwood, Paul. (2003, February). *Flashover—what do you know?* Retrieved January 5, 3013, from http://www.firetactics.com.

27. Miale, op cit.

28. Hall and Adams, op cit.

Courtesy of James D. Richardson.

axial loads, *p. 121*

compressive force, *p. 123*

compressive stress, *p. 122*

corbels, *p. 141*

dead-load, *p. 129*

eccentric loads, *p. 122*

gusset plates, *p. 128*

load-bearing, *p. 126*

non-load-bearing, *p. 126*

occupancy, *p. 120*

shear forces, *p. 139*

sole plate, *p. 147*

spalling, *p. 140*

tensile force, *p. 136*

tensile stress, *p. 122*

top plate, *p. 147*

torsional loads, *p. 122*

OBJECTIVES

After reading this chapter, you should be able to:

- Describe the five categories of construction recognized by the National Fire Protection Association.
- Explain how each construction category reacts under fire conditions.
- Identify the various construction elements of each construction category.
- Describe the potential hazards that each construction poses to firefighters.
- Identify the various forces and loads that affect building integrity and explain how these forces and loads can cause building failure under fire conditions.
- Identify construction elements that may contribute to building collapse under fire conditions and describe the warning signs that may alert firefighters to a possible building collapse.

Introduction

There is no such thing as an average fire. Each incident is unique. Different ignition sources, different fuels, and different construction types are a few of the variables that determine a fire's character. Although it is impossible to predict fully the course any fire will take, the more firefighters know about these variables, the more effective and safer they will be. In this chapter, we will consider building construction and how it affects fire behavior.

Building construction includes the design of a structure, the materials used to build it, and the configuration of structural elements in the design. Acquiring a comprehensive knowledge of building construction is one of the most challenging professional duties of today's firefighters. This requires knowing the properties of different building materials and how they react to fire. It also requires understanding basic principles of physics such as load and force.

Of course, in the heat of battle, firefighters cannot take the time to perform complicated calculations, but a basic knowledge of these principles can have immense benefits. It can help incident commanders decide whether to initiate offensive or defensive fire ground operations. It can teach firefighters to read the warning signs of life-threatening events such as rollovers, flashovers, and imminent building collapse. Firefighters may experience one or all of these phenomena in any type of structure.

Knowledge of building construction, coupled with all-important experience, can spell the difference between life and death.

Loads and Forces

A load is defined as "any of the forces that a structure is calculated to oppose, including any permanent force, any moving or temporary force, wind, or earthquake forces."[1] It is very important that firefighters understand how these forces affect a structure. Keep in mind that architects design and builders build with these forces in mind. They provide support elements adequate to withstand the forces that will be imposed on them. However, when these support structures are weakened by fire, the equations change, and the building can become unstable and subject to collapse. This scenario was played out in the early hours of February 14, 2000, when two members of the Houston (TX) Fire Department entered a burning fast food restaurant. They were unaware that there was fire in the concealed space overhead that was formed by the wooden roof trusses. The fire weakened the trusses to the point that they could no longer support the weight of the commercial equipment on the restaurant's roof. The roof collapsed, fatally trapping the two firefighters.

occupancy ■ General term for how a building structure or home will be used and what it will contain.

When calculating building design, architects and builders also consider how the building will be used and what it will contain. This is referred to as its **occupancy**. What forces will the contents of a building and the people who live or work there exert on the structure? Loads and stresses must be calculated and structural support elements provided to counter these forces. If the occupancy changes and the building is not renovated or is improperly renovated to support the new occupancy, new loads and stresses placed on the building may exceed the original load design. The economic downturn in 2009 created another problem for firefighters. Available income became scarce and individuals began performing their own repairs and remodeling, often without acquiring permits. This "do it yourself" (DIY) craze is a major concern for firefighters. Many homeowners have little or no experience in construction and are not knowledgeable in basic construction and design. They are only concerned with the esthetics and not the structural integrity. The newly renovated building may support the excess loads under normal conditions, but a premature collapse may occur when support elements are attacked by fire. On June 5, 1998, two New York firefighters were killed and four were seriously injured while fighting a fire in a commercial/residential wood-frame structure when the rear of the second floor collapsed. After the investigation, one recommendation made by National Institute of Occupational Safety and Health investigators was to "ensure that modifications/renovations to buildings are in compliance with applicable building codes, i.e., any renovations or remodeling does not decrease the structural integrity of the supporting members."[2]

LOAD TYPE

There are basically two types of loads—live loads and dead loads. A live load is any load that is not fixed or permanently attached to the structure, such as desks, occupants, and so on. Dead loads are permanent parts of the structure, floor, roof, walls, HVAC equipment, elevators, and so forth. Loads can be static or dynamic. Static loads are loads that are applied steadily or gradually, such as applying new materials for a roof. A dynamic load is a load imposed by motion, such as wind, water, moving equipment, and so on.

LOAD DISTRIBUTION

Load distribution can be categorized as uniform or concentrated. Uniformly distributed loads are applied over a large area. Roofing and flooring material would be an example of uniformly distributed loads. A concentrated load is applied to a small area. Bank safes or commercial copying machines are examples of concentrated loads. The type of load and the load distribution are important factors, but structural integrity relies on how the loads are imposed.

LOAD IMPOSITION

As we have seen, structural support members are designed to withstand designated load impositions. When these support elements are attacked by fire, they may be weakened or distorted, which causes the load imposition to change. Load imposition can be classified as axial, eccentric, or torsional.

AXIAL LOADS

Axial loads are proper loads—that is, the loads that structural elements were designed to support. They are evenly distributed and perpendicular to the central axis or plane of the supporting element. If load imposition is axial, the supporting elements are able to carry the maximum load capacity (see Figure 7-1).

axial loads ■ Loads that are placed perpendicular to the central axis of the support structure and utilize the full support capabilities of that structural component.

FIGURE 7-1 Axial loads take advantage of the full load-carrying capacity of the column. *Courtesy of James D. Richardson.*

FIGURE 7-2 Columns will bend under eccentric loads. *Courtesy of James D. Richardson.*

FIGURE 7-3 Torsional loads twist the column. *Courtesy of James D. Richardson.*

ECCENTRIC LOADS

eccentric loads ■ Loads that are placed perpendicular to the support structure but do not pass through the central axis, thus causing the support structure to bend.

tensile stress ■ Stress in a structure that tends to pull it apart.

compressive stress ■ The stress applied to materials that results in their decrease in volume.

torsional loads ■ Loads that are placed only on the longitudinal axis of the structural component causing it to twist.

Eccentric loads are improper loads. They are not evenly distributed over the central axis or plane of the supporting structure. Instead, they are perpendicular to that central axis or plane. This off-center position creates a bending force on the support structure. This places **tensile stress** on one side of the structure as it attempts to stretch and **compressive stress** on the opposite side as it contracts. The resulting imbalance weakens the support (see Figure 7-2).

TORSIONAL LOADS

Torsional loads are improper loads that are offset from the central axis of the support structure but are in the same plane as the cross section. These loads create a twisting motion in the supporting member (see Figure 7-3).[3]

Cataclysmic events, such as earthquakes or explosions, may subject a structure to both eccentric and torsional loads simultaneously. The resulting stresses are dynamic, three-dimensional, and highly intense. They radically increase the likelihood of rapid and serious damage.

STOP AND THINK 7-1

What are the most common loads and forces placed on a structure under fire conditions and why should firefighters be concerned?

Structural Components

Most buildings have the same basic structural components: a foundation or floor, exterior walls or supports, a ceiling support structure, and a roof. These components may be comprised of, or supported by beams, columns, and/or girders. It is essential to understand these structural components in order to make the best decisions in an emergency situation.

FOUNDATIONS

The foundation is the final recipient of all loads placed on a structure. It must be stable enough to accept the combined **compressive force** of the weight of the building and all loads delivered to the building. Today, most foundations are steel-reinforced concrete slabs. Concrete is a mixture of cement, sand or gravel, and water. It has a very high resistance to compression forces and continues to harden (cure) indefinitely.[4] The type of soil on which a building stands determines the structural design of the foundation, how much steel reinforcement is used, and the makeup of the concrete mixture (see Figure 7-4).

compressive force ■
A load that causes structural elements to shorten.

The foundations of structures that have basements are subject to a variety of forces. The walls of the basement must not only transfer the weight of the structure and all other loads to the ground, but also must withstand the horizontal forces exerted by the soil surrounding them. Soil pressure varies with the type of soil, its density, and its moisture content. An approximation of soil pressure on a structure can be computed by using the following formula:

Soil Pressure Formula:

$$P = \frac{Cwh}{2} \times \text{Height of soil} \times \text{length of wall}$$

$P = $ Pressure
$h = $ Depth of Soil
$w = $ Density of Soil
$C = $ Numerical Constant (based on physical properties of soil)[5]

FIGURE 7-4 Concrete is designed for compression loads.

What is the soil pressure on a basement wall that is 8 feet high and 10 feet long? The soil is 8 feet deep. The soil density is 0.4 and the numerical constant is 100.

$$P = \frac{100 \times 0.4 \times 8}{2} \times 8 \times 10$$

$$P = \frac{320}{2} \times 8 \times 10$$

$P = 160 \times 8 \times 10 = 12,800$ lbs. of horizontal force on the basement wall

Firefighters may encounter older structures with foundations built of wood joists and beams. This construction, sometimes referred to as *pier and beam,* consists of wooden floor joists supported by wooden girders, which are in turn supported by wood, concrete, or steel piers. The concealed space created by this type of construction provides a perfect environment for hidden fire to move around, over, and under firefighters, especially if the floor joists are of open-work truss construction. When fire weakens the support structures, they may well give way, resulting in floor or wall collapse (see Figure 7-5).

BEAMS

When a load is placed on a beam, the beam will be subjected to two different forces. The top of the beam will be compressed by the load while the bottom of the beam will be pulled by tensile forces. This creates a bending motion in the beam. The amount of weight that a beam will support is dependent on several characteristics of the beam: its width, depth, span, and composition. A change in any of these characteristics will greatly affect its load bearing capabilities. The characteristics of a beam are often changed during construction, renovation, and demolition. It is vitally important for firefighters to know if changes have been made. A continuous preincident planning program will allow firefighters to discover these changes and make proper adjustments to their firefighting strategy and tactics. The following three formulas and examples will demonstrate the changes in beam capacity by varying its depth, width, or span.

FIGURE 7-5 Old pier and beam foundations were made from tree trunks. This is very hazardous to firefighters when the piers are exposed to fire. *Courtesy of James D. Richardson.*

Depth

EXAMPLE:

If a 2 × 8 beam will support 2,000 lbs., how much weight would a 2 × 12 beam support?

$$\frac{\text{Load 1}}{\text{Load 2}} = \frac{\text{Depth 1}^2}{\text{Depth 2}^2}$$

$$\frac{2,000}{X} = \frac{64}{144}$$

$$\frac{288,000}{64X} = 4500 \text{ lbs.}$$

Width

EXAMPLE:

If a 2 × 8 beam will support 2,000 lbs., how much weight would a 4 × 8 beam support?

$$\frac{\text{Load 1}}{\text{Load 2}} = \frac{\text{Width 1}}{\text{Width 2}}$$

$$\frac{2,000}{X} = \frac{2}{4}$$

$$\frac{8,000}{2X} = 4,000 \text{ lbs.}$$

Span

EXAMPLE:

If a 4 × 12 inch beam spanning 12 feet can support 2,000 lbs. How much weight would a similar beam support if the span was 20 feet?

$$\frac{\text{Span 1}}{\text{Span 2}} = \frac{\text{Load 2}}{\text{Load 1}}$$

$$\frac{12}{20} = \frac{X}{2,000} = \frac{24,000}{20X} = 1,200 \text{ lbs.}$$

COLUMNS

Columns are designed to withstand compressive forces that are imposed perpendicular to the center axis of the column. If the load is imposed in an eccentric manner (off center from the axis), a bending motion will be placed on the column and failure could occur. There are several factors that determine the strength of a column: the material used to construct the column, its diameter, and its length. Example: A 10 foot column is supporting 1,000 pounds. How much weight would a 20 foot column of equal diameter and composition support?

Column Formula

$$\frac{\text{Length 1}^2}{\text{Length 2}^2} = \frac{\text{Weight of Load 2}}{\text{Weight of Load 1}}$$

$$\frac{10^2}{20^2} = \frac{X}{1,000}$$

$$\frac{100}{400} = \frac{X}{1,000} = \frac{100,000}{400X} = 250 \text{ pounds}$$

FIGURE 7-6 (a) Load-bearing walls transfer loads from other support structures to the foundation. (b) Non-load-bearing walls transfer only their own weight to the foundation. *Courtesy of James D. Richardson.*

Doubling the length of the column reduces its strength by a factor of four. Loss of strength of the longer column can be regained by bracing the column. If the same 20-foot column used in our example is braced at 10 feet, it will be able to support the same weight as the 10 foot column.[6]

WALLS

load-bearing ■ A component of a structure that is designed to bear the weight of the structure and loads and transmit this weight to the foundation.

non-load-bearing ■ A partition or wall that supports only itself or interior finish components.

Walls can be classified as load-bearing and non-load bearing. **Load-bearing** walls support the roof and any upper floors. All loads placed on the structure are transmitted through the load-bearing walls to the foundation. **Non-load-bearing** walls support only their own weight (see Figure 7-6). Structurally speaking, load-bearing walls function as columns. They are designed to accept vertical loads and to transmit those loads to the foundation. Bracing must be added to walls to give them resistance to lateral loads such as wind forces. There are a variety of brace configurations. The most common types are diagonal bracing, chevron bracing, and X bracing (see Figure 7-7). In some areas, builders incorporate $\frac{1}{2}$-inch thick plywood sheeting or $\frac{1}{2}$-inch oriented strand board for wind bracing in wood-frame construction by fastening the sheeting to the outside of the exterior walls. The 4×8 foot sheeting may be fastened at the corners of the structure, or the contractor may "wrap" the entire exterior of the structure with sheeting. This creates a very strong structure, but will create forcible entry problems if walls need to be breached.

Firefighters may underestimate the power of wind on the walls of a structure. Some architectural designs feature cantilevered walls (walls that are only supported on one end). This type of wall construction can become very unstable when subjected to wind forces while under fire conditions. Fire can also destroy bracing and weaken the structure. A building under construction may be temporarily braced with inferior materials and structures under demolition or renovation may have bracing removed. This makes walls very susceptible to failure from lateral forces. The pressure (P) exerted by wind forces can be calculated by multiplying .00256 (C, a numerical constant) by wind velocity squared (V^2).[7]

$$P = Cv^2$$

FIGURE 7-7 There are several configurations of lateral bracing for walls. *Courtesy of James D. Richardson.*

How much pressure would a 40-mile-per-hour wind place on a wall that is 10 feet high and 20 feet long?

$$P = .00256 \times 40^2$$
$$P = .00256 \times 1600$$
$$P = 4.096 \text{ psf on the wall}$$
$$20 \times 10 = 200 \times 4.096 = 819.20 \text{ pounds of force on the wall}$$

WOOD TRUSSES AND I-JOISTS

Today, wood trusses are designed by computer engineering programs. All of the angles and lengths of the components are cut with precision and precisely placed in the web, the area between the top and bottom cord. However, the actual strength of the wood truss does not rest completely with its components. It also depends on the connectors that tie the web components and cords together. A truss is an assemblage of building components that form a rigid framework. It consists of a top cord (top board), bottom cord (bottom board), and interconnecting members that create a web. The strength of the truss is achieved by configuring the web components in triangles, which provide greater stability than right-angled configurations when stress is imposed. Wooden trusses are replacing solid lumber joists and roof components in many wood-frame structures because they are lighter, easier, and faster to install. Also, truss construction for ceiling joists allows for spans as great as 80 feet without the need for support columns. Prior to 1952, truss components were connected with wood gussets, staples, nails, glue, or mortise-and-tenon joints. After that date, metal **gusset plates** came into use. Today's standard connectors are steel plates stamped from 16-, 18-, or 20-gauge structural steel and are coated with zinc to reduce the possibility of corrosion. One side of the plate has teeth that can range in length from 5/16 to 9/16 inch. The plate contains eight teeth per square inch and is sized according to the size of the wood components and the amount of stress that will be transferred (see Figure 7-8). Today, the connector plates are pressed into place by computer-operated machines. This creates a much stronger connection than hand-mounted connections. For example, a 3×4-inch connector plate, applied to both sides of a web joint, provides a working strength of 2,400 pounds, with a safety design load of 7,680 pounds.[8] This is very

gusset plates ■ Metal plates used to unite multiple structural members of a truss.

FIGURE 7-8 The teeth on the gusset plate vary in length from 5/16 to 9/16 of an inch. *Courtesy of James D. Richardson.*

impressive when viewed from a construction standpoint. However, from the firefighter's viewpoint, there is a caveat: under fire conditions, metal connectors will start to lose structural integrity when they are heated to 1,000°F. If the connectors fail, the entire truss is likely to collapse. Failure of the metal connectors due to heat exposure is not the only reason wood trusses collapse under fire conditions. The bowstring truss poses an additional threat to firefighters. This type of truss has an arched top cord and a straight bottom cord that are connected by the web. The truss resembles an archer's bow. Bowstring trusses are used to span large, open areas such as gymnasiums and public arenas. Spans can be over 100 feet. The arch shape of the top cord places an outward pressure on the structural element. The outward pressure is restrained by the straight, bottom cord. If any component of the truss fails, the tension in the truss is released and the truss will fail catastrophically. On March 4, 2012, one Wisconsin firefighter was killed and two injured when a bowstring truss roof collapsed during firefighting operations at a theater fire. Trusses may also be weakened by:

- Damage caused by pests (termites, carpenter ants, and so on)
- Corrosion of the metal connectors
- Excessive **dead-load** weight beyond the design load
- Illegal modifications to the truss (see Figure 7-9)

The wood I-beam construction component came into use in the 1990s and continues to increase in popularity in today's housing industry. It makes possible open spans of up to 60 feet without support columns. This is less than the 80-foot spans that truss construction allows. The I-beam was designed to take the place of solid-wood and wood-truss joists. In fact, construction professionals refer to it as an "I-joist."

Like the truss, the I-joist has three components: a top cord, bottom cord, and web. It offers a much higher strength-to-span ratio than dimensional lumber. This means that it can support more weight than a solid beam of equal length, width, and depth. Most I-joist components are held together by glue. However, wide I-joists (10 inches wide or more) may use a metal angle brace gusset for additional support. The cords are constructed of laminated veneer or solid lumber and range in dimensions from $1\frac{1}{2} \times 1\frac{1}{2}$ inches to $1\frac{1}{2} \times 3\frac{1}{2}$ inches. The I-joist web is composed of plywood, laminated veneer lumber (LVL), or oriented strandboard (OSB). Instead of using the triangle principle for strength, the I-joist web takes advantage of the 90° cross grain of the veneered layers in the plywood, LVL, and the cross strands of the OSB.[9] Engineered structural components such as LVL and OSB are becoming increasingly more common in the housing industry (see Figure 7-10).

dead-load ■ The structure and any equipment or building components that are permanently attached to the building.

FIGURE 7-9 Gusset plates tie the truss components together. Gussets may fail under fire exposure, which then causes the truss to fail. *courtesy of James D. Richardson.*

FIGURE 7-10 I-joists are engineered to perform equally as well as a solid wood beam of the same dimension. However, this lightweight support component will fail more often than a solid-wood component under fire conditions. *Courtesy of James D. Richardson.*

CEILING JOISTS

Ceiling joists are beams that support ceiling loads and are themselves supported by larger beams, girders, or load-bearing walls. Ceiling joists tie the exterior and interior walls together and double as floor joists for the attic or for a second story. When designed loads are imposed onto ceiling joists, the load is distributed horizontally to the ends of the joists resting on the load-bearing walls. The load is then transmitted vertically through the walls to the foundation and then to the ground. Any load-bearing wall or column under the joists will also assume part of the load. Ceiling joists may be constructed of lightweight steel trusses, wooden trusses, I-joists, I-beam steel, or solid-dimensional lumber. The length of the span and the load that must be carried determines the size of the joists. This is often specified in a jurisdiction's building code (see Figure 7-11).[10]

FIGURE 7-11 Ceiling joists are beams that connect parallel walls and support the interior ceiling covering. *Courtesy of James D. Richardson.*

It should be noted that wood trusses and I-joists, although stronger than solid dimensional lumber, have less mass and provide a larger exposed surface area. This increased surface area absorbs heat and will burn more readily than dimensional lumber, thus creating a shorter time span for collapse under fire conditions.

ROOF CONFIGURATIONS

There are two major kinds of roofs: flat or pitched.

Flat Roofs

Flat-roof construction utilizes the ceiling joists as support for the roof assembly. The ceiling joists in effect become the rafters. Truss construction is often found in flat-roofed buildings built since the 1950s and designed with large open areas. This type of construction provides large concealed spaces in the attic area in which fire can travel unnoticed by firefighters. Remember that the ceiling joists tie the exterior and interior walls together. If the trusses fail under fire conditions, not only will the roof collapse, but the walls that are tied together by these trusses may also fail.

Pitched Roofs

Pitched roofs can be of conventional or prefabricated truss construction. Conventional pitched roofs are shaped and assembled on the construction site and are built one rafter at a time. This type of construction is referred to as *stick frame*. The stick-framed roof may be constructed with a truss design, but the majority of truss roofs are prefabricated construction.

Prefabricated trusses are assembled at fabrication shops by cutting and assembling the components, using gusset plates to hold them together. The trusses are carried to the work site and lifted into place onto the support structure. Because the bottom cord of the truss is also the ceiling joist, failure of the truss during a fire will result in structural instability of the supporting walls (see Figures 7-12A and B). The arched roof is a form of pitched roof. It incorporates the truss design and is used to span large areas, such as gymnasiums. It is referred to as a *bowstring truss* because its design resembles an archer's bow.

Types of Pitched Roofs

Pitched roofs can be found in several configurations: gable, hip, lantern, gambrel, shed, sawtooth, butterfly, arch, and mansard.

FIGURE 7-12A The triangular shapes of their components give trusses their strength. *Courtesy of James D. Richardson.*

FIGURE 7-12B If one component of a truss fails, the entire truss will fail. *Courtesy of James D. Richardson.*

Components of a Truss Beam

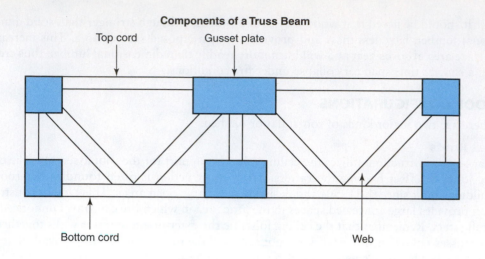

Top cord   Gusset plate

Bottom cord   Web

The gable roof is in the form of an "A." The rafters rest on the top plate of the exterior load-bearing walls and are fastened together over the center axis of the structure to form the A configuration. (See Figure 7-13.)

The hip roof is a roof with sloping ends and sides. All four exterior walls of the structure are load-bearing walls. (See Figure 7-14.)

The lantern roof combines the gable and shed roof designs. The appearance of the structure resembles a lantern. (See Figure 7-15.)

The gambrel roof (sometimes referred to as a barn roof) has two separate slopes on each side of the two sides. The lower slope is much steeper than the upper slope. (See Figure 7-16.)

The shed roof is a continuous structure that slopes from a wall to a lower parallel wall in order to allow drainage. (See Figure 7-17.)

The sawtooth roof is usually found in industrial locations. It is a series of shed roofs attached to a single structure. This gives the appearance of teeth on a saw blade. (See Figure 7-18.)

FIGURE 7-13 The gable roof forms a triangular-shaped structure. *Courtesy of James D. Richardson.*

FIGURE 7-15 The lantern roof combines the gable and shed roof designs. *Courtesy of James D. Richardson.*

FIGURE 7-16 The gambrel roof has two slopes on each side of the roof. *Courtesy of James D. Richardson.*

FIGURE 7-17 The shed roof slopes from front to rear of the structure. *Courtesy of James D. Richardson.*

FIGURE 7-18 The sawtooth roof is found in industrial structures.

The butterfly roof design incorporates two shed roofs that angle from the exterior support walls toward the center of the structure. (See Figure 7-19.)

The arch roof is a curved roof that extends in an arch from one wall to a parallel wall. It is used to span large areas, such as a gymnasium. (See Figure 7-20.)

FIGURE 7-19 The butterfly roof configuration utilizes the shed roof design that slopes toward the center of the structure. *Courtesy of James D. Richardson.*

The mansard roof slopes on all four sides of the structure and breaks near the top to form a slight slope or flat deck. (See Figure 7-21.)

Construction Types

Building design and construction are the major factors determining how a structure reacts under fire conditions. Firefighters knowledgeable about different building construction classifications have a definite edge when it comes to predicting a fire's behavior—knowledge that could save a life.

There are several model building codes that classify building construction: the International Building Code (IBC), the Uniform Building Code (UBC), the National Building Code (BOCA), the Southern Building Code (SBC), and the National Fire Protection Association (NFPA). The IBC, UBC, BOCA, and SBC are joined under the International Code

FIGURE 7-21 The mansard roof slopes on all four sides and breaks near the top. *Courtesy of James D. Richardson.*

Council (ICC). Each code defines five types of construction. Their differences vary slightly in fire-resistance requirements. For example, the IBC has a lower fire-resistive rating than the NFPA 5000 for Type I A and B and Type II A structural components. Firefighters should be familiar with the code that is used in their area of protection. This chapter will focus on the building construction types recognized by NFPA 220, *Standard on Types of Building Construction*, and the new NFPA 5000, *Building Construction and Safety Code*.

There are five basic building classifications. NFPA 220 lists these classifications as follows:

- Type I, Fire Resistive
- Type II, Noncombustible
- Type III, Ordinary
- Type IV, Heavy Timber
- Type V, Wood Frame

These five classifications are used in the new NFPA 5000 building code and the ICC. However, following the recommendations of the Board for the Coordination of the Model Codes (BCMC), the new building code removed the designations fire resistive, ordinary, and heavy timber, and so on, and simply designated Types I and II as noncombustible and Types III, IV, and V as combustible.[11]

In discussing these types of constructions, we will refer to fire-resistive ratings. A fire-resistive rating is based on the length of time (in hours) that a structural component can be expected to maintain structural integrity while being subjected to a standard fire exposure test, such as outlined in NFPA 251, *Standard Methods of Tests of Fire Endurance of Building Construction and Materials*. A fire rating of one hour means that the assembly will resist breakthrough for one hour under controlled test conditions.

TYPE I CONSTRUCTION

In Type I construction, the structural components are noncombustible and protected, usually by encasement in a heat-resistant material that provides temporary structural stability when the components are exposed to fire. Because the structural components are protected, they are given a fire-resistive rating. Although the structural components do not contribute fuel to a fire, some combustible materials are permitted for use in interior finishes, roof structures and coverings, doors and windows, and exterior trim.[12]

The new NFPA Building Code divides Type I construction into two subtypes: Type I-443 and Type I-332. Each numerical digit represents the number of hours of fire resistance required for certain components of the structure.

- The first digit identifies the fire-resistive rating for the load-bearing exterior walls as three or four hours.
- The second digit identifies the fire-resistive rating of interior load-bearing walls, columns, beams, girders, trusses, and arches as three or four hours.
- The third digit identifies the fire-resistive rating for floor and roof assemblies as two or three hours.

Although concrete block or masonry can be used in Type I construction, the most common structural materials used are steel-reinforced concrete or protected steel.[13]

Steel

tensile force ■ A force that causes the structural element to lengthen.

Steel is a very strong structural material, unique in its ability to withstand compressive and **tensile forces** equally. Alloying steel with other metals gives it versatility and allows it to be used for a variety of construction purposes. However, strong as it is, steel begins to lose its structural integrity when subjected to fire temperatures of 1,000°F, which is not unusual in structure fires. In fact, tests have shown that structural steel will fail when

FIGURE 7-22 Bolted or welded connections maintain structural integrity longer under fire conditions than supported connection joints. But when they fail structural collapse may occur. *Courtesy of James D. Richardson.*

heated above 1,200°F. The length of time between heat exposure and the structural failure of steel depends on the following factors:

- Mass of the Steel Component. The more density (mass) that a steel component has, the longer it will take to heat to 1,000°F. Steel components with less density absorb heat faster.
- Heat Release Rate of the Fire. The higher the heat release, the faster the steel is heated.
- Forces or Loads Exerted on the Steel Component. A steel component under stress will fail at a lower temperature than a component that is not load bearing.
- Type of Connections Joining the Steel Components. When structural steel members are welded or bolted together to form an assembly, they will maintain their structural integrity longer under fire conditions than if the components are simply supported (see Figure 7-22).[14]

Structural steel becomes very animated under fire conditions, especially if it is under a load or restrained at its ends. If the beam is restrained, it will distort by bending or twisting when fire conditions reach or exceed 1,000°F. This can create excessive load imposition on support structures and cause structural failure. Steel has a memory when it is placed under stress. The memory allows the steel to return to its original shape after the stress has been removed. This is referred to as *flexibility*. The point at which the steel component will not return to its original shape after the stress is removed is called the *point of strain*. When water streams are directed onto a distorted beam, it will return to its original shape if it has not passed its point of strain. If it has passed its point of strain, the beam will remain in its distorted shape when cooled. When buckled beams are cooled, they can retract and fracture the connections at the ends. This occurs because the safety rating for bolts and welds is higher than for structural steel. When the retracting steel is torn away from the bolts and welds, structural components are weakened and collapse is possible.

Unrestrained beams also pose dangers for firefighters. An unrestrained steel beam 100 feet in length will elongate 9.5 inches when heated to 1,000°F. However, when that same beam is cooled to a normal ambient temperature, it will return to its original length. Elongation can cause a wall or support to be displaced. If this has occurred, when the beam is then cooled and retracts to its original length, it may dislodge from the support and initiate a structural collapse (see Figure 7-23).[15]

There are several methods of protecting structural steel components from the adverse effects of fire exposure. These methods are not designed to make the steel components fireproof but to give them a fire-resistive rating. One method of protecting structural steel is encasement. The fire-resistive ratings for encasements depend on the moisture content and the thickness of the materials used. Fire-resistive materials can be made of tile, gypsum, plaster and lath, or concrete. Each material has advantages and disadvantages. Tile is a good insulator, as is gypsum, which has the added advantages of being lightweight and high in moisture content. Both materials are, however, easily damaged. If they should break away, the structural steel that they are meant to protect would be exposed to potential fire damage. Lath and plaster is used in conjunction with gypsum board to encase structural steel. It is a good substitute for concrete because it is much lighter, but concrete remains the best, and most permanent, fire-resistive material. Its main disadvantages are its weight and expense (see Figure 7-24).[16]

A second method of protecting structural steel is by spraying the steel components with a fire-resistive coating. It is difficult, however, to gauge the thickness of a sprayed coating or to maintain an even application on the steel component. Consequently, the

FIGURE 7-24 Encasement is a method of giving structural steel fire resistance. *Courtesy of James D. Richardson.*

FIGURE 7-25 Fire-resistive coatings can be sprayed on the steel structural components. *Courtesy of James D. Richardson.*

fire resistance of components treated in this way is not easy to determine. Moreover, although a cement-based spray coating is an excellent insulator, it is easily damaged and can be accidentally removed or damaged by maintenance personnel (see Figure 7-25).[17]

Concrete

Concrete is a mixture of cement, sand, and aggregate. Water is added to the mixture to transform it into a fluid state so that it can be molded into a desired configuration. Concrete building components are formed by pouring the liquefied mixture into a form or "falsework" and allowing it to harden or cure to a predetermined working strength. This material has a natural strength against compressive loads that makes it an ideal material for columns, beams, and floor construction. Concrete structural components can be made on the building site (cast in place) or at a remote location and delivered to the work site (precast).

There are several configurations in which concrete can be used to take advantage of its structural qualities:

- Plain Concrete. Composed of cement, sand, aggregate, and water, it is very strong against compressive forces. However, plain concrete is vulnerable to the tensile and **shear forces** (loads that cause the structural element to deform or fracture in a parallel direction to the force, by sliding one section of the element along others) that are placed on walls, beams, girders, columns, and floors.
- Steel-Reinforced Concrete. Steel bars or steel mesh are added to the concrete mixture when it is poured. As the concrete cures, it bonds to the steel. The strength characteristics of the two materials complement one another, creating a structural component that has compressive, shear, and tensile strength. (See Figure 7-26.)
- Prestressed (Pretensioned) Concrete. Prestressed structural components are cast at a concrete plant and subjected to a tensioning process. Prestressing cables, referred to as *tendons*, are placed in the falsework, or casting bed, and tightened to about 30,000 pounds of tension. The concrete is poured around the cables and allowed to bond with the steel as it cures. When the tension is released from the cable ends, the steel tries to retract, creating a compression force on the concrete component, which gives

shear forces ■ Loads that cause the structural element to deform or fracture in a direction parallel to the force, by sliding one section of the element along others.

FIGURE 7-26 Steel gives concrete tensile and shear strength. *Courtesy of James D. Richardson.*

it resistance against tensile forces. Prestressed components can be identified by the cut cables on the ends of the casted units.

■ Post-Tensioned Concrete. Post-tensioned structural components are cast in place. High-strength steel cables are placed in the forms prior to pouring the concrete. The cables are coated or placed inside tubes to prevent them from bonding with the concrete. After the concrete is cured to the desired strength, the cables are tensioned with hydraulic jacks placing compression forces on the component. Post-tensioning is used extensively for floors in high-rise buildings because it allows thinner structural members to carry heavier loads.

The moisture content of concrete makes it a very effective fire-resistive material, but moisture can contribute to its degradation under certain fire conditions. When it is exposed to intense or prolonged heat, the moisture in the pores of the concrete converts to steam and expands, causing the concrete to break apart. This phenomenon is called **spalling**. When spalling occurs, the concrete is broken away from reinforcement steel, exposing the steel to the fire environment. The heated steel, having bonded with the concrete, will try to expand and bend, causing more portions of the concrete to fail. Recently cured concrete will spall at a faster rate than older "aged" concrete because it has a higher moisture content.[18]

spalling ■ The process in which parts of a concrete structure break or chip off when the structure is exposed to extreme heat.

TYPE II CONSTRUCTION

Type II construction is described in the NFPA 5000 building code as noncombustible. The structural components are of noncombustible or limited-combustible materials. Type II construction is quite similar to Type I, but its fire-resistive rating is not as high, and protection is required in only two subtypes of this category: Type II-222 and Type II-111. These components must be protected and must have a two- or one-hour fire-resistive rating, respectively. They are considered adequate for residential, educational, institutional, business, and assembly occupancy structures. Structural components of Type II-000 are of unprotected noncombustible construction. These have no fire-resistive rating and do not provide structural stability under fire conditions. This type of construction is common and may be found on commercial and private properties in the form of business office buildings, garages, and barns (see Figure 7-27).

STOP AND THINK 7-2

What factors should firefighters be especially concerned about when fighting fires in structures of Type I or Type II construction?

TYPE III CONSTRUCTION

Type III construction is quite different from Types I and II. In Type III construction, the exterior walls are required to be constructed of noncombustible or limited-combustible building materials. They can be of stone or brick masonry, metal, concrete, or any combination of these. In old-style Type III buildings, the exterior bearing walls are usually of masonry construction. They are very thick at the lower levels and progressively decrease in thickness at each upper level. Ledges known as **corbels** are formed between stories and support the floor joists for the next floor (see Figure 7-28).

Most of these buildings were built in the 19th century. Their design poses a serious inherent threat to firefighters. Even though the exterior walls are noncombustible, the interior structural support components may be constructed partly, or entirely, of combustible material. Many of these buildings have ceiling and floor joists, floor and ceiling coverings, interior walls, girders, beams, and columns of wood. These components are not only combustible, but they are also assembled in a way that creates concealed spaces—for example, between floor and ceiling joists and between wall studs. Fire can move through these spaces unnoticed by firefighters, from room to room or from floor to floor.[19] Moreover, in Type III construction, the floors are structurally engineered to collapse under fire conditions to prevent collapse of the exterior walls. The floor joists, supported by the corbelled exterior walls, are "fire cut" (cut on an angle) on each end. This allows the ends of the joists to burn through faster and ultimately drop away from the wall (see Figure 7-29).

corbels ■ Pieces of stone jutting out of a wall to carry weight.

FIGURE 7-28 Corbels
are ledges constructed on
the inside of a masonry
wall to support the floor
joist of the upper level.
*Courtesy of James D.
Richardson.*

The fire-cut design was actually intended to save firefighters' lives. During the 19th century, fires were fought from outside buildings. An interior collapse was preferred over an exterior wall collapse. Exterior walls might still collapse, but it was fairly certain that the interior structures would go first. Modern firefighting strategy and tactics usually call for an aggressive interior attack. In older Type III buildings, such an attack could have fatal consequences. Firefighters must familiarize themselves with buildings of this construction category in their response area.

There are several indicators of danger that firefighters should look for in Type III buildings. It is critical that these be identified during preincident planning to avoid unpleasant surprises during an actual firefighting operation:

- Metal Stars or Plates on the Exterior Masonry Walls. These plates indicate that the walls of the building are supported by metal rods or cables connected to an opposite

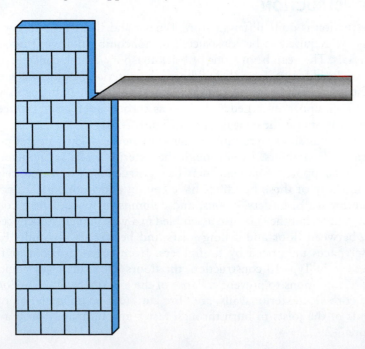

FIGURE 7-29 The fire cut allows the ends of the joist to burn faster and fall inward. *Courtesy of James D. Richardson.*

FIGURE 7-30 Metal stars on the outside of a building should be a warning to firefighters that the masonry wall is unstable and will collapse if the rods connecting the metal stars fail. *Courtesy of James D. Richardson.*

wall or a floor beam. The rods and cables are constructed of cold-drawn steel. Cold-drawn steel will fail when heated to 800°F. When the rods or cables fail, the wall will collapse. (See Figure 7-30.)

- Cracks in the Masonry and Mortar. In Type III construction, the load-bearing masonry wall is the support structure. Cracks in the masonry or mortar can be an indicator of wall movement, imposition of excessive load on the supporting elements, or foundation failure (see Figure 7-31).

- Sand Lime Mortar. Sand lime mortar is vulnerable to water erosion. Fire streams can cut the mortar from between the masonry and weaken a wall. Preincident inspection should identify deteriorating sand lime mortars.[20] Keep in mind that Type III buildings listed as historical structures are highly likely to have sand lime mortar, because historical preservation rules require that it be used in any restoration work (see Figure 7-32).

Like Type II construction, Type III construction has two subdivisions: protected and unprotected. In Type III-211 protected construction, the exterior load-bearing walls are required to have a fire-resistive rating of two hours. The combustible or limited-combustible interior structural support components must be protected and have a fire-resistive rating of

FIGURE 7-31 Cracks in a masonry wall may indicate foundation weakness. *Courtesy of Bob Hopkins, Weatherford (TX) Fire Department.*

FIGURE 7-32 Sand lime mortar is used in the restoration of historical buildings. *Courtesy of James D. Richardson.*

one hour. These components include all structural frames, columns, girders, and beams supporting loads of more than one floor. Floor construction must have a one-hour fire-resistive rating.

In Type III-200 unprotected construction, the exterior load-bearing walls are required to have a two-hour fire-resistive rating. However, the interior support structures—columns, beams, and girders supporting loads from more than one floor—and the floor assemblies are constructed of unprotected combustible materials. NFPA 220 categorizes Type III construction as ordinary construction. This description has been dropped by the new NFPA 5000 building code, which designates these buildings as combustible (see Figure 7-33).[21]

TYPE IV CONSTRUCTION

In the mid-19th century, insurance companies demanded that the textile industry build structures that would provide greater fire resistance than was offered by ordinary construction. In response, builders developed the heavy-timber construction method, now designated as Type IV. The exterior load-bearing walls were of solid masonry (just as in Type III construction), but interior support timbers had to be of larger dimensions than was standard. Concealed spaces between the floors and roofs were eliminated.

FIGURE 7-33 Unprotected Type III construction poses a threat to firefighters because it contains a large amount of concealed spaces in which fire can travel unseen. *Courtesy of James D. Richardson.*

Unfortunately, today there are different interpretations of what constitutes heavy-timber construction, which is also sometimes called *mill construction*. NFPA 220 uses the term "heavy timber" to identify Type IV construction, but due to the confusion, the new NFPA 5000 categorizes Type IV construction simply as combustible.

NFPA 220 defines Type IV construction as a structure with noncombustible or limited combustible exterior and interior walls. The interior structural components, such as columns, beams, arches, floors, and roofs, must be constructed with large dimensional solid or laminated wood and have no concealed spaces. The standard also requires that interior load-bearing walls have a fire-resistive rating of no less than one hour and exterior load-bearing walls have a fire-resistive rating of no less than two hours.[22] All wood columns used to support floor assemblies are required to be a minimum of 8 inches square (sawn posts) or 8 inches in diameter (round posts). The wood columns must be a minimum of 6 × 8 inches if they support a roof assembly. Floor decking must consist of 3-inch tongue-and-groove planks, and roof decking must be 2-inch tongue-and-groove planks. All wood beams and girders must be a minimum of 6 inches wide and 10 inches in depth if they support a floor assembly and 4 × 6 inches if they support a roof assembly. Even though the interior components are combustible, the large dimensions of the lumber give them a certain amount of fire resistance. In fact, wood over 6 inches thick can be more fire resistant than unprotected steel.[23] As the surface of the wood burns, it forms a layer (char) that insulates the unburned wood beneath. This process, called pyrolysis, reduces the burn rate of the wood and allows the wood to maintain structural integrity longer. In the United States, burn-rate testing of laminated and sawn timbers has established a char rate of 0.024 inch per minute.[24] For example, if a 2 × 4 inch solid wood board is exposed to heat and begins to pyrolyze, or char, at the point of heat contact, the 2 × 4 will be reduced in dimension by 0.024 inch per minute at the char site.

The load-carrying capability of structural components depends not only on the strength of the components but also on the connections that tie them together. The most stable way to connect two beams is to load the ends of the beams axially on top of a column or girder. When this is not possible due to height restrictions, beams may be tied to girders with mortise-and-tenon joints, or columns may be corbelled to accept the ends of the beams. It is recommended that the connections have higher fire resistance than the structural components. However, this may not always be the case. Mortise-and-tenon joints and corbels are not as solid as heavy-timber beams, girders, or columns. These connecting elements can fail under fire conditions and cause premature collapse, even though the structural elements they connect have maintained their strength. Firefighters should be aware of this during prefire planning (see Figure 7-34).

FIGURE 7-34 Used in post-and-beam construction, this type of connection is very strong but will readily fail under fire conditions if the tenon is exposed and burned. *Courtesy of James D. Richardson.*

TYPE V CONSTRUCTION

Wood is a popular building material because it is cheap and renewable. Its strength is determined by its density and moisture content. Unlike Type III and Type IV construction that have noncombustible components, Type V construction structural components are entirely of wood. Therefore, the building is a prime contributor to the fire load. The NFPA 220 wood-frame construction classification has been removed under the new NFPA 5000 and ICC building codes and replaced with the combustible classification. The NFPA places Type V construction under two subclassifications: Type V-111, which has a one-hour fire-resistive rating throughout the building, and Type V-000, which carries no resistive rating.[25]

Balloon-Frame Construction

Balloon-frame construction originated in the 1830s. It was a building method that required smaller dimensional lumber than previous forms of wood construction. Mortise-and-tenon joints were replaced with machine-made nails. The defining feature in balloon-frame construction is that the wall studs run continuously, two or more stories, from the sill to the top plate of the building. The studs and the first-floor joists rest on the sill. The floor joists for the upper stories are tied to the supporting exterior walls by a ribbon strip, which is a board that is nailed to the interior edge of the studs. The joists rest on the ribbon strip, much the same as beams rest on a girder, and are nailed to the inside face of the studs.

The natural inclination of wood is to expand and contract across the grain, not with the grain. In balloon construction, most of the structural components are assembled so that there is less cross-sectional wood grain framing than in other types of wood-frame construction. Because the exterior wall studs are continuous from floor to attic, the wood grain is in one direction and unbroken. Therefore, there is less expansion and contraction of structural components in a balloon-framed structure than in conventional wood-frame construction. This quality makes balloon framing ideal for two-story structures with brick veneer, stone veneer, or stucco exterior walls. It is also the construction method of choice for interior load-bearing components in Type III masonry structures. Because there is less cross-sectional grain construction, the dimensional changes in the walls and the settling that occurs between the exterior walls and interior support structures are minimized.

The biggest drawback to balloon-frame construction is the large concealed spaces and voids that are created by its design configurations. If the exterior walls of a balloon-framed structure are not provided with proper firestops, fire can easily travel from the basement to the attic between the studs and to opposite load-bearing walls between the ceiling or floor joists. This fire movement may be undetected by firefighters until it breaks out above or behind them. Firestops are blocks of wood placed between studs at each floor level. They may also be placed between floor or ceiling joists. Their sole purpose is to inhibit fire spread throughout the building. Unfortunately, many older structures with this type of construction are not properly firestopped. Firefighters should try to identify these structures during prefire planning. One way to identify balloon construction in older buildings is to observe the location of upper and lower level windows and doors. In these older buildings, the upper story windows are usually aligned vertically with the lower floor windows. Exterior wall studs must be cut to accommodate window and door openings, and vertical alignment of the openings reduces the number of studs that must be cut. The widespread use of balloon-frame construction declined in the early to middle 20th century because the long structural framing members required became more difficult to obtain. However, some custom builders still occasionally use this construction method (see Figure 7-35).[26]

Platform-Frame Construction

Platform framing gained popularity at the end of World War II. Platform framing allowed the use of shorter lengths of lumber, which reduced construction time and was easier to

FIGURE 7-35 Balloon-frame construction provides many concealed spaces and voids for fire to travel through. If all the upper level windows are aligned with the first-floor windows, it may indicate balloon-frame construction. *Courtesy of James D. Richardson.*

acquire at the time. This type of construction also reduced the amount of concealed spaces that were prevalent in balloon-frame construction. In platform-frame construction, wall studs extend only the height of each floor. The bottom of the first floor studs are nailed to a bottom board called a **sole plate**. The sole plate lies horizontally on a concrete foundation or on a wooden platform floor. The platform is constructed with floor joists covered by subflooring. The top of each stud on the first floor wall is nailed to a **top plate**. The top plate is constructed of two boards nailed together to form a beam. The top plate serves a dual purpose. It acts as a support structure for the ceiling or floor joists to rest on and as a firestop to prevent fire spread if a fire develops inside the wall. When a second story is added, another floor platform is constructed of wood joists and subflooring, and a separate wall is erected on this platform. Basically, platform framing consists of a structure on top of a structure. There is no structural continuity between the floors. This makes the structure very susceptible to wind forces (see Figure 7-36).

Platform construction allows for many variations of design and structural components. From the 1940s until the late 1950s, builders used solid-dimensional lumber for ceiling joists, usually measuring 2 × 6 or 2 × 8 inches depending on the span, and 2 × 12-inch

sole plate ■ The bottommost component of a wall assembly. The bottom of the wall studs are connected to this component, which in turn is connected to the foundation of the structure.

top plate ■ Horizontal part between where the roof and studs finish.

FIGURE 7-36 Platform framing is the most common form of wood construction today. *Courtesy of James D. Richardson.*

lumber for floor joists. To reduce cost and construction time, many builders have replaced solid-dimensional lumber with engineered-wood structural components. As mentioned earlier in this chapter, these lightweight components may be superior in strength under normal conditions, but they can readily fail under fire conditions. These structural components have less density, and because of their smaller dimensions, they have more surface area, which makes ignition easier.

Fire protection systems have been used for decades in all types of construction in commercial buildings. They have allowed larger, more open structures to be constructed with lightweight components. However, with the advent of lightweight engineered-wood components in residential structures, the need for firefighter safety has become paramount in fireground tactics and strategy. Some communities require the installation of residential sprinkler systems in all new residential construction. Residential sprinkler systems not only save lives, but also can delay structural collapse by slowing the fire's progression, or extinguishing the fire before severe damage occurs.

 ON SCENE

At 4:35 AM on February 6, 1997, firefighters of the Stockton (CA) Fire Department responded to a reported garage fire. On arrival at the scene, they found a single-family residence heavily involved with fire. It was reported that a victim was trapped in the residence. Search and firefighting teams entered the building, and a second alarm was ordered to increase the number of fire personnel and protect exposed residences in close proximity to the fire.

At 4:45 AM, the rear portion of the involved residence suddenly collapsed, killing two firefighters and severely injuring another. Firefighters later reported that there had been no warning signs of a potential collapse. The investigation that followed revealed that a two-story addition had been built onto the rear of the residence in the 1950s. This was a large open-area structure with no interior supporting walls. The exterior walls were the load-bearing supports for the second floor. When the structural integrity of the exterior walls was compromised by fire, the walls failed, and the entire second story collapsed to the first floor.

Further investigation revealed that the firefighters had not realized that the structure had a second level because the rear of the structure was obscured by smoke, fire, and darkness. They assumed that the structure was a one-story residence because there were no other two-story residences in the neighborhood.

1. What suggestions would you make to reduce the possibility that this type of incident would occur again?
2. How could this incident have been prevented?

Summary

Buildings are designed to carry specified loads and to endure specified forces. Renovations or changes in how a building is used (its occupancy) may alter the forces and loads in ways that exceed the building's capabilities. Firefighters must be able to identify the different types of loads and forces and to determine how a structure will react when subjected to a particular load or force.

A building consists of a foundation, floor, walls, and roof. The structural components are designed to transfer all loads to the foundation. The walls, columns, and foundation are designed to resist compression forces. Walls must be braced to strengthen them against lateral forces caused by wind. Beams and girders are designed to resist tensile and shear forces. Their function is to transfer loads horizontally to the support columns or walls.

The NFPA 220 and the NFPA 5000 building codes list five types of building construction. NFPA 220 designates Type I as fire resistive, Type II as noncombustible, Type III as ordinary, Type IV as heavy timber, and Type V as wood frame. NFPA 5000 eliminated these five designations and simply categorizes Types I and II construction as noncombustible and Types III, IV, and V as combustible.

Firefighters should study and be familiar with the characteristics of each construction category, focusing especially on those aspects of design and materials that would influence how a fire behaves. They should have a solid understanding of the structures in their jurisdiction and the special challenges these buildings may pose in the event of a fire. Keeping up-to-date on new, lightweight building components is also critical.

To put it simply: Firefighters should be ready for every possible situation. And then, after all their homework, painstaking preparation, and plans are in place, expect the unexpected!

Review Questions

1. List and categorize the types of construction designated by NFPA 220 and NFPA 5000.
2. What do the digits in the NFPA fire-resistive ratings represent?
3. What hazard does Type II construction pose to firefighters?
4. What types of forces are columns and foundations designed to withstand?
5. What hazard does old-style Type III construction pose to modern firefighting tactics and strategies?
6. What is the difference between prestressed and post-tensioned concrete?

Suggested Reading

Brannigan, Francis L., and Corbett, Glenn P. (2007). *Building construction for the fire service* (4th ed.). Quincy, MA: National Fire Protection Association.

Buchanan, Andy. (2005, September 13). *Fire resistance of solid timber structures*. Christchurch, New Zealand: University of Canterbury. Retrieved from http://www.branz.co.nz/branzltd/pdfs/AndyBuchanan.pdf.

Davis, Richard J. (2008). Structural integrity during fire. In *Fire protection handbook* (20th ed., Volume II). Quincy, MA: National Fire Protection Association.

Determine the construction classification number. (n.d.). *Firewise*. National Wildland/Urban Interface Fire Program. Retrieved September 13, 2005, from http://www.firewise.org/resources/operationWater/step1.htm.

Fisette, Paul. (2000). The evolution of engineered wood I-joists. *Building materials and wood technology*. Retrieved September 13, 2005, from the University of Massachusetts website: http://www.umass.edu/bmatwt/publications/articles/i_joist.html.

The history of wood trusses. (2003). *Wood truss design and manufacturing*. Weyerhaeuser. Retrieved September 13, 2005, from Truss-Frame.com website: http://www.truss-frame.com/truss-mfg.html.

Kodur, V. R., and Sultan, M. A. (1998, September). *Structural behaviour of high strength concrete columns exposed to fire* (NRCC-41736). National Research Council Canada, Institute for Research in Construction, International Symposium on High Performance and Reactive Powder Concrete. Sherbrooke, Quebec, Canada.

Miller, Mark, and Miller, Rex. (2010). *Carpentry and construction* (5th ed.). New York, NY: McGraw Hill.

Murnane, Lynne, and Fortney, Jeff. (2010). *Building construction related to the fire service.* Stillwater, OK: International Fire Service Training Association.

Naum, Christopher. (2010, December 3). *Operational safety considerations at ordinary and heavy timber construction: Predictability and performance of buildings and today's fireground.* Retrieved January 9, 2013, from http://commandsafety.com/tag/operational-safety-considerations-at-ordinary-and-heavy-timber-constructed-occupancies/.

Olson, Craig, and Smith, Laura. (1997, May 9). Post-tensioned concrete for today's market. *The Seattle Daily Journal of Commerce.* Retrieved from http://www.djc.com/special/concrete97/10024302.

Parkinson, David. (2002). *Performance based design of structural steel for fire conditions.* Paper submitted to the Faculty of the Worchester Polytechnic Institute. Retrieved from Worchester Polytechnic Institute website: http://www.wpi.edu/Pubs/ETD/Available/etd-0821102-115014/unrestricted/dparkinson.pdf.

Tide, R. H. R. (1998). Integrity of structural steel after exposure to fire. *Engineering Journal*, pp. 26–38.

Endnotes

1. Guralnik, David. (1982). *Webster's new world dictionary* (2nd college ed.). New York: Simon & Schuster.

2. National Institute for Occupational Safety and Health (NIOSH). (1998, June). *Fire fighter fatality investigation report F98-17.* Retrieved April 7, 2008, from the NIOSH website: http://www.cdc.gov.niosh/fire/reports/face9817.html.

3. Murnane, Lynne, and Fortney, Jeff. (2010). *Building construction related to the fire service.* (3rd ed.). Stillwater, OK: International Fire Service Training Association.

4. Brannigan, Francis L., and Corbett, Glenn P. (2007). *Building construction for the fire service* (4th ed.). Quincy, MA: National Fire Protection Association.

5. Murnane and Fortney, op cit.

6. Brannigan and Corbett, op cit.

7. Murnane and Fortney, op cit.

8. Davis, Richard J. (2008). Structural integrity during fire. In *Fire protection handbook* (20th ed., Volume II). Quincy, MA: National Fire Protection Association.

9. Miller, Mark, and Miller, Rex. (2010). *Carpentry and construction* (5th ed.). New York, NY: McGraw Hill.

10. Murnane and Fortney, op cit.

11. Fisette, Paul. (2000). The evolution of engineered wood I-joists. *Building materials and wood technology.* Retrieved September 13, 2005, from the University of Massachusetts website: http://www.umass.edu/bmatwt/publications/articles/i_joist.html.

12. Davis, op cit.

13. Murnane and Fortney, op cit.

14. Davis, op cit.

15. Ibid.

16. Tide, R. H. R. (1998). Integrity of structural steel after exposure to fire. *Engineering Journal.* Retrieved September 12, 2005, from http://www.steelstructures.com/IIW%20files/35_1_026.pdf.

17. Brannigan and Corbett, op cit.

18. Murnane and Fortney, op cit.

19. Brannigan and Corbett, op cit.

20. Kodur, V. R., & Sultan, M. A. (1998, September). *Structural behaviour of high strength concrete columns exposed to fire* (NRCC-41736). National Research Council Canada, Institute for Research in Construction, International Symposium on High Performance and Reactive Powder Concrete, Sherbrooke, Quebec, Canada. Retrieved from http://irc.nrc-cnrc.gc.ca/ircpubs.

21. Davis, op cit.

22. Brannigan and Corbett, op cit.

23. Davis, op cit.

24. Brannigan and Corbett, op cit.

25. Ibid.

Fire Prevention Codes and Ordinances

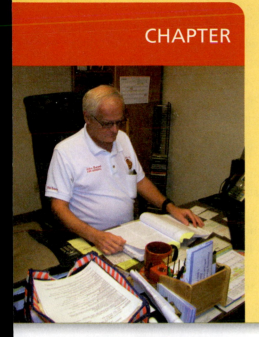

Courtesy of James D. Richardson

OBJECTIVES

After reading this chapter, you should be able to:

- Describe the relationship between fire and building codes.
- Identify the different model code organizations.
- Determine the legal basis for fire code enforcement.
- Identify the types of fire codes and explain how fire codes are adopted.
- Explain the legal right that a fire official has to issue warrants, summonses, citations, and injunctions.

Introduction

In Chapter 7, we looked at the resources that the fire service uses to protect the public from fires and, increasingly, from other kinds of emergencies as well. In this chapter, we will broaden our scope to look at ways in which the fire service cooperates with government, at the state and local level, to extend its protective reach. Specifically, we will look at codes and ordinances, the legal backbone of public safety. A *code* is a set of mandated standards set forth by a professional organization. When a standard is given legal status by a municipality or other local jurisdiction, it becomes an **ordinance**. We will define the function of codes and ordinances and examine how fire departments and other government agencies work together to administer them. We will consider some model codes and the organizations that create them. We will briefly trace the history of codes in the United States, explore some of the jurisdictional and civil liberties issues surrounding their implementation, detail how fire codes are adopted, and describe the enforcement process.

Major fires have played an all-too-frequent role in U.S. history. Almost every sizable city has suffered at least one major conflagration. It is important to remember that behind the rather dry language of codes and ordinances are real-life calamities involving tragic losses of life (see Table 8-1). For example, at 3:15 PM on December 30, 1903, a fire erupted in the Iroquois theater in Chicago. About 1,900 people filled the theater, which had a seating capacity of 1,724. Although there were 27 exits, most were locked, covered, or blocked with iron gates. A total of 602 perished in the fire. As a result of the tragedy, the fire codes were changed, requiring all theater doors to open outward and to have clearly marked exits. The new code also required that theater management conduct fire drills with theater employees.[1] In Boston on November 28, 1942, a fire started in the Cocoanut Grove nightclub. About 1,000 people filled the structure that had a licensed capacity of 500. A total of 492 people died as a result of locked exits and a revolving door. After the fire, new fire codes were written to require outward-swinging exit doors with push-down "panic hardware" on each side of a revolving door. Codes were also written limiting the use of flammable materials for interior decorations in nightclubs.[2] Codes and ordinances are written, and just as frequently revised, to prevent such disasters from recurring.

The writing of a code is only the beginning. In order to function, codes must be adopted as ordinances and rigorously enforced. This became evident on February 20, 2003, when the fourth deadliest nightclub in U.S. history occurred in West Warwick, Rhode Island. A total of 462 people were inside the 4,500 sq. ft. structure—100 people were killed and 230 were injured when stage pyrotechnics ignited polyurethane foam that was attached to the walls and ceiling around the stage. The fire spread rapidly and fully engulfed the building in less than 5 minutes. The owners of the club did not have permits for pyrotechnic displays and the foam, which had been installed, without permit, two years earlier, was not treated with fire retardant and violated the fire code. Enforcement means more than policing, however. First and foremost, it means informing the public about the importance of fire codes and explaining the reasoning behind code provisions and their importance. Public education is

TABLE 8-1	Historic Fires		
CITY	**DATE**	**OCCUPANCY**	**FATALITIES**
New York, New York	March 1911	Triangle Shirtwaist Factory	145
Chicago, Illinois	November 1942	Cocoanut Grove Night Club	492
Southgate, Kentucky	May 1977	Beverly Hills Supper Club	165
Las Vegas, Nevada	November 1980	MGM Grand Hotel	85
West Warwick, Rhode Island	February 2003	The Station Nightclub	100

the duty of all fire service organizations. Firefighters respond when the fire alarm goes off, but the primary responsibility for preventing a fire rests squarely on the shoulders of the informed citizen.

Building Codes and Fire Prevention Codes

There are two categories of codes that relate to public fire safety: building codes and fire prevention codes.

Building codes address fire and other safety issues related to building construction. They establish standards for the fire resistance of building materials and interior finishes; for means of egress; for vertical openings, such as stairwells, elevator, and utility shafts; and for other structural features. Typically, a municipality's building department is responsible for administering and enforcing building codes.

Fire prevention codes focus on fire safety issues that relate to how a building is used and maintained. These codes set standards for fire protection systems and equipment. They also define fire hazards, such as faulty electric appliances and improperly stored hazardous materials. Fire codes are usually administered and enforced by the fire department.

Building codes and fire prevention codes are different but are closely related, however, and a good working relationship between the building department and the fire department is the best recipe for protecting the public.[3] The relationship should be based on understanding that three basic ingredients are necessary for success: quality engineering, strict enforcement, and comprehensive public education programs. Cooperation is essential in all three areas.

Both the building department and the fire department should review new construction plans and make **engineering corrections** that enhance fire safety. These corrections may include the installation of fixed fire protection systems, such as automatic sprinklers. These sprinklers are designed to control or extinguish a fire in its early stage of development and provide a protected path of travel to exits, which will then reduce the amount of time required to evacuate a building. Coordinating efforts for code enforcement between the building and the fire departments will result in greater convenience for the public, reduce redundancy, and save time and money for both departments. Coordinated education programs will enhance public compliance with inspections and will foster better understanding on the part of both departments for changing community attitudes and needs.[4] Too often, unincorporated areas are ignored by these departments. Code enforcement and inspections are infrequent or nonexistent. Education programs should be expanded in order to bring building and fire code awareness to these areas.

History of Model Codes in the United States

When establishing legal standards on construction and fire safety, state authorities can choose a **model code**, which they can adopt or adapt, as the law allows, to coincide with the needs of their state. These codes are written, published, and revised (generally every three to five years) by professional membership organizations, some of which will be discussed later.

The origin of codes in the United States goes back to the 19th century. As a result of industrialization and population growth in the cities, urban fires became increasingly frequent and increasingly costly. By the middle of the century, major insurance companies were suffering such huge losses that some were forced into bankruptcy. After a major fire in Portland, Maine, in 1866, the insurance companies formed the **National Board of Fire Underwriters (NBFU)**. This new organization, which later became the American Insurance Association (AIA), was created to standardize rates and find ways to deter arson. Although it was never successful in standardizing rates, the NBFU did develop guidelines for fire apparatus and for municipal fire supplies. These guidelines became the nation's first fire codes.

building codes ▪ Laws adopted by states, counties, cities, or other government sources to regulate all aspects of building construction.

fire prevention codes ▪ Body of laws designed to enforce fire prevention and safety.

engineering corrections ▪ Corrections that enhance fire safety. These corrections may include the installation of fixed fire protection systems, such as automatic sprinklers.

model codes ▪ Codes developed and maintained by a standards organization separate from the governing body for building and fire codes.

National Board of Fire Underwriters (NBFU) ▪ Organized in 1866, the NBFU developed standards for fire safety in building construction and control of fire hazards.

In 1893, insurance companies became alarmed about the proposed Palace of Electricity at the World's Columbian Exposition in Chicago. The array of untested electrical connections, in close proximity to the flammable decorative façade, was considered so hazardous that the insurance companies threatened to pull their backing, which nearly ended the Exposition. A young Boston electrician, William Henry Merrill, saved the day. He reviewed the wiring, made certain that all proper safeguards were in place, and got the insurance companies back on board. With the backing of the NBFU, Merrill went on to form the Underwriters Electrical Bureau, which later became Underwriters Laboratories (UL).

In 1904, in response to the great Baltimore, Maryland, fire, the National Board of Fire Underwriters issued its Standard Grading Schedule for all municipalities. This standard was used to establish a municipality's basic insurance rate, which was contingent on the municipality's ability to protect its residents from fire. It factored in the number and types of fire apparatus available to the city's fire department, its water supply, and its average response time. The board also published the first National Building Code in 1905. This code set height and area restrictions and established standards for such safety features as means of egress, fire-resistive construction, and fire protection systems. It was offered free of charge to any state or local government that was willing to adopt the code. Fourteen editions of the National Building Code were published before it was discontinued in 1976.[5] The AIA also published fire and building codes until 1971.

BUILDING OFFICIALS AND CODE ADMINISTRATORS INTERNATIONAL

The Building Officials and Code Administrators International (BOCA) is a membership organization serving the northeastern and midwestern United States and Canada. It is comprised of state and local governments and agencies that are responsible for code enforcement, as well as members of the building construction disciplines. It was established in 1915 and published its Basic Building Code in 1950, which later became known as the BOCA National Building Code.[6]

INTERNATIONAL CONFERENCE OF BUILDING OFFICIALS

The International Conference of Building Officials (ICBO) was established in 1921. By 1927, it had published a number of codes pertaining to the building industry. One of these—the Uniform Building Code (UBC)—mandated construction design and building material requirements. This code was primarily used in the Pacific Coast, Rocky Mountain, and Great Plains states until publication of the code ceased in 2003.[7]

SOUTHERN BUILDING CODE CONGRESS INTERNATIONAL

The Southern Building Code Congress International (SBCCI) was established in 1940 and published its Standard Building Code in 1945. This code included special provision for dangers such as hurricanes that occur in the area that the SBCCI served—the southeastern and southwestern United States. It was similar to other model building codes in other respects.[8]

INTERNATIONAL CODE COUNCIL

In 1994, BOCA, ICBO, and SBCCI joined forces to found the International Code Council (ICC). Its mission was to develop a set of model international building codes. By 2000, the ICC's International Building, Fire, and Property Maintenance Codes were available for adoption. Three years later, BOCA, ICBO, and SBCCI consolidated all code development services into ICC and ceased publishing their own codes. ICC's work made standardized fire protection possible in the United States and, as an additional benefit, made it possible for building and construction inspectors to work in different parts of the country.[9] The ICC has developed many "I-codes," including building and residential construction, green construction, energy conservation, electrical, and fire codes. The council also offers training, education, and certification programs for safety professionals. Fifty states and the District of Columbia have adopted the ICC codes at the state or local level. Every three to five years the ICC code development cycle is initiated. The development cycle is a long process that begins when an agency or organization submits proposed changes. The changes are posted for public review. A code development hearing is held and the results from the public hearing are posted. Public comments are requested and distributed. A final action hearing is convened and the new code edition is published.[10]

NATIONAL FIRE PROTECTION ASSOCIATION

The National Fire Protection Association (NFPA) was established by the National Board of Fire Underwriters in 1896. The NFPA not only worked in developing national standards for firefighting education and apparatus, but also developed almost 300 fire prevention and building codes and standards that have been adopted by reference in many jurisdictions.

Among NFPA's landmark documents are its National Electric Code, first issued in 1911 and still in publication today (NFPA 70), and the NFPA 101, *Life-Safety Code*. The 2012 Life-Safety Code divides building occupancy into 16 classifications and establishes means of egress standards for new and existing structures. Although the term "occupancy" is commonly used to describe the number of people allowed in a structure, it is also used to identify the business operations being conducted within the structure. The 2012 Life-Safety Code occupancy classifications are assemblies, education, day care, health care, ambulatory health care, detention and correctional, one- and two-family dwellings, lodging and rooming houses, hotels and dormitories, apartments, residential boarding and care, mercantile, business, industrial, storage, and building rehabilitation. Because the code is so comprehensive in addressing life-safety requirements, many jurisdictions incorporate it into their fire and building codes.[11] In 2002, NFPA published an alternative to the ICC International Codes. This publication is known as NFPA 5000, *Building Construction and Safety Code*. The ICC International Codes and the NFPA 5000 are written by private organizations. The codes are recommendations, not mandates. Each state or province may choose (through legislation) which code to adopt when writing building and fire prevention ordinances. Table 8-2 describes the differences among the codes. There has been an attempt by the ICC to consolidate the codes into one International Code.

TABLE 8-2	Building and Fire Code Organizations
The Building Officials and Code Administrators International (BOCA)	Serves the northeastern and midwestern United States and Canada. It comprises state and local governments and agencies that are responsible for code enforcement, as well as members of the building construction disciplines.
The International Conference of Building Officials (ICBO)	Published a number of codes pertaining to the building industry, including the Uniform Building Code.
Southern Building Code Congress International (SBCCI)	Published the Standard Building Code in 1945. This code was very similar to the other model building codes, except it included special provisions for events, such as hurricanes, that affect the region it served. Membership consisted of representatives from the southeastern and southwestern United States.
International Code Council (ICC)	In 1994, BOCA, ICBO, and SBCCI joined forces and formed the International Code Council and published a set of model building codes called International Codes.
The National Fire Protection Association (NFPA)	Established by the National Board of Fire Underwriters in 1896. This organization has been instrumental in developing almost 300 fire and building codes and standards that are adopted by reference in many jurisdictions.

Historical Basis for the Legal Authority of Code Enforcement

In the United States, ordinances governing such matters as building codes and fire safety are enacted at the local level. However, the local government's discretion in these matters is limited by the state government, which establishes minimum standards, usually selecting the code that is to be used.

There is a history of controversy over the relative role of municipalities and states in legislating local standards (analogous to the national controversy between federal and states' rights). During the 150 years of America's colonial period, town governments enjoyed considerable autonomy in managing their local affairs. This concept, known as *home rule*, was based on principles inherited by the colonists from English law (the Carta Civibus Londoniarum of 1100 AD and the Magna Carta of 1215 AD).[12]

The Constitution of the United States, by contrast, does not provide for local government autonomy, instead granting to the individual states the rights and privileges that had traditionally belonged to towns. The Tenth Amendment to the Constitution reserves power to the states or the people that are not delegated to the federal government or prohibited to the states. The wording of the amendment is open to interpretation. Local jurisdictions in several states have argued that the phrase "or to the people" gives local governments the authority to develop their own codes and ordinances without interference from the state. The issue was brought before several state supreme courts before it was eventually decided by the U.S. Supreme Court in 1903.

DILLON'S RULE, THE COOLEY DOCTRINE, AND *ATKINS V. KANSAS*

In 1868, Iowa State Supreme Court Justice John Dillon wrote a landmark decision in the case of *City of Clinton v. Cedar Rapids and Missouri Railroad Company*. His opinion, known as Dillon's Rule, stated that local jurisdictions in Iowa could exercise only those powers that are expressly granted to them by the state.[13]

This strict constructionist decision was not welcomed in many states. Michigan Supreme Court Justice Thomas Cooley challenged Dillon's opinion by arguing that the Tenth Amendment reserved the right of home rule because it delegates power to the states "or to the people." Cooley argued that "to the people" meant "to the local government." Local governments should have powers to initiate policy. The Cooley Doctrine found support in the states of Indiana, Kentucky, Texas, and in some courts in Iowa. The home rule movement continued into the 20th century.[14]

In 1903, the U.S. Supreme Court heard the case of *Atkins v. Kansas*. The Court rejected Cooley's Doctrine and upheld Dillon's Rule, declaring that local governments can have only those powers specifically given to them by the state. In 1923, the Court further bolstered Dillon's Rule with a decision declaring that a state has the right to grant, withhold, or withdraw powers from local governments as the state deems necessary.[15] These decisions established the framework on which today's fire codes and ordinances are enforced.

Dillon's principle became the law of the land. However, in actual practice, states do sometimes grant a degree of decision-making autonomy to local government officials, and this may be reflected in the local codes and ordinances they enact.[16] In all cases, however, this power is granted at the discretion of the state.

WARRANTLESS ENTRY

Having explained the constitutional background for the authority granted to local governments by the states, we will now examine their use of this authority in enforcing codes and ordinances. Unquestionably, the most controversial aspect of code enforcement is the right granted to fire prevention authorities to enter premises either for inspections or to take emergency measures. Even though the state grants powers of code enforcement to local governments, and the local government in turn confers these rights to fire authorities, all authorities are legally bound by the U.S. Constitution and are limited in their ability to act. In most states, fire prevention codes are applicable to all premises, including single-family dwellings. However, what right does a representative of the government have to enter private property? The Fourth Amendment to the U.S. Constitution clearly states, "The right of the people to be secure in their persons, houses, papers, and effects, against unreasonable searches and seizures, shall not be violated, and no **Warrants** shall issue, but upon probable cause, supported by Oath or affirmation, and particularly describing the place to be searched, and the persons or things to be seized."[17]

The language of the Fourth Amendment would seem to forbid warrantless entry. The U.S. Supreme Court, however, has ruled that warrantless entry is legal under limited emergency circumstances. In three separate decisions—*Mapp v. Ohio, Frank v. Maryland,* and *Camara v. Municipal Court*—the Court ruled that warrantless entry is unconstitutional without the consent of the occupant or owner. It is justified if entry is made in response to an emergency. In another case, the Court ruled that a burning building is considered an emergency; therefore, warrantless entry is justified. In many states, the law recognizes that even if a building is burning, the right to privacy outweighs the police powers of the government if the burning building is so isolated that it poses no threat to other property or persons.[18]

warrants ■ Orders from a judge allowing fire officials the right to search or seize private property or arrest an individual.

RIGHT OF ENTRY

All model building codes contain clauses pertaining to the right of entry by fire officials, but these officials are bound to act within carefully delineated guidelines. Under normal circumstances, permission to enter a property must be granted by the owner or occupant, or a warrant must be issued. The fire official must present identification and the proper credentials. The purpose for entry must be related to code compliance issues, and the entry request should be scheduled during normal operating hours of the occupancy or at an hour that is considered reasonable. Fire officials may enter a property for the purpose of code compliance verification without a warrant or permission only if a life-safety condition exists or if the area being inspected is visible from a public way (meaning that the inspection can be conducted without actually setting foot on private property).

The right of entry to private property by a fire official often depends on the type of model used in establishing inspection priorities. The two models used for prioritizing inspections are the **permit model** and the inspection model.[19]

Permit Model

Under the permit model, businesses that have a high potential for loss of life, such as nightclubs, restaurants, and hazardous materials processing and storage facilities, require a permit to operate. They must prove that their facility meets code requirements before a permit is issued. The permit is usually site specific, which means that it identifies the hazard, location, time of operation, and purpose. It allows fire officials the opportunity to reference other codes, such as those that establish building and zoning requirements. The permit is considered the property of the issuing agency and does not belong to the individual receiving the permit. The issuance of a permit gives the jurisdiction the right of entry to the premises for the purpose of inspection. If after receiving a permit for operation, the occupant or owner of a business refuses to allow an inspection of the facility during a "reasonable hour," the permit can be revoked.[20]

The permit model does not give a fire official unlimited access to premises, however. In the U.S. Supreme Court case of *See v. The City of Seattle*, a Seattle businessman refused to allow fire inspectors into an area of his business, claiming that the area was used for private business matters and was not accessible to the general public. The Court ruled against the city, stipulating that a business owner can deny government officials entry to areas of a business that are not normally open to the public. If permission to enter that area is denied, the inspector must obtain a warrant before entering.[21]

permit model ■ Businesses that have a high potential for loss of life, such as nightclubs, restaurants, and hazardous materials processing and storage facilities, require a permit to operate.

Inspection Model

occupation classification ▪ Classification given to a structure by a model code used in that jurisdiction and pursuant to the given use of that structure.

statute ▪ A law established by a state or federal legislative body.

When using the inspection model, fire officials schedule inspections in their jurisdiction based on the occupation category of the buildings. The **occupation classification** of a building is based on the purposes it serves: residential, educational, mercantile, storage, and so on. Each occupancy type is designated for inspection during a set time period. For example, the fire inspector may decide that all fuel storage areas will be inspected during the month of May. Inspection schedules and procedures must be published, and the right of entry by the fire official will be based on the schedule. Any deviation from the schedules is discouraged, unless a life-safety hazard is found to exist. A **statute** establishes that a particular type of occupancy must be inspected a certain number of times each year and that fire officials have a duty to carry out those inspections. This reduces the possibility that a business owner will claim harassment.[22]

STOP AND THINK 8-1

Under what circumstances are unscheduled inspections allowed under the jurisdiction in your area?

Types of Codes and Code Adoption

The process by which fire codes are adopted is determined by state governments, but state governments can allow local jurisdictions discretion to adapt codes to meet particular local needs.

MINIMUM CODES VERSUS MINI-MAXI CODES

minimum codes ▪ Lowest acceptable standards.

mini-maxi code ▪ Establishes minimum and maximum building code requirements.

In some states, **minimum codes** are adopted by the state, and local jurisdictions are not allowed to implement regulations that have less restrictive requirements. They are empowered to implement more stringent code requirements if they see fit.

Under the **mini-maxi code** concept, local jurisdictions are not allowed to adapt or modify any regulation. The code is mandated by the state and is enforced statewide. The rationale behind the mini-maxi code is that it allows for uniformity in the application of the code. Builders, contractors, and architects support the concept because it simplifies their work—regulations are the same in all jurisdictions. All lobbying efforts for code changes are made at the state level. Fire officials are opposed to mini-maxi codes because they do not allow local jurisdictions the flexibility to write and enforce regulations that meet needs unique to their community. A positive aspect of the mini-maxi code concept is that it forces the fire departments in the state to work together and become proactive in code development.[23]

MODEL CODES

The code adoption process is defined by state law. Before a code can be adopted, notice must be published scheduling a public meeting so that citizens can voice their support or objections to the code. An effective date for the code must be written into the ordinance, and the agency responsible for code enforcement must be clearly identified. Model codes can be adopted by reference or by transcription.

Adoption by Reference or Transcription

When adopting by reference, the jurisdiction simply refers to an edition of the model code in the adopting ordinance. If the adoption is by transcription, the entire model code must be published in the adopting ordinance. The life expectancy of most model code editions is three to five years.

Prescription- or Performance-Based Models

States have the option to base their fire prevention efforts on one of two models: the prescription model or the performance model. These are different ways of viewing how codes work to protect the public. The choice of one over the other is strictly a matter of perspective.

Prescriptive Codes

Prescriptive codes establish specific requirements for building and fire safety. In the prescriptive model, fire safety is achieved through the use of mandated requirements in building materials, construction components, means of egress, and fire protection systems. These mandates are based on past fire experiences in which large losses of life or property occurred; the mandates are generally not revised until new lessons are learned, often as a result of another major fire loss. Prescriptive codes do not identify fire safety objectives; rather, the assumption is that if all of the mandated building requirements are followed, fire safety will be achieved.[24]

Performance Codes

Performance codes begin from the perspective of goals and objectives. This type of code does not mandate specific materials and so on, but relies on quantitative measures to determine the fire safety level of building construction assemblies and of the building as a whole. Code compliance can be verified by one of three methods:

Laboratory or On-Site Testing. The design or material is tested in a laboratory or on the building site to determine if it meets code requirements.

Verification that a Technical Solution Satisfies Performance Requirements. A standard unit of measurement is agreed on, so that test results can be validated in repeat tests.

Verification by a Technical Expert or a Certifying Entity. Performance is evaluated and certified by outside technical experts.[25]

Code Administration and Enforcement

The administration section of a model code identifies the enforcing authority and outlines the legal framework for its enforcement activities. It contains legal language that, like all legal language, calls for careful reading. The wording of the document is critical because it establishes the legal foundation for right of entry as well as for the issuance of orders for evacuations and other emergency measures intended to preserve life. Wording in this section should not only give the fire officials full authority to do their job effectively but should also specify the responsibility of the property owner for building maintenance and fire safety.[26] Wording in the code will also determine if the actions taken by a fire official are mandatory or discretionary. For example, if the code uses the word "shall," the action of the fire official is **mandatory authority** under the law. If the word "may" is used, the fire official is given **discretionary authority**. The fire official must be familiar with the wording in the code and follow the "letter of the law" for all mandated requirements. If the code mandates the inspection of certain occupancies, fire officials have a duty to inspect those facilities. Failure to do so may result in charges of **negligence**, resulting in legal action against the official.

DOCUMENTATION

To protect themselves legally, fire officials should document all inspections or visits to a facility. All actions taken must be recorded. If a violation of the code is discovered, a legal

mandatory authority
■ Makes the actions mandatory under the law.

discretionary authority
■ Allows the fire official to use discretion when enforcing the law.

negligence ■ Failure of a fire official to perform in the same manner as a prudent person would under the same circumstances.

notice of violation must be issued. The form this takes depends on the code that the jurisdiction is using, but there are some basic violation notice procedures that are common to all codes. First, the notice must be in writing, identifying the exact location (legal address) of the inspected facility, the date of inspection, and the name of the inspecting officer. Second, violations must be specifically enumerated, and the code sections supporting the notice must be cited. Third, any required corrective action must be clearly described. Finally, the notice must contain a date for a follow-up inspection to verify code compliance.[27]

STOP AND THINK 8-2

What are the legal ramifications if an inspector fails to perform a follow-up inspection when a code violation is documented?

SUMMONS AND CITATIONS

summons ■ A written document from a judge or a law enforcement officer ordering a code violator to appear in court on a certain date.

citation ■ Same as a summons except it is issued by a law enforcement officer instead of a judge.

injunction ■ A legal document from a judge that orders a person or entity to cease or perform certain activities.

Inspecting officials may be able to use other enforcement tools to ensure compliance with fire and building codes. The availability of these tools varies from jurisdiction to jurisdiction. Some jurisdictions give inspecting officials the power to issue a **summons** or **citation** (the two terms are synonymous). These documents are legal notices that effectively charge an individual with a criminal misdemeanor violation of the fire codes and direct that person to appear in a court of law on a certain date to answer the charges. Some codes require the inspector to issue a citation if a violation is discovered, whereas others may give officials discretionary powers to issue warnings and give individuals a chance to correct the violation before they face a criminal charge. In either case, a thorough and accurate report of the inspection must be submitted. If the owner of a facility does not correct a violation after receiving a summons or citation, the fire official can request that a court issue an **injunction** requiring the facility to cease operation until code requirements are met. An injunction is a court order. Failure to comply with stipulations stated in the injunction is considered contempt of court, and the guilty party can be held in jail until compliance with the code is achieved.[28]

ON SCENE

Case 1

A fire department responds to a building fire in a rural area. The building is approximately 1,000 feet from the road and several hundred feet from all other property lines. No exposures are threatened by the fire. On arrival, the first responding fire company is met by the landowner, who has blocked the entrance to the property and refuses to allow the fire department to enter.

1. What would you do if you were the fire officer in charge?
2. Does the owner have the legal right to refuse entry to his or her property in this situation?

Case 2

The Fourth Amendment to the U.S. Constitution protects citizens from unlawful search and seizure and limits the state's police powers. However, warrantless entry is allowed under certain circumstances.

1. What circumstances would allow nonconsensual entry?

Summary

When the public thinks of the fire service, they picture gleaming red engines, their sirens screaming, racing to the scene of a raging inferno. That picture is valid, certainly, but it is only the tip of the iceberg. The mission of today's fire service is not limited to emergency response. Fire prevention efforts are an equally vital part of that mission, and those efforts include providing public education and monitoring building design and construction. In addition to extinguishing fires, firefighters now have input at the grassroots level of community development and planning. If they have the enforcement authority and training to do their jobs effectively, firefighters can, in a sense, extinguish fires before they are ignited.

Codes and ordinances are the legal backbone of fire prevention. Considerable overlap exists between the fire safety principles that encourage intelligent building design and construction and the fire safety measures that prevent fires. Building codes, administered by a community's building department, and fire codes, administered by its fire department, are different but are closely related. Therefore, it is very important for municipal building and fire department personnel to work together to ensure compliance with all codes and to enhance overall fire safety.

Until 1994, there were a variety of model building codes and several building code organizations. In 1994, the Building Officials and Code Administrators International, the International Conference of Building Officials, and the Southern Building Code Congress International joined forces and formed the International Code Council. In 2000, ICC created the International Building, Fire, and Property Maintenance Codes and made them available for adoption. In 2002, the National Fire Protection Association published the NFPA 5000, *Building Construction and Safety Code.*

The legal authority for a fire prevention organization rests in its fire code. If the code mandates that a certain business be inspected, the fire prevention officer has a duty to inspect that business. Code violations can be enforced in several ways. The owner can be issued a citation or summons to appear in court to answer criminal (misdemeanor) charges. If the owner does not correct the violation after receiving a citation or summons, the inspector can seek a court-ordered injunction against the individual, requiring that all business operations cease until the violation is corrected. Failure to comply with the mandates in an injunction is considered contempt of court and can result in the individual's being incarcerated until code requirements are satisfied.

Full and accurate inspection reports must be kept on file. Fire officials are sometimes summoned to court several years after an inspection. A complete report will allow the inspector to "refresh" his or her memory before testifying. An inspection report is an official document and is considered public information. *Remember: If it is not documented, it never happened!*

Review Questions

1. What influence do building codes have on fire codes?
2. Explain the concepts of Dillon's Rule and the Cooley Doctrine.
3. How does the Fourth Amendment to the U.S. Constitution affect the duties of fire officials in enforcing codes and ordinances?
4. What is the difference between a minimum code and a mini-maxi code?
5. How are model codes adopted?
6. What is the difference between a prescription-based code and a performance-based code?

Suggested Reading

Crawford, Jim. (2002). *Fire prevention: A comprehensive approach.* Upper Saddle River, NJ: Prentice Hall Health.

Crawford, Jim. (2011). *Fire prevention organization and management.* Upper Saddle River, NJ: Pearson Education.

Diamantes, David. (2005). *Principles of fire prevention.* Clifton Park, NY: Thomson Delmar Learning.

Dillon's Rule: Law and legal definition. (n.d.). Retrieved April, 9, 2013, from USlegal.com website: http://definitions.uslegal.com/d/dillons-rule/.

Hammerschlag, Michael. (n.d.). *Inferno horror at the fire station.* Retrieved April 10, 2013, from http://members.surfbest.net/mikehammer/nightclub.htm

Robertson, James. (2009). *Introduction to fire prevention* (7th ed.). Upper Saddle River, NJ: Prentice Hall Health.

Sawyer, Steven. (2008). Fire prevention practices. In *The fire protection handbook* (20th ed.). Quincy, MA: National Fire Protection Association.

U.S. Constitution. Legal Information Institute. Retrieved December 20, 2005, from http://www.law.cornell.edu/constitution/constitution.billofrights.html.

Writ, Clay L. (1989, August). Dillon's Rule—Fairfax County Government. *Virginia Town and City Magazine, 24*(8).

Zimmerman, Joseph. (2010). State-local relations trends. In *The Book of States* (Chapter 2: Federalism and intergovernmental relations). Retrieved April 10, 2013, from http://www.csg.org/policy/publications/bookofthestates.aspx.

Endnotes

1. Eastland Memorial Society. (2008, March 24). *The Iroquois Theater fire.* Retrieved March 24, 2008, from the Eastland Memorial Society website: http://www.inficad.com/~ksup/iroquois.html.
2. *Cocoanut Grove fire Boston.* (n.d.). Retrieved March 24, 2008, from the CelebrateBoston website: http://www.celebrateboston.com/disasters/fires/cocoanutgrovefire.htm.
3. Robertson, James. (2009). *Introduction to fire prevention* (7th ed.). Upper Saddle River, NJ: Prentice Hall Health.
4. Crawford, Jim. (2002). *Fire prevention: A comprehensive approach.* Upper Saddle River, NJ: Prentice Hall Health.
5. Diamantes, David. (1998). *Fire prevention inspection and code enforcement.* Clifton Park, NY: Thompson Delmar Learning.
6. Sawyer, Steven F. (2008). Fire prevention practices. *Fire prevention handbook* (20th ed.). Quincy, MA: National Fire Protection Association.
7. Ibid.
8. Ibid.
9. Ibid.
10. International Code Council. (2012). Online at http://www.iccsafe.org.
11. Diamantes, op cit.
12. Writ, Clay L. (1989, August). Dillon's Rule—Fairfax County Government. *Virginia Town and City Magazine, 24*(8).
13. *Dillon's Rule: Law and legal definition.* (n. d.). Retrieved April, 9, 2013, from USlegal.com website: http://definitions.uslegal.com/d/dillons-rule/.
14. Zimmerman, Joseph. (2010). State-local relations trends. In *The Book of States* (Chapter 2: Federalism and intergovernmental relations). Retrieved April 10, 2013, from http://www.csg.org/policy/publications/bookofthestates.aspx.
15. Ibid.
16. Writ, op cit.
17. U.S. Constitution. Legal Information Institute. Retrieved December 20, 2005, from http://www.law.cornell.edu/constitution/constitution.billofrights.html.
18. Diamantes, op cit.
19. Ibid.
20. Ibid.
21. Sawyer, Steven. (2008). *The fire protection handbook* (20th ed., Volume I), Quincy, MA: National Fire Protection Association.
22. Diamantes, op cit.
23. Ibid.
24. Ibid.
25. Sawyer, Volume I, op cit.
26. Robertson, op cit.
27. Diamantes, op cit.
28. Ibid.

Courtesy of James D. Richardson.

KEY TERMS

branch lines, *p. 186*

clean extinguishing agent, *p. 199*

combination flame detectors, *p. 198*

combination system, *p. 166*

crossmains, *p. 186*

direct pumping system, *p. 166*

distribution grid, *p. 168*

distributor pipes, *p. 168*

encrustation, *p. 169*

enhanced 911, *p. 194*

flow pressure, *p. 179*

friction loss, *p. 180*

gravity system, *p. 165*

groundwater, *p. 164*

head pressure, *p. 165*

high-pressure systems, *p. 200*

indicating valve, *p. 183*

infrared flame detectors, *p. 198*

ionization detectors, *p. 197*

light-obscuration systems, *p. 197*

light-scattering systems, *p. 197*

local application systems, *p. 200*

low-pressure systems, *p. 200*

multipurpose dry chemical systems, *p. 201*

nonindicating valve, *p. 168*

nonpressurized dry hydrant system, *p. 174*

ordinary dry chemical systems, *p. 201*

photoelectric detectors, *p. 197*

pressurized dry hydrant system, *p. 174*

primary feeder mains, *p. 168*

residual pressure, *p. 170*

riser, *p. 177*

secondary feeder mains, *p. 168*

sedimentation, *p. 169*

static pressure, *p. 170*

subsidence, *p. 164*

surface water, *p. 164*

target areas, *p. 177*

total flooding systems, *p. 200*

tuberculation, *p. 169*

ultraviolet flame detectors, *p. 198*

watersheds, *p. 165*

wet chemical, *p. 202*

OBJECTIVES

After reading this chapter, you should be able to:

- Define the two sources of water supply.
- Describe the four components of a water system.
- List and describe the two categories of water supply systems and explain the advantages and disadvantages of each system.
- Explain the different types of pressures.

- List and describe the three basic types of water supply systems.
- Identify the three components of the water distribution grid.
- Identify and describe the three types of fire hydrants.
- Describe the various types of sprinkler systems.
- Understand the various types of detection and alarm systems.
- Describe the various types of special fire protection systems.
- List and describe the different types of extinguishing agents used in fire protection systems.

Introduction

This chapter is designed to introduce firefighters to fire protection systems. Topics covered include water supply and water distribution systems, fire hydrants, water-based sprinkler systems, standpipe systems, special systems, extinguishing agents, and detection and alarm systems. Entire courses and textbooks can be dedicated to each of these topics. The intent of this chapter is to present only the basic design features and operating principles of these systems, with the understanding that students will go into greater depth on these subjects in more advanced study.

The first part of this chapter is dedicated to water supplies and water distribution systems. Because the majority of fire protection systems encountered by firefighters use water as the extinguishing agent, it is important that the firefighter understand how this extinguishing agent (water) is supplied. Inadequately designed or antiquated distribution systems can wreak havoc at emergency fire scenes and endanger firefighters' lives. The remainder of the chapter is dedicated to the various sprinkler systems, standpipe systems, and special systems found in private sector construction.

Water Supplies and Distribution Systems

Why are water supplies being discussed in a chapter dedicated to fire protection systems? The answer is simple. When firefighters connect their hoses or apparatus to a fire hydrant, the system that supplies that hydrant becomes a fire protection system. The volume of water and the amount of pressure the hydrant provides is directly related to the size of the mains, the design of the distribution grid, the type of system, and the volume and availability of the water supply. It is the responsibility of all firefighters to acquire a working knowledge of the types of systems that serve their community.

SOURCES OF WATER SUPPLY

There are two basic sources of water supply. **Groundwater** is found in flowing springs or in underground geologic formations known as aquifers. **Surface water** comes from rivers, lakes, streams, reservoirs, or various kinds of storage tanks.

Groundwater

The amount of water that can be drawn from an aquifer depends on the stability of the aquifer formation and the amount of rainfall in any particular season. In most cases, aquifers in limestone or other rock strata can withstand fluctuations of the water table during times of high demand and low recharge without extensive damage to the aquifer formation. In some areas of the country, the aquifers are in strata of sand and clay. If water is drawn from these aquifers faster than the water is replenished, the formation has a tendency to collapse. In some of the coastal areas of the country, collapse of these formations has caused a phenomenon known as **subsidence**, in which the ground sinks as the aquifer collapses. Municipalities in some affected areas have had to convert from relying on

groundwater ▪ Water located beneath the ground surface.

surface water ▪ Water collected on the ground from a stream, lake, ocean, or wetland.

subsidence ▪ A phenomenon in which the surface of the ground sinks as water is drawn from the underground aquifer faster that it can be replenished.

FIGURE 9-1 A flowing spring is classified as groundwater. Water flows from the aquifer to the surface. Aquifers are replenished by rainfall in the recharge zone. *Courtesy of J.D. Richardson.*

groundwater to using surface water. Houston relied heavily on groundwater from large wells. In the 1970s, Houston experienced a very large increase in population. The aquifers under Houston and the surrounding areas could not support the increased flow demand. As the water was drawn from the ground, the aquifers were unable to replenish fast enough to meet demand and began to collapse. Low-lying areas around Houston began to sink. Areas near the Gulf Coast dropped below sea level and were flooded by Gulf waters. The city and county governments realized that Houston could no longer rely on groundwater as a primary supply source. Large lakes were constructed around the Houston area, and the use of groundwater ceased (see Figure 9-1).

Surface Water

Like groundwater supply, surface supply is dependent on rainfall or runoff during spring thaws in some areas. Surface supplies are more readily recharged than groundwater supplies, especially if they have large recharge zones, also known as **watersheds**. It is possible to funnel the water that drains into a watershed into reservoirs constructed to store the water to supplement supply needs during periods of drought. Prolonged droughts can make this source of water supply unreliable (see Figure 9-2).

WATER SYSTEM COMPONENTS

In addition to a water supply source, all water systems require three additional components: a way to transport the water from the source to its final destination, a water treatment facility, and a distribution grid.

Water Transport Systems

Water may be supplied by the gravity system, the direct pumping system, or by some combination of the two.

The **gravity system** is the most efficient. In this system, water is moved through pipes from an elevated water source to the treatment facility and distribution system. There is no need for pumping apparatus. The height of the surface of the water source determines the amount of pressure in the system. This is called **head pressure**. Head pressure increases by 0.434 pound per square inch (psi) with each foot of elevation. Therefore, the higher the elevation of the water source, the higher the pressure in the system. For example, if the surface of a water reservoir was 1,000 feet above sea level and the area being served was at sea level, the pressure in the system at sea level would theoretically be 434 psi (see Figure 9-3).

watersheds ■ Geographic areas around a body of water that have a topography that allows all rainfall in that area to flow into and replenish that body of water.

gravity system ■ A water supply system that uses only gravity to move water.

head pressure ■ The height of the surface of the water supply above the discharge orifice is equal to 0.434 pound per square inch (psi) per foot of elevation.

FIGURE 9-2 Surface water is replenished by rainfall in the watershed. *Courtesy of James D. Richardson.*

A **direct pumping system** relies completely on mechanical apparatus to move the water. Pumps are used to draw water from groundwater supplies or from surface supplies that cannot supply elevation head pressure. The pumps are usually located at or near the water supply, direct the water to a treatment facility, and then on into the distribution system (see Figure 9-4).

A **combination system** uses features of both the gravity and pumping systems. Pumps move the water from the source to the distribution system. Some of this water is directed into elevated storage tanks located along the distribution path. These storage tanks not only act as a supplementary water supply but also are used to temporarily maintain pressure should the system pumps fail (see Figure 9-5).

FIGURE 9-3 The gravity system is the most reliable distribution system. *Courtesy of James D. Richardson.*

Water Treatment Systems

Regardless of the type of system used, all public water systems must have a means to filter and purify the water so that it is safe for human consumption. The complexity of the treatment system depends on the size of the distribution area and the amount of purification needed. Groundwater is often naturally filtered, to some degree, as it moves through the aquifer and may not require as much purification as surface water. Some communities use chlorine injection to treat against coliform bacteria contamination. Others inject both chlorine and fluoride to provide protection against dental decay. Firefighters responding to an emergency in a water treatment facility should be aware that these chemicals are toxic and pose a severe health hazard to anyone not wearing the proper personal protective equipment. Treatment facilities can also hold millions of gallons of water in large vessels. These vessels may collapse if subjected to seismic or extreme environmental forces (see Figure 9-6).

FIGURE 9-5 The combination system utilizes mechanical pumps and gravity to move water through the system. *Courtesy of James D. Richardson.*

FIGURE 9-6 Water must be treated to ensure that it is safe for human consumption. *Courtesy of James D. Richardson.*

Distribution Grids

Water is distributed to the individual customer through a configuration of pipes, referred to as a **distribution grid**. The size of the pipe used for each section of the grid depends on water demand and the size of the area covered by the grid. The grid consists of three sections:

- **Primary feeder mains** are very large pipes that are capable of carrying large volumes of water into a given area. Primary feeders form the outside borders of the grid.
- **Secondary feeder mains** are slightly smaller in diameter than the primary feeder pipes. These secondary mains interconnect portions of the primary feeders so that the water supply will flow from two directions. These pipes supply the distributors.
- **Distributor pipes** supply water to street fire hydrants and to the individual customers up and down the block.

A common practice is to place control valves in the grid mains so that no more than 500 feet of main will be shut down at any one time. Because the valves are within the grid structure, they may be buried many feet underground, depending on the depth of the frost line. Concrete or metal shafts, leading from the valve to the surface of the ground, are constructed to allow access to the valves. The shafts have metal or polymer lids identifying the location of the valve. A special long shaft "key" tool is needed to operate the valve. Such a valve is classified as a **nonindicating valve**, which means that the valve's operating position (open or closed) cannot be determined by visual inspection. The valve has to be physically operated to determine if it is open or closed (see Figure 9-7).

PIPE SIZE REQUIREMENTS

The National Fire Protection Association (NFPA) recommends that pipes used to supply firefighting needs should be no less than 6 inches in diameter. Six-inch pipe, however, should be used only in a looped grid in which no leg is longer than 600 feet. A *looped grid* is a grid that is constructed in the configuration of a square or rectangle, which allows the water from the water source inlet to flow in two directions inside the loop. In congested areas, distributors should be no less than 8 inches in diameter—or no less than 12 inches in diameter if they form part of a long grid line. It is recommended that for fire protection purposes, pipe diameters should be increased one size over the required minimum. Increasing the pipe diameter by one size can dramatically increase water flow.

---

distribution grid ■ The system in which power travels from a power plant to individual customers.

primary feeder mains ■ Large-diameter pipes that form the outside perimeter of the water distribution grid and receive water from the water source.

secondary feeder mains ■ Piping that interconnects with the primary feeder main.

distributor pipes ■ Piping located inside a building.

nonindicating valve ■ A valve that has to be physically operated to determine if it is open or closed.

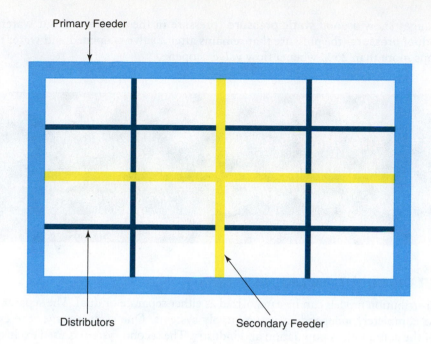

Primary Feeder

Distributors

Secondary Feeder

FIGURE 9-7 The distribution grid must be designed to provide adequate water supplies and flow to the community. *Courtesy of James D. Richardson.*

Flow ratio can be determined by using this simple formula:

$$\frac{\text{LD}^2}{\text{SD}^2} = \text{Flow Rate}$$

LD = Large-diameter pipe
SD = Small-diameter pipe

To compute the increased flow ratio, divide the square of the diameter of the SD into the square of the diameter of the LD.

EXAMPLE:

If the distributor size is increased from 8 inches to 10 inches, the flow increase can be determined as follows:

$$\frac{10^2 = 100}{8^2 = 64} = 1.56$$

By increasing the pipe by one size, the amount of flow is increased 1.56 times.

The condition of the interior lining of a pipe also determines the amount of flow that it can provide. Water pipes can be constructed of cast iron, steel, reinforced concrete, ductile iron, or polyvinyl chloride. Steel and cast iron pipes are subject to damage by various phenomena, such as tuberculation, encrustation, and sedimentation. **Tuberculation** is a buildup of rust on the interior lining of metal pipe. **Encrustation** is caused by chemical deposits or biological growth inside the pipe walls. **Sedimentation** is the accumulation of mud, sand, and other debris. All of these reduce the interior diameter of pipes, which reduces water flow. Concrete and polyvinyl chloride pipes are not subject to tuberculation because they do not rust, but they are affected by encrustation and sedimentation. Water distribution systems must be routinely flushed out and flow-tested to ensure that flow rates have not been reduced due to these problems. One indication of a restricted flow is

tuberculation ■ The buildup of rust on the inside surface of a metal water pipe, which reduces the inside diameter of the pipe.

encrustation ■ The buildup of chemical deposits and biological growth on the inside walls of a water pipe, which reduces the inside diameter of the pipe.

sedimentation ■ The accumulation of mud, sand, and other debris inside a water pipe, which reduces the inside diameter of the pipe.

static pressure ■ Energy that is available to force water through pipes and fittings.

residual pressure ■ The pressure that remains in the system when water is being discharged.

when gauges show a good **static pressure** (pressure in the system without water flow), but **residual pressure** (the pressure that remains after a valve is opened and water is flowing) drops more than 25% when a flow valve is opened.

STOP AND THINK 9-1

While conducting flow tests in your district, you find a hydrant that has a static pressure of 85 psi, but when you open the hydrant valve, the residual pressure drops to only 40 psi. What information should you seek? How will this situation affect you and your fire crew?

WATER DISTRIBUTION MODELS

Water distribution models can be categorized as either separate or dual. The separate model has two completely independent water supply systems. One system serves the everyday needs of the community's population and industry. The second system is used exclusively by the fire service and is independent of the community water distribution system, having its own water supply, pumps, and distribution mains.[1] San Francisco, California, has the only high-pressure separate system in the United States. Known as the San Francisco Auxiliary Water Supply System, it is capable of supplying enough water to fill a 100,000 square feet area with 25 feet of water in one day. The system has three reservoirs with a total capacity of 11,750,000 gallons. The reservoirs are normally supplied from the domestic water system. However, if the fresh water system should fail or be depleted, two pumps will supply salt water from San Francisco Bay. Each pump has a 10,000 GPM pump capacity and can provide 300 psi pressure. If this system should fail, two fire boats can connect to the system and supply a total flow capacity of 33,600 GPM.[2] The dual model has one system that supplies water for all domestic, industrial, and fire protection needs. Both options have advantages and disadvantages. These are outlined in Tables 9-1 and 9-2.

TABLE 9-1	Advantages and Disadvantages of a Separate System
ADVANTAGES	**DISADVANTAGES**
The system is under complete control of the local fire department.	The community must pay for the construction of two distribution systems.
Cross-contamination of the local drinking water is avoided.	Firefighters must travel to a pump station to start and operate the pumps during an emergency. This reduces the available staffing for fire ground operations.
Nonpotable water can be used, and the water supply does not require treatment.	The pumps and other apparatus are in a nonoperational mode and may fail to start during an emergency if they have not been subjected to a regular schedule of maintenance and operation.
The water supply for firefighting is not impacted by population growth or by periods of heavy demand.	
The system can be custom-designed to meet all firefighting requirements.	

TABLE 9-2	Advantages and Disadvantages of a Dual System	
ADVANTAGES	**DISADVANTAGES**	
The pumps and other apparatus at the source of supply are in operation 24 hours a day, seven days a week.	There is a possibility of cross-contamination of the community's drinking water.	
There are redundant systems that can be put into operation in case of equipment failure. Personnel are available for maintenance and repair at all times.	A common water source is used for domestic, industrial, and firefighting operations. Therefore, the water supply for firefighting is affected by population growth and periods of heavy demand.	

Standards for installation and maintenance of various components of these systems are provided by organizations such as the National Fire Protection Association and the American Water Works Association. These standards are supported through testing by Underwriters Laboratories and Underwriters' Laboratories of Canada.[3]

Fire Hydrants

The humble fire hydrant is the most visible component of the public fire protection system. As previously mentioned, the ability of a hydrant to support firefighting operations depends on the ability of the distribution system to support the hydrant. The first water distribution systems in the United States consisted of buried water mains constructed of hollowed wooden logs. To acquire water in an emergency, firefighters would dig a hole in the street to expose the main and then drill into the main. The hole would fill with water, becoming a cistern from which pumpers could be supplied. In 1803, a Philadelphia engineer named Frederick Graff designed a standpipe that could be permanently inserted into the wooden main. A control valve on the standpipe allowed the water to be turned off so that the pipe could be drained to protect the system during freezing temperatures. The visible part of this early hydrant was a wooden or metal box that protruded about 2 feet above the ground encasing the standpipe. It would be over 50 years before this design would be improved. (Ironically it is not possible to verify that Mr. Graff held the first U.S. patent for a fire hydrant, because the Patent Office burned to the ground in 1836, destroying all U.S. patent records!)

There are three classifications of fire hydrants: the base-valve dry-barrel hydrant (often referred to as a dry-barrel hydrant), the wet-barrel hydrant (also known as the California hydrant), and the dry hydrant. Base-valve dry-barrel and wet-barrel hydrants are linked to a distribution grid that provides a pressurized water supply. They are commonly found in towns and urban areas. The dry hydrant does not rely on a municipal distribution system and is adapted for use in rural areas.

THE BASE-VALVE DRY-BARREL HYDRANT

Samuel R. C. Mathews was a Detroit inventor who worked for a rotary pump fire engine manufacturing company. In 1858, he patented the design for a base-valve dry-barrel fire hydrant. This type of hydrant is constructed of cast iron and designed to be free standing (no wooden or metal box is needed for protection). The barrel of the hydrant is dry because the base valve is installed below the frost line. Water does not enter the barrel until the base valve is opened. To open this valve, it is necessary to turn a nut that is located on top of the hydrant bonnet for easy access. When the valve is closed, drain ports open, which allow water to drain from the hydrant barrel. This feature is of great benefit in areas where freezing temperatures occur (see Figure 9-8). When the barrel is filled, water is made available to all hydrant outlets. The main disadvantage of this design is that

FIGURE 9-8 The base-valve dry-barrel hydrant is designed to ensure proper operation during freezing weather.
Courtesy of James D. Richardson.

when more lines need to be connected to the hydrant, the hydrant must be closed so that the outlet caps can be removed.

Base-valve dry-barrel hydrants are the most common hydrants used in the fire service today. They are manufactured of three basic materials. The barrel, caps, and bonnet are usually cast iron; all internal moving parts are bronze; and the compression valve face is neoprene or rubber.

Standards for base-valve dry-barrel hydrants are found in the American Water Works Association Standard C502-94, *Standard for Dry Barrel Fire Hydrants*, and Underwriters Laboratory Standard UL 246, *Hydrants for Fire Protection Service*.

THE WET-BARREL (CALIFORNIA) HYDRANT

A young immigrant from France named Morris Greenburg invented the wet-barrel hydrant in San Francisco in 1851. It is constructed of the same materials as the dry-barrel hydrant, but pressurized water is supplied to the barrel of the hydrant at all times—hence the name wet-barrel. Wet-barrel hydrants do not require a base valve. Individual valves control each outlet on the hydrant. This simple design allows additional hose lines to be attached without disrupting the water flow. Hydrants of this type are used in areas that are not subject to freezing temperatures[4] (see Figures 9-9A and B).

Standards for wet-barrel hydrants are found in the American Water Works Association Standard C503-88, *Standard for Wet Barrel Fire Hydrants*, and Underwriters Laboratory Standard UL 246, *Hydrants for Fire Protection Service*.

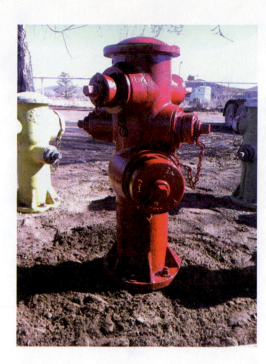

FIGURE 9-9A The wet barrel hydrant has a control valve on each outlet.
Courtesy of Willis Lamm

Dry-Barrel Hydrant

Stem Nut

Operating Stem

Drain Hole

Valve

Wet-Barrel Hydrant

Operating Stem

Hose Outlet and Valve Seat

Automatic Check

FIGURE 9-9B Cut-away view of a dry barrel and wet barrel hydrant

DRY HYDRANTS

Dry hydrants are generally found in rural areas where there are no water distribution systems or where the existing systems cannot adequately support firefighting needs. These hydrants generally do not directly supply water to the firefighting apparatus, but are used to fill tenders that transport the water to the fire ground for use by the firefighting apparatus. They can be designed as either nonpressurized or pressurized. The **nonpressurized dry hydrant system** is installed at a static water source, such as a pond, lake, river, or bay. Construction consists of a pipe that extends into the water source on one end and has a threaded connection for attaching fire apparatus at the other end. The apparatus connects to the hydrant and drafts water from the water supply. The maximum vertical lift from the water surface to the pump should not exceed 15 feet, and an all-weather road should be provided to allow 24-hour access to the hydrant (see Figure 9-10).

The Natural Resource Conservation Service (NRCS), an office of the U.S. Department of Agriculture, provides recommendations to rural communities for dry hydrant operations. NRCS suggests that a dry hydrant water supply have a minimum capacity of 30,000 gallons.

A unique style of dry hydrant system is the **pressurized dry hydrant**. This system may be found in communities located higher than the water distribution grid. The system consists of a piping grid connecting a fire department connection and one or more fire hydrants. The fire department connection is located at the lower level, close to a water distribution hydrant. During firefighting operations, a pumping apparatus connects to the supply hydrant and transfers water to the dry hydrant system through the fire department connection, filling the system and providing a minimum flow rate of 1,000 GPM at 150 psi. The fire department connection (FDC) may be a standard FDC manifold intake, with two 2 to $2\frac{1}{2}$ inch intake ports, a base- valve dry-barrel, or wet hydrant (pumper hydrant) to accommodate large-diameter hose. If a pumper hydrant is used, it must be clearly marked as an FDC.[5] An example of this type of system can be found at the Moraga-Orinda Fire District in Moraga-Orinda, California (see Figures 9-11A and B).

HYDRANT MAINTENANCE AND FLOW TESTING

The NFPA suggests that private fire hydrants should be visually inspected once each month, lubricated twice yearly, and opened and closed once yearly. Graphite is the

nonpressurized dry hydrant system ■ A hydrant system that is used to draft water from a static water source.

pressurized dry hydrant system ■ A piping grid that connects several dry hydrants. A fire department connection is placed at the base of the grid near a water source. During fire-fighting operations, a pumping apparatus connects to the water supply and supplies water to the dry hydrant system.

FIGURE 9-11A The pressurized dry hydrant FDC can be a standard FDC manifold or a fire hydrant. *Courtesy of Willis Lamm and the Moraga-Orinda Fire Department.*

FIGURE 9-11B Pressurized dry hydrant pumping operation. *Courtesy of the Moraga-Orinda Fire Department.*

preferred lubricant. Petroleum-based products can cause dirt and other foreign matter to collect on the outlet threads, which could impede proper hose connection. The NFPA has also published standards for flow-testing and marking fire hydrants.

There are three types of pressure that must be considered for hydrant flow testing: static pressure, flow pressure, and residual pressure. Static pressure is the pressure in the system when the water is at rest (not flowing). Flow pressure is pressure at the discharge outlet when water is flowing. Residual pressure is the pressure that is left in the system when water is flowing. The amount of drop in residual pressure when a hydrant is flowing will determine the amount of water that can be supplied to a pumping apparatus.

There are several formulas for determining hydrant flow. The first method determines the natural flow of a single hydrant with no pumping apparatus connected to it. The gallon per minute flow can be determined by removing a cap on one of the $2\frac{1}{2}$ inch hydrant outlets, fully opening the hydrant valve, and measuring the discharge pressure by placing a pitot gauge into the water stream. When the flow pressure is determined, the following formula is used to establish the gallon per minute flow of the hydrant:

$$GPM = 29.7 \times d^2 \times \sqrt{P} \times C$$

C is the coefficient of the outlet based on its design. The coefficient can be 0.7, 0.8, or 0.9. Information on the coefficient can be obtained for the community's water department, or the utility responsible for hydrant installation.

EXAMPLE:

A hydrant has a flow pressure of 85 psi and an outlet coefficient of 0.8.

$$GPM = 29.7 \times 2.5^2 \times \sqrt{85} \times 0.8$$
$$GPM = 29.7 \times 6.25 \times 9.2 \times 0.8$$
$$GPM = 1,366.2$$

This would be the expected flow if a hand hose line were connected directly to hydrant. When a pumping apparatus is to be connected to the hydrant, a quantity residual formula must be used to determine an accurate flow calculation. Most fire departments recommend that pumping apparatus operate with a desired residual pressure of 20 psi. In the first formula, only flow pressure was used to calculate GPM. With the quantity residual formula, static and residual pressures, and GPM are used in the calculation. Values for the formula are found in the H factor chart found in the NFPA handbook or NFPA 291.[7]

$$Q_R = Q_F \times \frac{H_R}{H_F}$$

Q_R = Flow at desired residual pressure
Q_F = Flow of hydrant during test
H_R = Pressure drop to desired residual pressure
H_F = Pressure drop during test

EXAMPLE:

Using the hydrant from the first formula:

A hydrant is flowing 1,366 GPM and the static pressure is 85 psi. The residual pressure is 60. The desired residual pressure is 20 psi.

$$Q_R = 1,366 \times \frac{\text{Static Pressure} - \text{Desired Residual Pressure} = 85 - 20 = 65 = 9.53 \ (H \text{ chart value})}{\text{Static Pressure} - \text{Residual Pressure} = 85 - 60 = 25 = 5.69 \ (H \text{ chart value})}$$

$$Q_R = 1,366 \times \frac{9.53}{5.69} = 2287 \text{ GPM}$$

By bringing the residual pressure down to 20 psi, this hydrant's flow was increased by 921 gallons per minute.

The flow capacity of an entire grid section can be determined by flowing one hydrant and recording the static pressure and residual pressure drop on a second hydrant that is on the same main. This will allow calculations to be made to determine the amount of flow that can be expected from the main if multiple hydrants are needed simultaneously. Multiple hydrant testing is required by the Insurance Service Organization (ISO) that

TABLE 9-3	NFPA 291, Fire Hydrant Classification and Color-Code Recommendation[6]	
CLASS	**FLOW RATE**	**COLOR OF BONNET AND CAPS**
AA	1,500 GPM or greater	Light Blue
A	1,000 to 1,499 GPM	Green
B	500 to 999 GPM	Orange
C	Less than 500 GPM	Red

grades a community's firefighting capabilities to determine the fire insurance rates its citizens will pay. NFPA 291 recommends that the hydrants be classified and color-coded so that firefighters can visually determine if the hydrants are capable of supplying their needs. NFPA standards are not laws; they are only recommendations. Many jurisdictions use their own color-code system. It is the responsibility of all firefighters to familiarize themselves with the color-code system used in their response areas (see Table 9-3).

For NFPA standards for maintenance, inspection, and flow testing of private fire hydrants, see NFPA 25, *Standard for the Inspection, Testing, and Maintenance of Water-Based Fire Protection Systems,* and NFPA 291, *Recommended Practice for Fire Flow Testing and Marking of Hydrants.*

The NFPA recommends that the distance between fire hydrants should not exceed 800 feet. On average, the standard engine company apparatus carries 1,000 feet of water supply hose. This would allow the engine to maintain a 200 feet reserve when laying a supply line into a fire. In closely built or **target areas**, fire hydrants should be spaced at an interval of 500 feet or less.[8]

target areas ■ Areas that are of high value or where there is a potential for a large loss of life.

Standpipe Systems

The NFPA defines standpipe systems as "fixed piping systems and associated equipment that provides a means to transport water from a reliable water supply to designated areas of buildings where hoses can be deployed for fire fighting."[9] These systems are found in multistory buildings and other areas, such as large warehouses and wharves, which cannot be reached by hose lines from hydrants or fire apparatus. They are intended for manual firefighting and allow the firefighter to bring the hydrant to the fire.

The insurance industry has taken the lead in lobbying for local or state legislation and for the passage of building codes that mandate standpipe systems. These can vary from the very simple to the very complex. The simplest design consists of a water supply (feeder main), a check valve (backflow preventer), a **riser** (the piping), and a hose valve station. The NFPA recognizes three classes of systems: Classes I, II, and III.

riser ■ The largest pipe in a sprinkler system. It is connected directly to the main water supply line and contains the sprinkler alarm valve. It supplies water to the crossmain.

NFPA 14, *Standard for the Installation of Standpipe and Hose Systems.*

FIGURE 9-12 Only trained and experience personnel should use a Class I standpipe system. *Courtesy of Pickett's Video & Training.*

CLASS I SYSTEM

The Class I system is designed to be used only by trained firefighting professionals for full-scale firefighting. It is required to have one or more $2\frac{1}{2}$-inch hose connections. The system is also required to have an outside fire department connection that allows water to be supplied by fire department apparatus. If the system is in a high-rise building with several fire zones, a fire department connection must be provided for each zone.

In buildings higher than three stories, a Class I standpipe system may be required even if the building is equipped with sprinklers (see Figure 9-12).

CLASS II SYSTEM

The Class II system is designed for use by building occupants. A $1\frac{1}{2}$-inch outlet and pre-connected $1\frac{1}{2}$-inch hose and nozzle are provided. They are usually located in wall-mounted cabinets in the building hallways.

Class II systems were once required in large unsprinklered buildings, but today their use is discouraged for three principal reasons. First, the hose lines in these systems can carry as much as 100 GPM. This can create safety issues when such heavy lines are manned by individuals not trained in the strategy and tactics needed to use them effectively. Second, a delay in reporting the fire may result while occupants are trying to extinguish the fire. Third, energy spent on firefighting might be put to better use in evacuating the building. Individuals not trained in the use of hose lines can do little against a fire once it has passed the incipient stage. Once it has entered the free-burning state, heat and smoke will rapidly increase in intensity, and unprotected occupants should abandon hose lines and evacuate the building immediately (see Figure 9-13).

STOP AND THINK 9-2

You and your crew advance to the fifth floor of an old hotel and find heavy smoke and one room heavily involved in fire. The building has no sprinkler system and only a Class II standpipe system. You see a charged hose line from the standpipe extending down the hallway to the involved room. What actions should you take to initiate extinguishment operations?

CLASS III SYSTEM

The Class III system is a combination of the Class I and II systems. A $2\frac{1}{2}$-inch outlet is provided for full-scale firefighting and a $1\frac{1}{2}$-inch outlet with preconnected hose and nozzle for occupant use. This system is declining in use and is discouraged for the same reasons as those for Class II systems.

TYPES OF STANDPIPE SYSTEMS

In 1993, NFPA 14 classified standpipe systems into five categories or types: automatic wet systems, automatic dry systems, semiautomatic dry systems, manual dry systems, and manual wet systems.

Automatic Wet System

The automatic wet system is the most common standpipe system. The piping is filled with pressurized water from the water supply. When the hose valve is opened at the standpipe station, water is immediately available for firefighting operations.

Automatic Dry System

The automatic dry system is used in areas that are subject to freezing temperatures. The system is filled with pressurized air that is controlled by a dry pipe valve. When the hose valve is opened at the standpipe station, air is expelled, and the dry pipe valve opens, allowing the water supply to enter the system.

FIGURE 9-13 Class II systems delay evacuation and expose untrained and unprotected individuals to potentially deadly hazards. *Courtesy of Pickett's Video & Training.*

Semiautomatic Dry System

The piping in the semiautomatic dry system is also filled with pressurized air. When a remote device is activated at the hose station, the air is exhausted from the system. A deluge valve opens and allows the water supply to fill the piping.

Manual Dry System

The manual dry system is not connected to a permanent water supply. The piping is filled with air at normal atmospheric pressure. Water supplies must be manually provided by the fire department through a fire department connection.

Manual Wet System

The manual wet system is not connected to a permanent water supply. The piping is filled with pressurized water from the domestic water supply. This supply is inadequate for firefighting. In the event of a fire, the fire department must supply water manually to the system through the fire department connection.

STANDPIPE FLOW RATE REQUIREMENTS

NFPA 14 requires that Class I and Class III systems be capable of delivering 500 GPM from the first standpipe and 250 GPM from each additional standpipe, up to a maximum total of 1,250 GPM. The water supply must be adequate to sustain this flow rate for a minimum of 30 minutes. These systems must also maintain a residual pressure of 100 psi while flowing 250 GPM at the two topmost outlets, plus 250 GPM from each additional standpipe, up to the maximum flow rate of 1,250 GPM. Class I and Class III systems must provide a minimum **flow pressure** (the pressure at which water is discharged from the outlet valve) of 100 psi to each outlet of the system. Class II systems must provide 100 GPM for 30 minutes and maintain a residual pressure of 65 psi at the topmost outlet while flowing 100 GPM.

flow pressure ■ The pressure created by the forward velocity of a fluid at the discharge orifice.

In Class I and Class III standpipe systems, the risers (pipes that carry water to ceilings and upper stories), must meet the following requirements:

- If the building's height is 100 feet or less, the standpipe riser diameter must be a minimum of 4 inches.
- If the building's height exceeds 100 feet, the standpipe riser diameter must be a minimum of 6 inches; however, the top 100 feet of the standpipe can be 4 inches in diameter.

Standpipe risers for Class II systems must meet the following requirements:

- If the building height is less than 50 feet, the riser must be 2 inches in diameter.
- If the building height exceeds 50 feet, the minimum riser diameter is $2\frac{1}{2}$ inches.
- If the building height is over 275 feet, the system must be divided into zones.

PRESSURE-REGULATING DEVICES

As the height of the standpipe increases, the pressures at the lower outlets increase. This is caused by head pressure, the pressure of the column of water inside the riser. The fourth principle of pressure states that pressure in an open vessel is proportional to the depth of the vessel. To reduce the dangers of high pressure on lower outlets, standpipe systems must be divided into zones in buildings over 275 feet in height. At this height, the head pressure at the lowest outlet will be approximately 120 psi ($275 \times 0.434 = 119.35$ psi). It is possible to install pressure-reducing devices that will lower outlet pressures to 100 psi or less. If such devices are installed in a standpipe system, the zone height requirement can be extended from 275 feet to 400 feet. A major disadvantage of working with a pressure-reducing device is that most fire departments use fog nozzles on $2\frac{1}{2}$-inch hose lines. These nozzles are designed to be effective and reach maximum flow rates at 100 psi nozzle pressure. If the outlet is regulated to a pressure of 100 psi, there is no compensation for **friction loss** in the fire department hose line. Friction loss is the pressure that is lost in a fire hose or standpipe as a result of water turbulence and resistance caused by the water moving against the sides of the conduit. Friction loss reduces the flow rate and maximum effectiveness of the hose stream. Some fire departments extend maximum outlet pressures to 120 psi to compensate for friction loss.

The entire standpipe system should be tested annually to ensure proper installation and adjustment.[10]

friction loss ■ Pressure loss caused by turbulence as water flows through a conduit.

Automatic Sprinkler Systems

John Carey invented the first automatic sprinkler system in England in 1806. The ingenious system consisted of perforated pipes suspended from the ceiling. Water supply was controlled by a set of valves held closed by counterweights. Strings suspended the counterweights. In the event of a fire, the strings would burn through and the counterweights would drop, opening the valves and sending water through the perforated pipe. Water would be discharged over the entire area, not just on the fire. As this system evolved, solid pipes with open-style sprinkler heads replaced the perforated pipes. In 1812, British inventor Sir William Congreve designed a perforated pipe system that utilized a manually controlled valve. Mounted on the exterior of the building, the valve could be operated by anyone in the event of a fire.

Henry Parmelee, a Connecticut piano maker, is given credit for inventing the first practical automatic sprinkler in 1874. To prevent extensive water damage to the expensive wood that he used in his craft, he soldered brass caps over the open sprinkler heads. The system could be filled with water because the caps would hold the water back. In the event of fire, heat would melt the solder holding the caps, and water pressure would blow them away from the sprinkler heads, allowing water to be directed onto the fire. Areas not affected by the fire would remain dry.

(A)

(B)

FIGURE 9-14 The basic operation of these devices has changed little since their inception. (A) Sprinkler alarm valve; (B) sprinkler head. *Courtesy of James D. Richardson.*

Frederick Grinnell improved the Parmelee Sprinkler by designing a sprinkler head with a more sensitive, heat-actuated release mechanism that could be calibrated to open at a predetermined temperature and a valve seat that eliminated leakage due to water pressure fluctuations. The improved Parmelee Sprinkler later incorporated a variable-pressure alarm valve as well, a device invented by an Englishman named John Taylor. These innovations are still used in modern sprinkler systems (see Figures 9-14A and B).

Today, the automatic sprinkler system is considered the most effective method of controlling fires in their incipient stage. Automatic sprinklers were primarily installed for property conservation purposes, but their life-safety benefits were soon realized. With the addition of John Taylor's alarm valve, the system not only attacked the fire in its early stages, but it also sounded an alarm to warn occupants of fire danger. The NFPA stopped collecting data on sprinkler performance and effectiveness in 1970. However, each year, the Fire Analysis and Research Division of the NFPA publishes a report entitled, "U.S. Experience with Sprinklers and Other Fire Extinguishing Equipment." The division gathered data on sprinkler operation in structure fires from 1999 to 2002. The report listed the percentage of fires extinguished or controlled compared to number of sprinkler heads that activated. The study included wet pipe, dry pipe, and "other" type sprinkler systems. With wet pipe systems, 65% of the fires were controlled or extinguished with the activation of only one sprinkler head. The percentage for dry pipe and other type systems were 41% and 55%, respectively. However, 97% of wet pipe systems, 86% of dry pipe systems, and 100% of other type systems controlled or extinguished the fire with the activation of 10 or fewer sprinkler heads. Sprinkler systems are so effective in controlling and extinguishing fires that the NFPA Life-Safety Code 101 allows increased travel distances to exits and reduced fire ratings on wall assemblies and interior finishes in building with complete sprinkler coverage. The exemplary performance of automatic sprinklers in supporting life safety was verified in 1995 when the NFPA's Fire Analysis and Research Division released a report entitled "U.S. Experience with Sprinklers: Who Has Them? How Well Do They Work?" It states, "NFPA has no record of a fire killing more than two people in a completely sprinklered building where the system was working properly."[11] Sprinkler reliability is dependent on proper design, installation, and maintenance. The NFPA published data listing reasons for unsatisfactory sprinkler performance. These were[12]

1. Water shut off—35.4%
2. Inadequate system of level of hazard—18%
3. Inadequate water supplies—9.9%

4. Inadequate maintenance—8.4%
5. Obstruction to water distribution—8.2%
6. Partial sprinkler protection—8.1%
7. Faulty building construction—6%
8. System components defective or damaged—5.6%
9. Exposure fire—1.7%
10. System frozen—1.4%
11. Other—1.9%

The NFPA published the first codes for automatic sprinkler installation in 1896. Since then, it has established three standards and two recommendations dealing with these systems. It is important that all firefighters familiarize themselves with these standards and recommendations, especially with NFPA 13.[13]

NFPA 13, *Standard for the Installation of Sprinkler System;* NFPA 13D, *Standard for the Installation of Sprinkler Systems in One- and Two-Family Dwellings and Mobile Homes;* NFPA 13R, *Standard for the Installation of Sprinkler Systems in Residential Occupancies up to Four Stories in Height;* NFPA 13A, *Recommended Practice for the Inspection, Testing, and Maintenance of Sprinkler Systems;* and NFPA 13E, *Recommendations for Fire Department Operations in Properties Protected by Sprinkler and Stand-pipe Systems.*

AUTOMATIC SPRINKLER SYSTEM COMPONENTS

An automatic sprinkler system consists of the following components: a water supply main, a backflow preventer (check valve), a water flow control valve (shut off valve), an alarm valve, a fire department connection, a local audible alarm, a riser, a crossmain, branch lines, and sprinkler heads.

Water Supply Main

The water supply main should be at least as large as the riser, a minimum of 6 inches in diameter. It must be able to deliver the required fire flow to the highest sprinkler with a residual pressure of 15 psi. If the system is designed by hydraulic calculation, the residual pressure to the highest sprinkler can be lowered to 7 psi. When the water supply is delivered from an elevated storage tank, the bottom of the tank must be a minimum of 35 feet above the highest sprinkler. This will maintain a minimum residual pressure of 15 psi.

Backflow Preventer

If the sprinkler system is being supplied from a public water distribution system or a private system that is also for public use, a backflow preventer must be installed between the public supply main and the water flow control valve. A backflow preventer, or check valve, is a one-way valve that allows water to flow toward the sprinkler system but closes when pressure in the system is reversed, thus preventing water from flowing water back into the public water supply. Fire protection systems may sit idle for many years. If they are not flow tested regularly, the water in the pipes may become stagnant. If this water were allowed to contaminate the public water supply, the health quality of the system would be compromised. When inspecting the installation of a sprinkler system, the backflow preventer should be examined to ensure that it is properly installed. An arrow imprinted on the side of the valve housing indicates the direction of flow (see Figure 9-15).

Water Flow Control Valves

The water flow control valves, or shut off valves, are located in the riser between the water supply main and the fire department connection. There are several types of control valves.

Outside Stem and Yoke (OS&Y) Valve.   This is the most common control valve used on sprinkler systems and is classified as a *gate valve*. The yoke controls a threaded stem that opens or closes the valve as the yoke is turned. The OS&Y is classified as an **indicating valve**, which means that the position of the valve can be determined by visual inspection. If the threaded stem protrudes from the yoke, the valve is open; if the stem is flush with the top of the yoke and the threads are not visible, the valve is closed (see Figure 9-16).

Post Indicator Valve.   When sprinkler control valves are installed underground, a post indicator is provided so that the valve can be operated. The post indicator is a hollow post that houses a valve stem, which is connected to the operating

indicating valve
■ A valve that can be visually inspected to determine whether it is open or closed.

FIGURE 9-16
If the stem of the OS&Y valve is extended out from the operating wheel, the valve is open. *Courtesy of James D. Richardson.*

yoke of the underground valve. That valve is opened and closed by turning a valve stem nut on top of the post. Indicators are attached to the stem. When the valve is closed, the words "Shut" or "Closed" will appear in a window opening on the side of the post. When the valve is opened, the word "Open" will appear in the window (see Figure 9-17).

Wall Post Indicator Valve. This valve is similar in operation to the post indicator valve, but it is located inside the building, next to an exterior wall. The valve stem extends through the exterior wall of the building and is connected to an operating wheel or lever mounted on an indicating post outside of the exterior wall. The post has an indicating window that shows the words "Shut" or "Open" to indicate the position of the valve gate (see Figure 9-18).

Post Indicator Valve Assembly. This valve can be wall-mounted or placed in-line between the water supply and the alarm valve inside the alarm valve room. The post indicator valve assembly is the only one of the four valve types that operates as a butterfly valve. A crank handle or wheel is mounted on the side of the valve housing. When turned, a worm gear rotates the butterfly 90° inside the valve housing waterway. The indicator is a pointer mounted on top of the valve housing that points to the words "Open" or "Closed."

FIGURE 9-18
The wall post indicator valve is an operating device for a gate valve inside the building.
Courtesy of James D. Richardson.

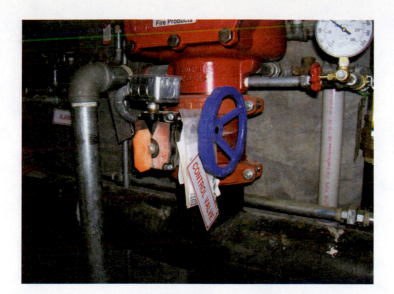

Note: All water flow control valves on sprinkler systems are held in the "Open" position by a chain or lock, and most are electronically monitored to signal a valve shutdown (see Figure 9-19).

Fire Department Connection

FDCs are required on most sprinkler and standpipe systems. Partial, or limited area, systems may not have an FDC. This can be determined by preincident planning. The FDC allows fire department pumping apparatus to attach hose lines and supply additional pressure and water into the system. The connection can be found on the exterior wall of the building or mounted as a manifold outside of the building. There are several variations of fire connections. The most common consists of two $2\frac{1}{2}$-inch inlets that "siamese" into a 4-inch pipe. For multiple or large systems, the connection may be set up as a manifold, having three or more $2\frac{1}{2}$-inch or larger inlets to accommodate a large-diameter hose. All inlets have female threads. If the system has only one riser, the fire connection pipe will connect to the sprinkler side of the system, above the alarm valve. If more than one riser is attached to the supply main, the pipe will attach to the supply side of the alarm valve (see Figure 9-20).

FIGURE 9-20
The fire department connection. *Courtesy of James D. Richardson.*

Alarm Valves

The alarm valve is a clapper valve that can be designed to serve several functions. It can be operated as a wet valve, a dry valve, or a deluge valve. Different configurations enable the valve to be operated hydraulically, pneumatically, or electrically. The purpose of the valve is to protect against false alarms and direct the flow of water to the sprinklers and alarm signaling system (see Figure 9-21).

Often referred to as a "water gong," a local audible alarm is attached to the alarm valve by piping. As water flows through the valve, a portion of the water is diverted through the pipe and travels to the gong. The force of the water passing through a chamber in the gong turns a wheel that causes a clapper to strike a gong. The gong will sound as long as water is flowing through the sprinkler system. After the water passes through the gong, it is discharged outside the building (see Figure 9-22).

Risers, Crossmains, and Branch Lines

The riser is the largest pipe in the sprinkler system. It is the portion of pipe that carries the water from the water supply to the ceiling and upper levels. The water supply main, alarm valve, crossmains, and fire department connection (on single-riser systems) are connected to the riser. **Crossmains** are slightly smaller pipes that extend from, and run perpendicular to, the riser. They carry water from the riser to other areas of a building. **Branch lines** are

crossmains ■ Pipes that supply the branch lines from the main riser. They are slightly smaller than the riser but are larger than the branch lines.

branch lines ■ The smallest pipes of a sprinkler system on which the individual sprinkler heads are mounted.

FIGURE 9-22
Bees have been known to nest in water gongs and make them inoperable. *Courtesy of James D. Richardson.*

(A)

FIGURE 9-23
(A) The riser supplies water to the upper levels and the crossmain. (B) The crossmain delivers water from the riser to different areas of the building being protected. (C) The branch lines are supplied by the crossmain and are the mains on which the sprinkler heads are mounted. *Courtesy of James D. Richardson.*

(B)

(C)

the smallest pipes in the system. They extend from the crossmain into specific areas or rooms. The sprinkler heads are attached to the branch lines (see Figures 9-23A, B, and C).

Sprinkler Heads

The sprinkler head serves a dual purpose. It is both a heat detector and a systems activation device. The heat detection device on the sprinkler can be designed as a fusible link, frangible bulb, chemical pellet, or fusible pellet. The fusible link, chemical pellet, and fusible pellet operate by melting after reaching a predetermined temperature. This releases the mechanism holding the nozzle cap and allows water pressure to blow the cap away from the sprinkler's discharge orifice. Temperature activation settings are determined by the amount of surface area, thickness, and type of material used for the release mechanism. Fusible links are generally constructed from an alloy of cadmium, bismuth, lead, and tin.

The frangible bulb sprinkler utilizes a liquid-filled glass bulb to hold the nozzle cap in place. Heat causes the liquid inside the bulb to expand until the glass bulb ruptures, releasing the nozzle cap and allowing water flow. The activation temperature is determined by the amount of liquid in the glass bulb (see Figure 9-24).[14]

Sprinkler heads are designed for installation in either pendent, upright, or wall configuration, and they cannot be installed in any configuration except that for which they were designed (see Figure 9-25).

FIGURE 9-25
Sprinkler heads are designed to be mounted in only one configuration. For example, pendent sprinkler heads cannot be mounted in an upright configuration. *Courtesy of Pickett's Video & Training.*

The sprinkler must be able to provide the required flow rate for the area being protected. The flow rate depends on the size of the discharge orifice and the amount of discharge pressure being provided. Sprinkler discharge orifice diameters vary from $\frac{3}{8}$ inch to $\frac{3}{4}$ inch. The most common orifice diameter is $\frac{1}{2}$ inch. The formula used to calculate sprinkler flow rate is

$$Q = K\sqrt{P}$$

Q = Quantity of flow
K = Discharge coefficient based on the orifice size and surface characteristics
P = Discharge pressure at the sprinkler

The K factor coefficient for a $\frac{1}{2}$-inch sprinkler ranges from 5.3 to 5.8, depending on the condition of the interior surface of the discharge orifice. The minimum discharge pressure for a nonhydraulically calculated sprinkler is 15 psi. If the discharge coefficient of a sprinkler is 5.5 and the discharge pressure is 15 psi, the flow rate can be calculated by multiplying 5.5 times the square root of the discharge pressure (15 psi).

$$5.5 \times \sqrt{15} = \text{GPM}$$

$$5.5 \times \sqrt{3.87} = 21.28 \text{ GPM}$$

Flow rate is adjusted by varying the pressure or changing the diameter of the sprinkler orifice.[15]

TYPES OF AUTOMATIC SPRINKLER SYSTEMS

Four categories of automatic sprinkler systems are generally recognized: wet-pipe systems, dry-pipe systems, preaction systems, and deluge systems. The type of alarm valve used in a system determines its category. These systems can be complicated in design, engineered to protect against very specific types of hazards, such as electrical transformer vaults or high-value building contents. No matter how sophisticated the system, they all share the same basic components.

Wet-Pipe System

The basic design of the wet-pipe system has changed little since its inception in the 1870s. Because of its simple design and steadfast reliability, it is the most common sprinkler system in use today. Wet-pipe systems are normally installed in areas that are not subjected to freezing temperatures. However, portions of a wet-pipe system can be used to protect cold areas. If such a system is connected to a public water supply, the water is removed from the sprinkler piping and replaced with U.S. Pharmacopeia grade glycerin or propylene glycol. Special designs and safety features prevent the antifreeze from entering other portions of the sprinkler system.

When a sprinkler head on a wet-pipe system is activated, the water pressure in the pipe above the alarm valve drops. The clapper (a one-way valve that covers the opening of the water supply pipe) in the alarm valve is thrown open by the force of the pressurized water supply. The water in the piping is immediately discharged from the sprinkler head. As the water flows through the alarm valve, a portion of the water is diverted through another pipe leading from the alarm valve. This pipe diverts water to the retard chamber; after the chamber is filled, the water travels through piping and activates an electronic signaling device and a water-flow-operated bell (water gong), which sounds a local alarm. The water is then discharged onto the ground under the water gong.

Water distribution systems are subject to a phenomenon known as *pressure surge* that is caused by the constant use of the water system (the opening and closing of water valves). During a pressure surge, the clapper in the wet-pipe alarm valve is lifted off its seat by the force of the surge. This allows water to enter the alarm valve piping. To prevent false alarms, the wet-pipe system is equipped with a retard chamber. This appliance is placed in the piping that connects the alarm valve to the water gong. The chamber has a capacity of approximately 2 gallons of water. Water must fill the chamber completely before it can reach the alarm devices. Water flow during a pressure surge is of short duration and is not adequate to fill the retard chamber, thus preventing the false alarm. The amount of water that does enter the chamber is released through a calibrated drain valve, mounted on the bottom of the device, after the alarm valve clapper reseats. When sprinklers are activated, and the clapper is locked in the open position, the volume of water is high enough to override the drain valve, fill the retard chamber, and activate the alarm devices.

Dry-Pipe System

The dry-pipe system, more complicated than the wet-pipe system, is found in areas where temperatures can fall to 40°F or below. The riser, crossmains, and branch lines are filled with pressurized air instead of water. Air pressure in the piping is maintained by an air compressor or compressed air system. The clapper in the dry-pipe valve is held closed by utilizing the pressure differential principle: a small amount of air pressure above the valve holds back a larger amount of water pressure on the water supply side of the valve.

There are two basic designs that utilize this differential principle: the double-seated clapper and the single clapper. The double-seated clapper valve has two clappers: the upper air seat clapper and the lower water seat clapper. The upper air seat clapper is substantially larger than the water seat clapper and is held closed by the air pressure in the piping. This creates the pressure differential needed to hold the lower water clapper closed. The single clapper valve simply incorporates the features of the double clapper valve into a single clapper. The top of the clapper has a much larger surface area than the water side of the clapper, thus creating the pressure differential. The normal pressure differential in a standard dry-pipe system is approximately 6 to 1. With this design, the dry-pipe system is not affected by pressure surges and does not require the installation of a retard chamber. The pressure of the water supply will determine the amount of air pressure required in the system. For example, if the water supply pressure is 100 psi, the dry-pipe system will require a minimum air pressure of approximately 17 psi to hold the dry-pipe valve closed: 100/6 = 16.66 psi. This is referred to as the *trip pressure*. The air pressure in the system should be 15 to 20 psi above the trip pressure. Therefore, the air pressure in our system would actually be set at 32 to 37 psi.

One disadvantage of a dry-pipe system is the time delay between sprinkler activation and water discharge. It takes time for the pressurized air to be purged from the system to allow water to flow into the piping. NFPA 13 requires a quick-opening device on any dry-pipe system that has a pipe capacity of 500 gallons or more. The two quick-opening devices currently used are accelerators and exhausters.

The accelerator is used in conjunction with the double-seated dry-pipe valve. It is designed to sense the first 1 or 2 psi of pressure drop when a sprinkler activates. It opens a valve that diverts the system's air pressure into a chamber under the upper air clapper. When the air pressure equalizes on both sides of the air clapper, the valve is tripped open by water pressure. This process takes about 10 to 15 seconds.

The exhauster can be utilized with either double-seated or single clapper valves. When it senses a drop in air pressure in the system, it opens a large valve to evacuate the pressurized air rapidly to the outside, thus allowing the dry-pipe valve to trip. When trip pressure is reached, the exhauster closes to prevent the loss of water (Figure 9-26).

Preaction System

A preaction system is a dry-pipe system with a preaction valve and a supplemental detection system. The detection system is the central operating component. When the detection system senses a problem, it opens the preaction valve and fills the piping with water. Therefore, when the sprinkler head fuses, the water is ready for discharge. What happens, essentially, is that a dry-pipe system is converted into a wet-pipe system prior to sprinkler operation. A positive feature of this system is that the preaction valve will not open until the detection system activates. A loss of air pressure, sprinkler activation, or pipe failure will not open the preaction valve. The detection system must sense a problem and send a signal to open the valve. This type of system is practical in situations where inadvertent water damage is a major concern. If the preaction system is configured as a double-interlock system, both the sprinkler and the detection system must activate before the preaction valve will open

Deluge System

The deluge system is a nonpressurized dry-pipe system with a detection system. It is used to protect an area that requires rapid application of large quantities of water, such as electrical transformer rooms or outside transformer pads. High-speed, rapid-response deluge systems may be installed in facilities where processes create explosive atmospheres, such as flammable gas cylinder filling facilities. The risers, crossmains, and branch lines in a deluge system contain air at atmospheric pressure. Deluge systems utilize open nozzles or sprinkler heads that do not contain operating elements. When the deluge valve is opened, water fills the piping and discharges from all of the nozzles or sprinkler heads simultaneously. The detection system controls the operation of the deluge valve. The detectors can be of the smoke-, heat-, or flame-detection variety, and the deluge valve can be configured to operate hydraulically, electronically, or pneumatically. A hydraulically operated system can use any kind of detection system. An electronically operated deluge valve requires a detection system capable of transmitting an electrical signal to the valve. A pneumatically operated deluge valve is designed to use detectors that operate by sensing the rate of increase in air pressure inside the detector that is caused by exposure to heat. Fire pumps are used to supply the large volumes of water that are required by the deluge system.[16]

Special Note: It should be mentioned that most commercial sprinkler systems utilize fire pumps to maintain pressure in the system when it is activated. This is to supplement the fire flow provided by fire apparatus connected to the fire department connection. In high-rise buildings, sprinkler systems are zoned off in areas, usually 12 stories for each zone. Pressure to the zones is provided by jockey pumps that are placed at each zone to maintain pressures to upper floors. A firefighter should be stationed in the sprinkler operation room of each activated zone, to monitor the pumps.

Residential Sprinkler Systems

Residential sprinkler systems have become very popular with the fire service. Many communities have adopted fire codes requiring sprinkler system installation in all new residential construction. This has been a controversy in many areas. Building contractors complain of the higher cost of construction that may disqualify some potential buyers. The fire service and insurance companies lament the life-safety and loss prevention features of the systems. Studies have shown that residential sprinklers not only save money, but also lives. One study was conducted in 1986 when Scottsdale, Arizona, passed legislation requiring the installation of residential sprinkler systems in all new single-family homes. In 1996, the results of the legislation showed that the average per square foot installation cost for residential sprinklers dropped 45% from $1.14 to 59 cents. Over 90% of the fires were controlled by the activation of one or two sprinkler heads with an average water flow of 299 gallons per fire. In 1990, Prince George's County in Maryland passed an ordinance requiring the installation of sprinkler systems in all new residential structures. Over a 10-year period, there were 117 fires involving sprinklers. The total fire loss was $401,220. It was estimated that the loss would have been $38,230,000 if the structures had not be sprinklered. The study also reported that 154 lives were saved by the sprinkler system with only seven minor injuries. There are, truly, more pros than cons in the argument for residential sprinklers.

Residential sprinkler systems can be very complex or very simple. The average unit does not resemble the commercial system, but its functions are very similar. The unit does not have an alarm valve, but has an in-stream alarm switch that activates when water begins to flow through the piping. The main water shut-off valve is usually a simple ball valve that can be operated with one-quarter turn of a handle. The sprinkler heads are fast response and designed to activate in a much shorter time span than commercial system sprinkler heads. The residential sprinkler head flow pattern is more uniform than those used in other systems. They are designed to direct water high on the walls to protect contents on the perimeter of the room. With the advent of lightweight, engineered building components, residential sprinklers have proven valuable in reducing fire spread and delaying building collapse. But most importantly, these systems greatly increase occupant survivability. Residential sprinklers save lives and property. Installation should be in compliance with NFPA 13R.[17]

Signaling Systems

A signaling system is classified according to the function the system is expected to perform. All signaling systems, regardless of function, have six basic features.

Control Unit. This unit acts as the "brains" of the system. Its primary function is to receive a signal from the detection device and route the signal to a visual and audible alarm and, if designed to do so, to send an alarm to a monitoring agency.

Primary Power Supply. This is usually supplied by the public utility that serves the building.

Secondary Backup Power Supply. This can be a system of batteries or engine-driven generators. The secondary power supply is required to provide full signaling and detection system operational power within 30 seconds of primary power supply failure.

Trouble Signal Power Supply. This power supply can be from the building's public utility connection or from a separate source.

One or More Initiating Device Circuits. The initiating device may be activated by manual operation, heat, smoke, or flame (light).

One or More Alarm-Indicating Device Circuits. When the control unit receives a signal from an indicating device, it must activate audible and visual alarm devices.

Note: Some control units are able to activate other life-safety systems in the building, such as automatic ventilation or extinguishing systems. They may also send signals that close fire doors and return all elevators to the ground floor.

To ensure that operational standards are met, all components of a signaling and detection system should be recognized by a testing laboratory, such as Underwriters Laboratory or Factory Mutual Research Corporation. Installation, maintenance, and use of the system should conform to the standards set forth in NFPA 70, *The National Electrical Code*, and NFPA 72, *The National Fire Alarm Code*.

TYPES OF SIGNALING SYSTEMS

The type of signaling system used depends on several factors, including the kinds of hazard that the system expects to handle, jurisdictional requirements, and the design and construction of the building. During a preincident survey, the firefighters should be able to recognize the type of system in use and understand how it operates. They will then be able to judge whether it is providing adequate protection to the occupants of the building and their property.

The following are basic descriptions of the most common signaling systems encountered by fire service personnel. We will look at local alarm systems, auxiliary systems, remote station systems, and central station systems.

Local Alarm System

A local system is designed to activate a visual and audible alarm and alert occupants on the immediate premises to evacuate the building. The system does not transmit a signal outside of the property being protected. The four most common types of local systems are the noncoded system, the zone-noncoded system, the master-coded system, and the zone-coded system.

Noncoded System. This is the simplest of all local alarm designs. When the control unit receives a detection signal, it initiates all audible and visual alarms throughout the protected property. The unit is not designed to indicate the location of the problem. Firefighters must search the entire property to locate the cause of the alarm. Audible and visual alarms usually remain activated until the system is reset at the control unit.

Zone-Noncoded System. In this system, the protected property is divided into zones. Each zone is wired as a separate circuit to the control unit and is identified by a corresponding indication lamp on the unit. When an initiating device is activated in a zone, all visual and audible alarms on the premises will activate, and the indication lamp on the control unit assigned to that zone will illuminate. By viewing the control unit, firefighters can determine the general location of the problem.

Master-Coded System. This system utilizes audible and visual notification devices that are a normal part of a building's everyday operations, such as the bells that signal a lunch break. When the control unit receives a signal from a fire detection device, it initiates a different sound pattern from that normally used in the everyday operation. For example, if the signaling device normally sounds one long tone to indicate a break for lunch, the fire alarm control unit will cause the same device to sound several short tones. This system does not indicate the location of the trouble because all signaling devices throughout the premises are activated at once.

Zone-Coded System. This system utilizes the same zone circuitry and indicating lamp features as the zone-noncoded system. However, a coded audible circuit is added to each zone's circuitry at the control unit. When the control unit receives a signal from an indicating device, the zone indicator lamp is illuminated on the control unit, and all visual and audible alarms sound throughout the entire premises. The signals are coded, however, to indicate where the trouble is located. The pattern of the sound signal directs fire personnel to the affected zone.

Auxiliary System

The auxiliary system enables a building's alarm system to interface with a public alarm system. It can be used only in areas that utilize fire alarm boxes. When the building's control unit receives indication of an activated detector, it sends a signal to a municipal fire alarm box located on a nearby street. The signal activates the fire alarm box, which, in turn, sends a coded signal via dedicated phone line or radio signal to the fire department dispatch office. The code identifies the location of the fire alarm box. A fire-location indicator panel may be located inside the fire alarm box if it services more than one alarm system or if the system is zoned. The auxiliary system can be activated manually, by automatic fire detectors, or by water flow indicating devices.

Remote Station System

The remote station system operates in a similar manner to the auxiliary system in that it sends an alarm notification directly to the fire department dispatch office. This system, however, does not utilize a fire alarm box for signal transmission. Instead, the signal is sent directly from the protected premises through leased telephone lines to the telephone exchange, and then on to the dispatch officer. If multiple premises are being protected, the system must be coded. There are alternatives for communities that do not have a fire dispatch office. NFPA states, "If the public agency is unwilling to receive the remote station fire alarm signals, or if that agency is willing to allow another organization to receive those signals, then the signals may be received at a location acceptable to the authority having jurisdiction that is attended by trained personnel 24 hours a day."[18]

enhanced 911 ■ An emergency phone system that automatically gives a dispatcher the caller's location.

Another type of remote system is the public safety answering point (also referred to as the public safety access point; PSAP). This is used for 911 and **enhanced 911** (E911) services. The PSAP is usually established by a municipal or county agency that is responsible for receiving 911 calls and passing them on to public or private agencies. Trained PSAP operators are on duty 24 hours a day, seven days a week. Calls can be received by land-line or wireless systems. Most PSAP operations have E911 systems that can identify the location of the caller and route the call to the closest PSAP to decrease response time. When a land-line call is received, the system identifies the name, address, and phone number of the caller. If a wireless call is received, the system identifies the address of the wireless tower that transmitted the call to the PSAP, the mobile number, and an approximate location from where the call was initiated.[19]

Proprietary System

The proprietary system is designed to protect large commercial and industrial complexes. It can also be found in high-rise buildings and in multiple buildings that are concentrated in one location. Each building or protected area has its own system. When the control unit receives a signal from an activated detector, it transmits that signal to an alarm panel that is located in a separate building on the premises or in a separate room within the structure that is protected from hazardous operations. NFPA 72 requires that the building or room housing the alarm panel, which is also called the *supervising station*, be staffed 24 hours a day by employees or representatives of the occupant. These individuals must be trained in the system's operations and in response procedures if the system is activated. They must also be able to notify the fire department through an automatic or manual notification system.

The control units of proprietary systems are required to transmit alarm, supervisory, and trouble signals to the alarm panel in the supervising station. NFPA 72 also requires that a runner be dispatched from the supervising station to investigate any such signal. If a fire alarm signal is received, supervising personnel must retransmit that signal to the fire department communication center. It must be emphasized that all personnel staffing the supervising station are employees or representatives of the protected entity.

Central Station System

A central station system is very similar to a proprietary system. The main difference is that the alarm panel and supervising station is not located on the protected property and is not supervised by employees or representatives of the occupant. Instead, the alarm panel and supervising station is located off premises and constantly staffed by employees of a contract service company whose sole purpose is providing central station service. Underwriters Laboratory (UL) or Factory Mutual Research Corporation (FMRC) must list the contract service company. The installation of a central station system must be certified by UL or FMRC and should meet the standards set forth by NFPA 71, *Standard for the Installation, Maintenance, and Use of Proprietary Central Station Signaling System.*

When a fire alarm signal is received, the central station operators notify both the fire service communications center and designated personnel, either at the protected location or in their homes. They must also dispatch a runner or technician to the location to reset the system. Runners must be able to reach the location within one hour. Runners must also be dispatched when supervisory or trouble alarms are transmitted.[20]

Automatic Detection Systems

We described the different kinds of signaling systems that are used to alert building occupants and firefighters to danger. We will now look at the automatic detection devices that activate those signaling systems.

Automatic fire detectors are designed to sense one or more of the products of combustion: smoke, heat, or light (flames). The operations of these devices range from the very simple to the very complex.

HEAT DETECTORS

Heat detectors are the oldest automatic fire detection devices, dating back to the 1860s, about the time that the first automatic sprinkler heads came into use. They were used with early automatic sprinkler heads. Heat detectors that are designed to operate when a predetermined temperature has been reached are called *fixed-temperature* detectors. Others, called *rate-of-rise* detectors, respond to rapid temperature increases. Detectors that provide both fixed temperature and rate-of-rise capability are referred to as *combination* heat detectors. Heat detection systems are relatively inexpensive compared with more technologically sophisticated systems. They have the lowest false alarm rate of all automatic fire detection devices, but they are the slowest to activate under fire conditions.

Heat detectors should be located on or near the ceiling. To ensure optimum performance, they should be installed in areas where heat is expected to build rapidly, such as a small confined space, where environmental conditions would not allow the use of other types of fire detection systems, and where speed of detection is not a primary concern. Heat detectors may be used in environments that have dust or other suspended particles in the air. Smoke and light detectors would easily be activated in this type of atmosphere.

Fixed-Temperature Heat Detectors

The operating element of a fixed-temperature heat detector can be a fusible link, frangible bulb, bimetallic, or continuous line type. Fusible link, frangible bulb, and bimetallic detectors are referred to as *spot types*. They respond to environmental changes in the specific area where they are located. Fusible link and frangible bulb components are most often associated with automatic sprinklers.

When the fusible link is used in a heat detector, the eutectic (easily melted) metal alloy of cadmium, bismuth, lead, and tin acts as a solder to hold open a spring-loaded contact. When the temperature reaches the melting point of the solder, the contact is released, closing the circuit and initiating a signal.

The function of the frangible glass bulb in a heat detector is to hold two circuit contacts apart. When heat causes the liquid in the bulb to expand and rupture the glass, the contacts close and complete the circuit, thus initiating a signal. Fusible link and frangible bulb detectors do not reset when normal environmental conditions return and must be replaced after they have been activated.

The contacts in a bimetallic detector are held apart by a bimetallic strip or bimetallic snap disc. These elements are constructed by bonding two metals that have different heat expansion coefficients. When heat is applied to the bimetallic strip or disc, the bimetallic operating element will bend in the direction of the contact point, closing the circuit and initiating a signal. This type of detector automatically resets itself when the ambient temperature drops below its activation point.

Unlike spot-type detectors, *continuous line* detectors are designed to sense temperature changes over a large area. Two basic versions are available. In one design, two steel wires are separated by heat-sensitive insulation. When the ambient temperature reaches a predetermined level, the insulation melts from around the wires and allows them to make contact with each other. This closes a circuit and initiates an alarm signal. The fused segment of the line must be replaced in order to restore the system to normal operation. The other continuous line detector has a coaxial design. It consists of a conductive metal wire housed inside a stainless steel tube. A ceramic semiconductor separates the conductive metal wire and the stainless steel tubing. A minute electrical current flows through the circuit. Under normal conditions, this current is not strong enough to trip the alarm signal. However, when the stainless steel exterior tubing is heated, the electrical resistance of the ceramic semiconductor is lowered, allowing a stronger current to pass through the system and activate the alarm.

A fixed-temperature device activates when the ambient temperature at the ceiling exceeds its predetermined setting. At the time of activation, the actual ambient temperature at the ceiling, however, will be much higher than the activation setting of the device. This delay, called a *thermal lag*, is caused by the time needed for the operating element to absorb the heat necessary for activation. Thermal lags are an inevitable shortcoming of fixed-temperature heat detection systems.

Rate-of-Rise Heat Detectors

The rate-of-rise system eliminates thermal lag. When a fire occurs, the temperature of the air directly above the fire increases rapidly. The rate-of-rise detector is designed to sense this rapid increase in temperature and signal an alarm if the rate exceeds 12°F to 15°F per minute. These systems may be of the line or spot type, but both operate under the same principles.

The line-type rate-of-rise detector consists of metal tubing mounted on the ceiling in a loop configuration. This tubing is connected to a chamber that contains a diaphragm. There is an air chamber on one side of the diaphragm and a contact point on the other. When the air inside the tubing is heated, its pressure increases. This, in turn, brings increased pressure to bear on the air chamber, which forces the diaphragm outward and closes the contact points to initiate a signal. The detector is equipped with a vent hole in the air chamber. If the rate of temperature rise is less than 12°F to 15°F per minute, the vent is capable of maintaining pressure equilibrium in the chamber to prevent the diaphragm from moving. This prevents false alarms. The NFPA recommends that each circuit be limited to no more than 1,000 feet of tubing.

The spot-type rate-of-rise detector also works on a pneumatic principle. The main difference between the two is that the spot-type detector has a single air container instead of tubing.

Combination Heat Detectors

The combination heat detector employs both rate-of-rise and fixed-temperature principles in its design. It has an air chamber and diaphragm, as well as a contact that is held open by a fusible link. The rate-of-rise element gives it the capability to respond swiftly to rapidly developing fires, whereas the fixed-temperature element is more responsive to fires that develop slowly.

SMOKE DETECTORS

The smoke detectors that firefighters most commonly encounter operate on either the ionization principle or the photoelectric principle. **Ionization detectors** respond faster to high-energy or open-flame fires, whereas **photoelectric detectors** respond faster to low-energy or smoldering fires.

The ionization detector contains a small amount of radioactive material that is used to ionize air in a sensing chamber. When this occurs, the air becomes conductive, allowing a small electric current to flow between two charged electrodes. When smoke particles enter the chamber, they reduce the conductance of the air and disrupt the flow of current. This closes a circuit and activates the alarm (see Figure 9-27).

Photoelectric detectors work on either the light-obscuration or light-scattering principle. **Light-obscuration systems** consist of a light source, a light transmitter, and a light-sensitive device. A light beam is directed onto a light-sensitive receiver. When smoke particles enter the beam, the light is blocked (or obscured) from reaching the sensing device. This reduction in light causes the initiation of an alarm. This type of system is used to provide protection for large open areas (see Figure 9-28).

Light-scattering systems are usually of the spot type. They consist of a chamber that houses a light source and a light-sensitive receiver. When smoke particles enter the chamber, the light beam becomes scattered. Some of this light is refracted onto the light-sensitive receiving device, and the alarm is activated (see Figure 9-29).

FLAME (LIGHT) DETECTORS

Flame detectors are designed to respond to either ultraviolet or infrared light waves. Because of their rapid detection capabilities, they are used in locations with a high hazard potential, such as areas where explosive gases are present or where rapid fire spread might occur. **Infrared flame detectors** are best suited for protecting areas housing hydrocarbons. Because of their sensitivity to solar radiation, they should be sheltered from the sun's rays or provided with infrared filters. **Ultraviolet flame detectors** are sensitive to fires involving hydrocarbons, metals, sulfur, hydrogen, and ammonia. They do not respond to sunlight. **Combination flame detectors** are designed to sense both ultraviolet and infrared light waves (see Figure 9-30).[21]

FIGURE 9-27 Ionization smoke detector. *Courtesy of James D. Richardson.*

ionization detectors
■ Smoke detectors that use a finite amount of radioactive material to make the air within a sensing chamber conduct electricity.

photoelectric detectors
■ Smoke detectors that use a small light source to detect smoke.

light-obscuration systems ■ Smoke detection systems that are activated when smoke particles block a light beam from a photoelectric light-sensitive cell.

light-scattering systems
■ Smoke detection systems that are activated when light beams are scattered by smoke particles and strike a light-sensitive photoelectric cell.

Light Obscuration Smoke Detector

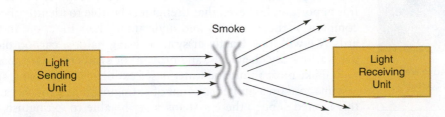

FIGURE 9-28
Light-obscuration smoke detection systems are used to protect large open areas. *Courtesy of James D. Richardson.*

FIGURE 9-29
Light-scattering smoke detection systems are also used to protect large open areas. *Courtesy of James D. Richardson.*

Light Scattering Smoke Detector

Light Sending Unit

Light Receiving Unit

Smoke

Light Sending Unit

Light Receiving Unit

infrared flame detectors ■ Flame detectors that operate in extreme environmental conditions.

ultraviolet flame detectors ■ Fire detection devices that are activated when ultraviolet light waves are detected.

combination flame detectors ■ Alarm devices that can detect abnormal conditions by several means.

CARBON MONOXIDE DETECTORS

Carbon monoxide (CO) is a toxic, colorless, odorless gas formed by incomplete combustion of a fuel. It is responsible for several hundred deaths each year. A great majority of these deaths occur in dwellings. In 1997, the NFPA published NFPA 720 *Recommended Practice for the Installation of Household Carbon Monoxide Warning Equipment*. This was not a standard, but a recommendation and did not include commercial properties. The focus of the recommendation was for family dwellings. In 2005, the recommendation was converted to NFPA 720 *Standard for the Installation of Carbon Monoxide (CO) Warning Equipment in Dwelling Units*. Like the recommendation, hotels, congregate residential buildings (assisted living), and other commercial nonresidential buildings are not included in this standard. CO detection units are recommended for any family dwelling that contains fossil burning fuel equipment, fireplaces, or has an attached garage. NFPA Standard 720 requires CO detection equipment to be interconnected to home fire or fire/burglary alarm system. To differentiate CO alarm signals from fire or smoke alarm signals, the CO unit must have a "single-tone, temporal 4 four cycle audible pattern of 100 milliseconds 'on' and 100 milliseconds 'off,' followed by 5 seconds 'off.' After 4 minutes, the "off" cycle can be extended to 60 seconds. It must have enough power to continually operate for 12 hours. Each state has set regulations for the installation and maintenance of CO detectors and alarm systems.[22] These statutes can be obtained from the National Conference of State Legislatures website: www.ncsl.org.[23]

Special Extinguishing Systems

For some kinds of fires, water is not the most effective extinguishing agent. In areas where fires of this kind are apt to break out, special systems must be installed. A great many such systems are in use today and discussing all of them is beyond the scope of this chapter. It is important, however, that firefighters be able to identify the most common of these systems and understand their function and hazards. We will look briefly at systems using halons, carbon dioxide gas, various dry chemicals, and protein- or synthetic-based foams.

Unlike most automatic water-based extinguishing systems, special systems contain a limited amount of extinguishing agent. Because of this limited supply, these systems must be able to extinguish a fire

FIGURE 9-30 Flame detectors are used in areas where very rapid detection is required. *Courtesy of Tyco/Ansul Corporation.*

effectively in a short duration of time. Automatic water-based extinguishing systems are rated as successful if they control the fire; special systems are rated successful only if they extinguish the fire. Some special extinguishing agents have the ability to extinguish a fire by cooling or disrupting the chemical chain reaction of combustion. However, they all have one common characteristic—they all separate the oxygen from the fuel.

HALON SYSTEMS

Halogenated hydrocarbon extinguishing agents, or halons, date back to the 19th century. They are derived from a process by which one or more of the hydrogen atoms in a hydrocarbon are replaced by atoms of the halogens fluorine, chlorine, bromine, or iodine. Halons are used to protect areas containing high-valued contents, such as electronic or computer rooms, records and document rooms, museum storerooms, or other materials that would be damaged by other extinguishing agents that leave extinguishing residues. The two most likely halon agents that a firefighter may encounter today are Halon 1211 and Halon 1301. The numbers indicate the number of carbon and halogen atoms in each molecule. The first digit is the number of carbon atoms; the second digit, fluorine; the third digit, chlorine; and the fourth digit, bromine. Halon 1211 is best suited for use in portable extinguishers because it is normally expelled as a liquid. It will rapidly vaporize, however, when it comes in contact with the fire. Halon 1301 is best suited for use in fixed extinguishing systems because it is normally discharged as a vapor. These two agents are classified as **clean extinguishing agents** because they do not leave any residue or cause collateral damage after extinguishing a fire, but they are dangerous. Firefighters working with these extinguishers should wear full protective clothing and use self-contained breathing apparatus (SCBA) systems in unoccupied areas or areas that are only temporarily occupied. The system should be equipped with a predischarge alarm to allow occupants time to evacuate before the extinguishing agent is released.

When Halon 1301 is decomposed by heat, the halogens combine with hydrogen formed by the combustion process to create toxic byproducts: hydrogen fluoride, hydrogen bromide, and bromine. Decomposition by-products of Halon 1211 include all of the above, plus hydrogen chloride and chlorine. Halon 1211, therefore, is the more toxic of the two agents. Toxicity is not the only drawback of halons, however. Because of their harmful effects on the ozone layer, the production of halons was discontinued at the end of 1993. The Montreal Protocol, an international treaty, mandated that all existing halon systems be decommissioned by December 31, 2003. These systems, however, are still being used by the military and could still be encountered by firefighters.

clean extinguishing agent ■ An extinguishing agent that leaves no residue or collateral damage resulting from the discharge of the agent.

Recommendations for installation, use, and maintenance of these systems are published in NFPA 12A, *Standard on Halon 1301 Fire Extinguishing Systems*, and NFPA 12B, *Standard on Halon 1211 Fire Extinguishing Systems*.

Alternative agents have been developed to replace halons. They are listed under several commercial names, including FM200, FE-25, and FE-13. These agents are ozone friendly but pose various levels of toxicity hazards to exposed personnel. The first halon replacement agent was FM200 and is the most widely used replacement agent. It is stored as a liquid in cylinders and discharged as a gas. It must be discharged in 10 seconds or less. This requires the cylinders to be located close to the protected area. Some of the alternative agents, such as FE-25 or HFC-125, are similar to Halon 1301 and can be used as "drop-in" replacements in existing Halon systems. Only minimum modifications to the piping and nozzles are required to accommodate the higher vapor pressures of the replacement

FIGURE 9-31 Halon replacement agents are toxic and require that full protective equipment be worn. *Courtesy of Tyco/Ansul Corporation.*

agents. If the agent is not compatible with the existing system, a new system, designed for the new agents, must be installed[24] (Figure 9-31).

CARBON DIOXIDE SYSTEMS

Carbon dioxide (CO_2) is an inert gas that is produced naturally in our atmosphere. It is colorless, odorless, nonconductive, and nontoxic. It is also nonflammable and $1\frac{1}{2}$ times heavier than air. These last two characteristics allow CO_2 to displace oxygen-rich air on a burning surface, creating an atmosphere that inhibits combustion. Carbon dioxide has long been used in the commercial arena for refrigeration and in other industrial processes. It was first used as a fire-extinguishing agent in 1914, when the Bell Telephone Company developed the portable CO_2 fire extinguisher to protect its electrical wiring and equipment. Automatic fixed CO_2 extinguishing systems were developed in the 1920s, and the NFPA published a standard on their proper installation, operation, and maintenance in 1929 (NFPA 12, *Standard on Carbon Dioxide Extinguishing Systems*).

CO_2 systems are classified according to their method of application: total flooding or local application.

Total flooding systems are designed to fill the protected area with CO_2, creating a CO_2-rich atmosphere in which combustion cannot continue. The system consists of a storage tank or cylinder, piping, and nozzles (referred to as *horns*). It can be activated by an automatic fire detection system, by normal manual operation, or by emergency manual operation. If the system is activated by an automatic fire detection system or by normal manual operation, it must be equipped with a predischarge alarm that will alert occupants and give them time to evacuate the area before the system activates. Systems with an emergency manual operation feature must be equipped with a discharge alarm but do not require a predischarge alarm. When this operation system is used, the occupants must evacuate while the system is discharging because no predischarge warning is sounded. Therefore, the emergency manual operation feature should be activated only when other initiating devices fail.

Local application systems are designed to direct the CO_2 onto the object being protected instead of flooding the entire area. The system components are the same as in a total flooding system: a storage tank, piping, discharge nozzles, and a fire detection system. When CO_2 is discharged in a local application system, dry-ice particles are produced. These particles have the potential to carry a charge of static electricity. To eliminate shock hazards to personnel operating the equipment and to prevent static discharge in a volatile atmosphere, the discharge nozzles must be properly grounded (Figures 9-32A and B).[25]

There are two ways in which the CO_2 can be stored in extinguishing systems. In **high-pressure systems**, the carbon dioxide is stored at room temperature in standard Department of Transportation–approved cylinders at a pressure of approximately 850 pounds per square inch. The maximum capacity of a high-pressure cylinder is 100 pounds of CO_2. If more agent is needed, additional cylinders must be connected to the system through a manifold. In **low-pressure systems**, liquefied carbon dioxide at 0°F is contained in large refrigerated storage tanks pressurized to 300 pounds per square inch. Low-pressure storage tank capacities range from $1\frac{1}{4}$ to 60 tons. The effectiveness of a CO_2 system depends on the structural integrity of the area being protected. If the area is tight, the CO_2 that is discharged will be retained longer. If the area has openings, especially on the sides and bottom, the heavier-than-air carbon dioxide may leak out before effectively extinguishing the fire.

total flooding systems
■ Extinguishing systems that are designed to completely fill the protected area with extinguishing agent when activated.

local application systems
■ Permanent suppression systems that are required to cover a protected area with 2 feet of foam depth within 2 minutes of system activation.

high-pressure systems
■ Carbon dioxide is stored at room temperature in standard Department of Transportation–approved cylinders at a pressure of approximately 850 pounds per square inch.

low-pressure systems
■ Liquefied carbon dioxide at 0°F is contained in large refrigerated storage tanks pressurized to 300 pounds per square inch.

FIGURE 9-32
(A) Total flooding system;
(B) local application system. These systems
require a preactivation
alarm to allow occupants
time to evacuate the area.
*Courtesy of Tyco/Ansul
Incorporated.*

(A)

(B)

STOP AND THINK 9-3

Your fire crew has been dispatched to an activation alarm at the local newspaper printing room. What hazards would you anticipate facing on arrival? How can you protect yourself and your crew?

DRY CHEMICAL SYSTEMS

Dry chemical systems are characterized as either ordinary or multipurpose. **Ordinary dry chemical systems** are designed primarily to extinguish flammable liquid fires (Class B fires). They may also be used on electrical equipment that is subject to flammable-liquid fires, such as transformers, and on energized electrical fires (Class C fires). **Multipurpose dry chemical systems** are rated for Class A (ordinary combustibles), Class B, and Class C fires. Neither one of these systems should be used in areas such as computer rooms and

ordinary dry chemical systems ■ Dry chemical systems designed primarily to extinguish flammable liquid fires (Class B fires).

multipurpose dry chemical systems ■ Dry chemical systems that use extinguishing agents that are rated to be effective on more than one type of fuel.

FIGURE 9-33 Dry chemical agents are nontoxic but can be highly irritating.
Courtesy of Tyco/Ansul Corporation

switchboards where there is electronic or electrical equipment, because dry chemical residue may be harmful to sensitive circuitry.

The basic chemicals used to produce dry chemical agents are sodium bicarbonate, potassium bicarbonate, potassium chloride, urea–potassium bicarbonate, and monoammonium phosphate. Additives, such as metallic stearates, tricalcium phosphate, and silicone, are used to increase flow characteristics, shelf life, and water resistance. These chemicals in various combinations give dry chemical systems four fire-extinguishing properties:

- **Smothering.** When monoammonium phosphate decomposes, it leaves a residue (metaphosphoric acid) on the surface of the fuel. This residue smothers the fire by sealing off oxygen from the fuel.
- **Cooling.** Although this is not a primary extinguishing property of dry chemical agents, it has been proven that these compounds absorb large amounts of heat when they decompose. This adds to the extinguishing ability of the agents, albeit only in a minor way.
- **Radiation Shielding.** When dry chemical agents are discharged, a large dense cloud of powder is generated between the fuel and the flame. This prevents heat from being radiated back to the fuel from the flame, thus aiding in the extinguishing process.
- **Breaking the Chain Reaction Process.** This is the most powerful of the four extinguishing properties of dry chemical agents. When dry chemical agents are exposed to heat, they decompose. Particles from the decomposition process interfere with the chain reaction of combustion, reducing the number of free radicals in the flame. Preventing these reactive particles from coming together breaks the chain reaction needed to continue the combustion process.

Dry chemical agents are nontoxic and nonconductive, but they can irritate the eyes, mucous membranes, and the respiratory system. Some agents may cause chemical burns on moist skin. Personnel must always wear full protective equipment, including SCBA, when entering areas that have been extinguished with dry chemical agents (see Figure 9-33).[26]

Dry chemical systems can be either fixed systems or handheld hose line systems and must meet the requirements of NFPA 17, *Standard for Dry Chemical Extinguishing Systems*.

wet chemical
- Potassium carbonate, potassium acetate, potassium citrate, or a combination of these chemicals, and phenol-phthalein or phosphoric acid, mixed with water. Used for extinguishing cooking oil fires.

It should be noted that although dry chemical agents are effective on most flammable liquid fires, they have a low performance rating on cooking oil fires (Class K fires). Dry chemical systems for cooking areas have been replaced with wet chemical systems.

WET CHEMICAL SYSTEMS

Wet chemical agents consist of potassium carbonate, potassium acetate, potassium citrate, or a combination of these chemicals, phenolphthalein, or phosphoric acid, mixed with water. Wet chemicals will react with any material that is reactive with water. They are not considered toxic, but many cause skin or respiratory irritation. Because wet chemical agents are alkaline solutions, they react with the fatty acids in cooking oils. This reaction is called *saponification*.

Saponification is a process in which the fatty acids in the cooking oil reacts with the alkaline agent and is converted into a soapy foam. The foam covers the surface area of the liquid, separating the oxygen from the fuel, smothering the fire. Water vaporization aids in cooling the fuel below its ignition temperature[27] (see Figure 9-34).

FOAM SYSTEMS

Foam extinguishing systems are especially effective in fighting fires fueled by hydrocarbons (methane, propane, benzene, naphthalene, and various polymers such as polyethylene and polypropylene). The foams work by forming a blanket over the burning fuel, separating it from combustion vapors and oxygen.

FIGURE 9-34 Wet chemical extinguishing agents are used over deep frying areas in kitchens. *Courtesy of J.D. Richardson.*

Firefighting foams are classified as low-expansion, medium-expansion, and high-expansion. Low-expansion foam has a 20:1 expansion ratio. This means that 1 gallon of foam concentrate will expand to 20 gallons of foam when mixed with water and agitated with air. Medium-expansion foam has an expansion ratio of 20:1 to 200:1, and high-expansion foam expansion ratios range from 200:1 to 1000:1.

Foam concentrates can be synthetic- or protein-based. Synthetic foams are detergent-based and are very effective on liquid hydrocarbon fires. They can be premixed in water storage tanks. An effective foam blanket, with a 20:1 expansion ratio, can be applied with a standard fog nozzle; an aerating nozzle is not required. Synthetic alcohol-resistant foams are also available. Protein foams are made from plant or animal matter. These foams are not often used today.

There are six basic types of foam systems:

- *Self-Contained Apparatus.* Found on fire apparatus, this system has all of the components needed to develop foam: a water storage tank, a foam storage tank, piping, a foam inductor, and the equipment needed to agitate the foam and water mixture.
- *Fixed.* This system is a permanently installed unit. It has a central foam station that delivers foam through piping to a discharge nozzle in the protected area. This unit may be of the total flooding or local application type.
- *Semi-Fixed Type A.* This system has installed piping and nozzles but no permanent source of foam. An apparatus must be connected to the system to pump the foam solution into the piping. This arrangement is similar to the dry standpipe system.
- *Semi-Fixed Type B.* This system provides foam protection by piping the foam solution through piping to foam hydrants located throughout the protected property. This system does not discharge the foam but makes the foam available for use. This can be compared to a fire hydrant on a water distribution system.
- *High-Expansion Foam.* The high-expansion foam system consists of an activation device (automatic or manual), a foam generator, a water supply, a foam supply, and piping. It can be used for local application or total flooding application, but in either case, it must supply enough high-expansion foam to cover the hazard with a minimum of 2 feet of extinguishing agent. High-expansion foam systems are used to protect aircraft hangers, basement areas, and confined spaces where flammable

FIGURE 9-35 Foam extinguishing agents are highly effective in fighting hydrocarbon fuel fires. *Courtesy of Tyco/Ansul Incorporated.*

atmospheres or processes may exist. The high expansion ratio (200:1 to 1000:1) allows large areas to be filled with agent with very little water damage.

- *Foam and Water.* This system is a deluge system that discharges foam. The piping contains a foam induction system and aerating sprinkler heads, which are mounted at the end of the piping. This configuration allows the foam to expand five or six times when discharged.[28]

- *Compressed Air Foam (CAFS).* These systems can be fixed or self-contained. They consist of a fire pump, a rotary air compressor, and a foam-proportioning system. The foam is generated by injecting compressed air into the foam solution as it travels through piping. The compressed air injection increases the momentum of the foam solution, allowing the

stream to be projected farther than a water stream, which results in better fire penetration. A CAFS uses less foam solution and less water than other types of foam induction systems. This greatly reduces property damage caused by the extinguishing process. The expansion ratios for CAFS-generated foam range from 1:4 to 1:20. However, tests have shown that foam solutions with an expansion ratio exceeding 1:10 are too dry to be very effective in fire suppression[29] (see Figure 9-35).

For a foam application to be effective, it must be applied correctly. A foam stream should not be applied directly onto the fuel. The agitation of the fuel will prevent the foam from forming a protective blanket between the fuel and the vapors it is generating. If the fuel is burning, the agitation will cause the release of more vapors and the fire will intensify. The heat will cause the foam to break down and reduce, or nullify its effectiveness. There are three basic techniques used to apply foam. The roll-on method is used by applying the foam on the ground at the edge of the liquid and rolling it over the surface. For unignited fuels or burning Class B, fuels the bank-down method can be used. This method is used when the fuel surrounds an object. The foam stream can be banked off the object and deflected onto the fuel. The rain-down method is most effective when using Aqueous Film Forming Foam (AFFF) foam. When using this method, a fog foam stream is directed into the air over the fuel. The droplets come down like rain onto the liquid and gently form a film over the fuel when the foam bubble burst.

 ON SCENE

The fire chief of a small city was informed that a developer had been given a permit to construct over 1,000 residential structures on a large parcel of land inside the city limits. It was suggested that this growth would be good for the city's economic growth. The fire chief is responsible for approving any new water distribution systems. Groundwater is the major water source for the city and the 28 municipalities bordering it. Groundwater also supplies several large military installations in the county.

1. What issues face this fire chief?
2. What construction requirements must he or she demand to ensure that an adequate water supply will be maintained now and in the future?
3. What questions must he or she ask the developer and city administration?

Summary

This chapter introduced much new terminology and presented a wealth of technical information. Read the text carefully, study the illustrations, and begin to use the proper vocabulary in talking about water systems, hydrants, sprinklers, and extinguishers.

Full mastery of the topics covered in this chapter will come with future in-depth study and with all-important on-the-job experience. If the student has gained an appreciation for the sheer volume of what there is to know and respect for the importance of this information, the chapter will have served its purpose. Two main points can stand repeating:

■ Firefighters must know the firefighting needs of their jurisdiction and be able to identify and understand the equipment (water systems, hydrants, standpipes, alarms, fire detectors, and extinguishing equipment) used in buildings they serve. They must be aware not only of the abilities of each system but also of the dangers these systems may pose for responding personnel.

■ Communities and their needs change. When the population of a rural district grows or new industry comes into a community, the fire service needs of the area change. Firefighters must keep abreast of these developments and be vigorously proactive in ensuring that the local water distribution system can handle the increased demand and that updated and expanded fire protection systems are in place before they are needed.

Review Questions

1. List and describe the four components of a water distribution system.
2. What methods are used to move water in a distribution system?
3. How are water distribution systems categorized?
4. Describe the differences between a base-valve dry-barrel hydrant and a wet-barrel hydrant.
5. The NFPA requires that Class I and Class III standpipe systems be capable of delivering _____ GPM from the first standpipe and _____ GPM from each additional standpipe.
6. _____ is the standard for the installation of sprinkler systems.
7. List the four types of sprinkler control valves.
8. Which of the four sprinkler control valves operates as a butterfly valve?
9. How are heat detectors categorized?
10. Describe the two methods used to initiate alarms with photoelectric smoke detectors.
11. Describe the hazards involved with the following special extinguishing agents:
    a. Halon 1301
    b. Halon 1211
    c. Carbon dioxide
    d. Dry chemicals

Suggested Reading

Ashtabula Soil and Water Conservation District. (2004). *Dry fire hydrants*. Retrieved April 10, 2013, from http://www.ashtabulaswcd.org.

Bachtler, R., and Brennan, T. (1995). *The fire chief's handbook* (5th ed.). Saddle Brook, NJ: Fire Engineering Books and Videos.

Beranek, Joe. (2007, June). An overview of pre-engineered fire suppression systems and industry changes that may affect your property. *Firewatch*. Retrieved April 10, 2013, from http://www.nafed.org/Files/Publications/Firewatch/FW_2007June_PreEngineered.pdf.

A brief history of the hydrant. (2003). Retrieved from http://www.firehydrant.org/pictures/hydrant_history.html.

Browmann, Mark. (2004). Fire protection: Usage of common standpipes. *PM Engineer*, Issue 4/04. Retrieved from *PM Engineer* website: http://www.pmengineer.com/CDA/ArticleInformation/features/BNP_Features_Item/0,273.

Casey, James F. (1970). *Fire service hydraulics*. (2nd ed.). New York, NY: Reuben H. Donnelley Corp.

Colletti, Dominic. (1998). *Class A foam—Best practice for structure firefighters* (1st ed.). Royersford, PA: Lyon's Publishing.

DiNenno, Philip J., and Taylor, Gary M. (2008). Halon and Halon replacement agents and systems. In *Fire protection handbook*. (20th ed., Volume II). Quincy, MA: National Fire Protection Association.

Beattie, Walter S. (2011). Evolution of the fire sprinkler. *Fireline*. Retrieved April 10, 2013, from http://waltbeattie.files.wordpress.com/2010/08/evolution-of-as-beattie.pdf.

Grant, Casey C. (2005). The birth of NFPA. *NFPA Journal*. Retrieved from http://www.nfpa.org/displaycontent.asp?categoryid=500.

Hague, David R. (2008). Standpipe and hose systems. In *Fire protection handbook* (20th ed., Volume II). Quincy, MA: National Fire Protection Association.

Huggins, Roland. (2008). Automatic sprinkler systems. In *Fire protection handbook* (20th ed., Volume II). Quincy, MA: National Fire Protection Association.

Jensen, John D. (2008). Water distribution. In *Fire protection handbook* (20th ed., Volume II). Quincy, MA: National Fire Protection Association.

Kim, A. K., and Crampton, G. P. (2000). *A new compressed-air-foam technology*. National Research Council of Canada, Halon Options Technical Working Conference, Albuquerque, NM. Washington, DC: National Institute of Standards and Technology.

Kroll, Karen. (2007, May). Fighting fires in data centers. *Facilitiesnet*. Retrieved January 12, 2013, from http://www.facilitiesnet.com/datacenters/article/Fighting-Fires-in-Data-Centers--6657.

Lake, James D. (2008), Chemical extinguishing agents and application systems. In *Fire protection handbook* (20th ed., Volume II). Quincy, MA: National Fire Protection Association.

Lamm, Willis. (2001). *History of California fire hydrants*. Retrieved from FireHydrant.org website: http://www.firehydrant.org/info/history1.html.

Lamm, Willis. (2001). *Pressurized dry hydrant systems. Retrieved* from FireHydrant.org website: http://www.firehydrant.org/info/dryhsys1.html.

Long, Thomas, Jr., Wu, Neil P., and Blum, Andrew, F. (2010, October 1). Lessons learned from unsatisfactory sprinkler performance: An update on trends and a root cause discussion from the investigating engineer's perspective. *Fire Protection Engineering* Retrieved January 12, 2013 from http://magazine.sfpe.org/sprinklers/lessons-learned-unsatisfactory-sprinkler-performance-update-trends-and-root-cause-discuss.

Scheffey, Joseph L. (2008). Foam extinguishing agents and systems. In *Fire protection handbook* (20th ed., Volume II). Quincy, MA: National Fire Protection Association.

Weldon, Corey. (2004, May). Commercially available clean agents in the U.S.A. *International Fire Protection Magazine*. Retrieved from http://www.ifpmag.com.

Wieder, M., Smith, Carol, and Brakhage, Cynthia. (2005). *Fire detection and suppression systems*. International Fire Service Training Association. Stillwater, OK: Fire Protection Publications, Oklahoma State University.

Wysocki, Thomas J. (2008). Carbon dioxide and application systems. In *Fire protection handbook* (20th ed., Volume II). Quincy, MA: National Fire Protection Association.

Endnotes

1. Wieder, M., Smith, Carol, and Brakhage, Cynthia. (1998). *Private fire protection systems* (2nd ed.). Stillwater, OK: Fire Protection Publications, Oklahoma State University.
2. Van Dyke, Steve. (n.d.). San Francisco Fire Department water supply system. *Virtual Museum of the City of San Francisco*. Retrieved January 12, 2013, from http://www.sfmuseum.org/quake/awss2.html.
3. Wieder et al., op cit.
4. Ibid.
5. Lamb, Willis. (2001). *Pressurized dry hydrants*. Retrieved April 16, 2008, from Pressurized Dry Hydrants website: http://www.firehydrant.org/info/dryhsys1.html.
6. Reproduced with permission from NFPA 291-2013, *Recommended Practice for Fire Flow Testing and Marking of Hydrants,* Copyright 2012, National Fire Protection Association, Quincy, MA. This reprinted material is not the complete and official position of the NFPA on the referenced subject, which is represented only by the standard in its entirety.
7. Ibid.
8. Ibid.
9. Ibid.
10. Shapiro, Jeffrey M. (1997). *Standpipe and hose systems. NFPA* (18th ed.). Quincy, MA: National Fire Protection Association.
11. Huggins, Roland. (2008). Automatic sprinkler systems. In *Fire protection handbook* (20th ed., Volume II). Quincy, MA: National Fire Protection Association.
12. Ibid.
13. Long, Thomas, Jr., Wu, Neil P., and Blum, Andrew F. (2010). Lessons learned from unsatisfactory sprinkler performance: An update on trends and a root cause discussion from the investigating engineer's perspective. *INNOVATION: Victaulic Fire Protection Products and Services*. Retrieved January 12, 2013, from http://magazine.sfpe.org/sprinklers/lessons-learned-unsatisfactory-sprinkler-performance-update-trends-and-root-cause-discuss.

14. Isman, Kenneth E. (2008). Automatic sprinklers. In *Fire protection handbook* (20th ed., Volume II). Quincy, MA: National Fire Protection Association.

15. Wieder et al., op cit.

16. Isman, op cit.

17. Huggins, op cit.

18. Madrzykowski, Daniel, and Fleming, Russell P. (2008). Residential sprinkler systems. In *Fire protection handbook* (20th ed., Volume II). Quincy MA: National Fire Protection Association.

19. Wilson, Dean K. (1997). *Fire alarm systems*. *NFPA* (18th ed.). Quincy, MA: National Fire Protection Association.

20. Janssen, Cory. (2012). Public safety answering point (PSAP). *Techopedia*. Retrieved January 17, 2013, from *Tecopedia* website: http://www.techopedia.com/definition/2969/public-safety-answering-point.

21. Wieder et al., op cit.

22. Black, Art. (2008). Carbon monoxide detection in residential occupancies. In *Fire protection handbook* (20th ed., Volume II). Quincy, MA: National Fire Protection Association.

23. Carbon Monoxide Detectors State Statutes. (2012). National Conference of State Legislatures. Retrieved January 12, 2013, from National Conference of State Legislature website: http://www.ncsl.org/issues-research/env-res/carbon-monoxide-detectors-state-statutes.aspx.

24. DiNenno, Philip J. (1997). *Direct halon replacement agents and systems*. *NFPA* (18th ed.). Quincy, MA: National Fire Protection Association.

25. Kroll, Karen. (2007, May). Fighting fires in data centers. *Facilitiesnet*. Retrieved January 12, 2013, from http://www.facilitiesnet.com/datacenters/article/Fighting-Fires-in-Data-Centers--6657.

26. Makowka, Norb. (2005). *Why carbon dioxide (CO_2) in fire suppression systems?* Retrieved from National Association of Fire Equipment Distributors website: http://www.nafed.org.

27. Beranek, Joe. (2007, June). *An overview of pre-engineered fire suppression systems and industry changes that may affect your property*. Retrieved April 10, 2013, from http://www.nafed.org/Files/Publications/Firewatch/FW_2007June_PreEngineered.pdf

28. Lake, James D. (2008). Chemical extinguishing agents and application systems. In *Fire protection handbook* (20th ed., Volume II). Quincy, MA: National Fire Protection Association.

29. Kim, A. D., and Crampton, G. E. (2000). *A new compressed-air-foam technology*. National Research Council of Canada, Halon Options Technical Working Conference, Albuquerque, NM. Washington, DC: National Institute of Standards and Technology.

10
Organizational Structure and Emergency Incident Management Systems

Courtesy of James D. Richardson.

KEY TERMS

compensation and claims unit, *p. 218*

cost unit, *p. 218*

Federal Emergency Management Agency (FEMA), *p. 218*

fire ground command system (FGC), *p. 216*

FIRESCOPE, *p. 215*

incident command system (ICS), *p. 211*

incident effectiveness, *p. 209*

Incident Management System (IMS), *p. 216*

line operation, *p. 211*

National Incident Management System (NIMS), *p. 215*

National Interagency Incident Management System (NIIMS), *p. 219*

paramilitary organization, *p. 209*

procurement unit, *p. 218*

public safety department, *p. 210*

span of control, *p. 222*

staff operation, *p. 211*

time unit, *p. 218*

unity of command, *p. 222*

OBJECTIVES

After reading this chapter, you should be able to:

- Describe the principles of command used in fire and emergency services.
- Describe a fire department's chain of command.
- Identify the different ranks and their general responsibilities.
- List and describe the four types of fire departments.
- Explain the history of the incident command system.
- Describe the differences among the FIRESCOPE, FGC, IMS, NIIMS, and NIMS ICS structures.
- Identify the components of the incident command structure.
- Describe the areas of responsibility of the ICS components.
- Explain why a national incident management system is needed.

Introduction

To work effectively in a fire department, personnel need to understand how the department is organized. Who reports to whom? Who is responsible for what? How does each member fit in? This chapter describes the different kinds of fire departments, how they are structured, and how structure affects function. We look at the day-to-day operation of the department as well as how it operates in an emergency. We will also trace the development of the Incident Management System (IMS), which culminated in the Department of Homeland Security's National Incident Management System (NIMS).

Fire Department Basics

In a certain sense, a fire department is like any other organization of people working together to achieve a common purpose. A critical difference, however, is the urgency and importance of the fire department's mission. When human lives are on the line, there can be zero tolerance for mistakes and misunderstandings. In an emergency, priorities must be set, objectives established, and an action plan initiated without debate and immediately. Lines of command need to be clear, and communications among all parties must be kept open at all times. To achieve this, fire departments and emergency services typically adopt a structure similar to that of a **paramilitary organization**. This organizational style is hierarchical. Personnel usually wear uniforms and carry insignia denoting their rank. It is very clear who makes the decisions and who gives the orders. Keep in mind, however, the importance of rank does not outweigh the importance of teamwork. When teamwork is compromised, injuries and death take a tragic toll.

paramilitary organization ■ A group of civilians organized in a military fashion.

TYPES OF FIRE DEPARTMENTS

There are five basic types of public fire departments in the United States: volunteer, combination, public safety, full-time paid, and paid on-call. Industrial fire brigades are specialized and are usually private, not public organizations. We will devote a brief discussion to the features that differentiate these department types. All, however, have certain features in common. Almost without exception, they adhere to the one-department model. Standard equipment is supplied by the department to all employees, and all employees follow the same guidelines. All personnel have in common the desire for **incident effectiveness**. This refers to the ability of the department's personnel to perform their functions safely and effectively during emergencies. It demands that physical fitness, skills training, and equipment training are department priorities. Finally, all departments share a commitment to that most important ingredient in the formula for success: team building.

incident effectiveness ■ The ability of the department's personnel to perform their functions safely and effectively during an emergency. This requires physical fitness, training, and equipment.

Volunteer Fire Departments

In October 2012, the NFPA published a report entitled "U.S. Fire Department Profile." The document reported that of the estimated 30,145 fire departments in the United States, 20,200 were all volunteer departments and another 5,530 departments were mostly volunteer.[1]

Volunteer fire departments tend to be small, ranging from 15 to 50 members, and generally serve either small towns or rural areas. Ninety-four percent of volunteer fire departments serve communities with populations of 25,000 or less.[2] There are exceptions to this rule. For example, Pasadena, Texas, has the largest municipal volunteer fire department in the United States. Its 200 members protect 140,000 citizens. In the United States, the members of the first fire departments were all volunteers, and volunteer departments clearly are still a major factor in public safety and a focus of life in the communities they serve. Because the firefighters are not paid, these departments are able to control costs and can use the money they bring in from donations and fund-raising events to purchase equipment and fund training opportunities. The National Fire Academy (NFA) has numerous training programs designed for volunteer firefighters.

Combination Fire Departments

The combination department has some paid employees and some volunteers. The demographics of rural communities are changing. The quaint, once-quiet townships are becoming busy satellite communities to the urban areas they surround. With increased population and business growth, they find themselves unable to provide adequate fire protection with an all-volunteer department. However, they also do not have the adequate revenues to fund a fully paid fire department. Many of these communities have chosen to form combination fire departments. Frequently, in combination departments, the officers, drivers, and apparatus operators are paid, whereas the line firefighters are volunteers. This arrangement obviously results in cost savings. The volunteers in these departments are sometimes referred to as reserves. The reserves receive the same training and experience as the fully paid personnel.

Public Safety Departments

public safety department ▪ A department that employs a person who is triple-certified as a police officer, emergency medical technician (Basic or Advanced level), and firefighter.

In some jurisdictions, a **public safety department** takes the place of separate fire, police, and emergency medical services (EMS) departments. A public safety department officer, often referred to as a *public safety officer* (PSO), must be triple-certified as a police officer, EMT (Basic or Advanced level), and firefighter. Public safety departments are most often found in communities that experience seasonal population fluctuations, such as winter ski and summer lake recreational areas. Public safety departments save money in personnel costs, but they also have their disadvantages. The same employee can handle all three types of calls, but he or she cannot be expected to have the same level of skills as the professional who specializes in only one of these highly demanding jobs.

Fully Paid Fire Departments

Fully paid (career) fire departments are the rule in large metropolitan areas where service demands require full-time professionals. The jurisdiction normally has more control over these employees than in other kinds of fire departments. Expert management is necessary, and prerequisites for employment at all levels tend to be high. Generally, firefighting experience is not enough. College hours, advanced certifications, and even officer experience or advanced degrees may be required.

Paid On-Call

On-call fire departments share characteristics of both volunteer and fully paid departments. Like volunteer departments, personnel do not stay at the fire station. When an emergency arises, they are contacted by pager, mobile telephone, or other communication devices. On-call personnel then respond to the fire station, or the scene of the emergency. On-call firefighters are compensated either hourly or for each response. The department shares characteristics with a fully paid career fire department because most of the funding comes from the local community being protected.[3]

Specialized Fire Brigades

A specialized fire brigade is a private fire department organized by a business, usually an industrial concern, such as an oil refinery or a manufacturing plant. These fire brigades consist of personnel hired and paid by the company. Most fire brigade personnel have regular job-related assigned duties and only respond in case of an emergency. Unlike public fire service organizations, fire brigade personnel are not required to be trained and/or certified by the state. They generally receive training on the work site and focus on the particular hazards and operations of that work area. Employers must comply with OSHA regulation 29 CFR 1910.156, which requires fire service education and training for all fire brigade personnel. Some private organizations choose to have their fire brigade personnel trained by local public fire training organizations through contract courses.

Rank Structure and Organization

RANK STRUCTURE

To further break down these fire department units, a rank structure has been established. Each rank has assigned duties and areas of responsibility. In an emergency, their roles are coordinated by an overarching organizational structure referred to as the **incident command system (ICS)**. Incident command systems will be described later in this chapter. For now, let's look at the functions of the different ranks, from line firefighter up to fire chief.

incident command system (ICS) ■ A blueprint for organizing the response to an emergency situation.

Firefighter

The firefighter is responsible for performing all physical tasks associated with emergency scene operations. This is the entry-level fire service position. Some fire departments have a probationary firefighter rank for new members. Probationary firefighters must achieve certain levels of proficiency during a designated period that ranges from three months to two years. On successful completion of the probationary period, they are granted the rank of firefighter.

Driver

A fire company's driver is responsible for delivering personnel and equipment to the emergency scene and returning them safely to the fire station. He must be proficient in apparatus operations, such as operating pumps and aerial equipment. The driver is also responsible for ensuring the operational readiness of the apparatus, for inventory control, and for keeping track of tools and equipment taken off apparatus during emergency operations. The driver may be referred to as fire apparatus operator (FAO), engineer operator, driver/operator, driver/engineer, or chauffeur. In some fire departments, this position is a promotional rank from firefighter. In other departments, being a driver is considered part of the firefighter's normal duties.

Lieutenant

The lieutenant is the lowest of the supervisory ranks. He or she is usually in charge of an engine or rescue company, although specific duties may vary based on departmental policy. If there are no higher ranking officers at the station, the lieutenant would be assigned as station or shift officer, assuming all duties required for the station's operation, including personnel management, discipline, supplies, and so on. If a lieutenant is the first officer arriving at the emergency scene, the lieutenant assumes command and implements the tactics and strategies deemed necessary to conduct a successful operation. The lieutenant will remain in command until relieved by an officer of higher rank.

Captain

The captain is the second level in the supervisory hierarchy. When a captain is assigned to a fire station, he is considered to be the station or shift officer. The captain is responsible for supervising all officers and firefighters under his or her command during emergency and nonemergency operations. Captains must also be experienced and knowledgeable in tactical and strategic planning. They, like lieutenants, assume command of an emergency scene until a higher ranking officer arrives.

Battalion or District Chief

The position of battalion or district chief may be in either the **line operation** or **staff operation** of a fire service department. The line battalion or district chief supervises multiple fire stations and their assigned companies. Line battalion or district chiefs are responsible for ensuring that the personnel under their command are adequately trained and equipped to respond to any emergency. Staff battalion or district chiefs are strictly administrative officers, who concentrate on planning, logistics, and finance.

line operation ■ Those activities that involve the direct physical tasks necessary for the completion of the mission.

staff operation ■ The administrative activities necessary to accomplish day-to-day operations.

Deputy or Assistant Chief

In larger departments, the rank of deputy or assistant chief is predominantly a staff position. However, some fire departments have line deputy chiefs, who supervise several battalions or districts and respond to large, multiple-alarm fires. The deputy chief has the discretion to assume command from the battalion chief or simply to observe the operation.[4] The organizational structures of fire departments vary. In some geographic areas, the deputy chief and assistant chief are separate ranks. In some fire departments, the deputy chief is the line chief over the districts (battalions) and district chiefs (battalion chiefs). The assistant chief is one rank below the fire chief. In other departments, the assistant chief supervises the battalions and battalion chiefs, and the deputy chief is subordinate to the fire chief. Individuals seeking a profession in the fire service should become acquainted with the organizational structures of the departments in their geographic area.

Fire Chief

The fire chief is the top administrator of the department. In paid, combination, or public service departments, the position is appointed by the mayor or city manager and approved by the city council. In most volunteer departments, all officers, including the fire chief, are elected by the department's members. The fire chief is responsible for all aspects of the department's operations. Duties include, but are not limited to, establishing an annual budget, ensuring that standard operating procedures are reviewed and updated, maintaining labor relations, overseeing disciplinary proceedings, and providing the resources that the department needs to operate effectively and efficiently. In some communities, the fire chief is also assigned the role of emergency manager and is responsible for disaster planning and operations.

ORGANIZATION

Most fire and emergency service organizations use a scalar military-type organizational structure. It is scalar because there is a distinct line of authority or chain of command that moves from the top downward to subordinate sections. The scalar organization is usually divided into two functional areas referred to as line and staff. In private industry, line and staff are considered two separate hierarchies. In the emergency services, and especially the fire service, the line and staff organizations have two separate functions, but one hierarchical structure.

The basic fire service organizational chart has the fire chief at the top. The organization is divided into line and staff functional areas of operations, support (logistics), prevention (planning), and finance (see Figure 10-1).

Line

The line organization is responsible for all day-to-day activities that deal with emergency response. The line organization in the fire service begins with the fire company.

Fire Company

The fire company is the smallest unit in the organizational structure. It consists of a fire apparatus and crew and is named for the type of apparatus, such as engine company, ladder company, or rescue company. The company is usually composed of four personnel: an officer with a rank of lieutenant or captain, an engineer/operator, and two firefighters. This unit is responsible for responding to emergencies and performing initial operations as directed by the incident commander. A fire station may house several fire companies, such as an engine company and a ladder company. These companies respond together in an emergency. In areas where fire departments are small, companies from several departments may join forces to respond to a multiple-alarm incident.

FIGURE 10-1 The basic fire department structure. *Courtesy of James D. Richardson.*

Battalion or District

The battalion or district organizational unit is found in large cities and communities that have many fire stations. This type of unit is usually comprised of multiple fire stations and is supervised by a battalion or district chief. Line battalion chiefs are responsible for personnel administration in their respective district and assume the role of incident commander at emergency scenes within the battalion's area of responsibility.[5]

Staff

The staff organization is responsible for day-to-day nonemergency administrative activities. This would include the duties of the fire chief, assistant chief, training, planning, logistics, finance, human resources, and so on. The function of the staff organization is to support the line organization. The two distinct functions of line and staff are combined into one hierarchical organization with the fire chief at the top. The assignments of line and staff duties rely on the fire departments' organizational structure. This organization is based on a concept known as the principles of command. The five principles are guidelines for use during every day, nonemergency operations and emergency response incidents. They do not act independently of each other, but interact closely to form an effective command structure.

Basic Principles of Command

Unity of Command: Everyone must know whom to report to and who is to report to them. This is known as the one boss principle. Personnel are more effective and efficient when they are supervised by only one person. Multiple supervisors can create conflicting orders that may result in mission failure or greatly reduced effectiveness.

Chain of Command: Chain of command places everyone in the organization into an unbroken line of authority that defines who reports to whom. This principle is related very closely to unity of command and identifies the reporting levels in the organization.

Span of Control: There is a limit to the number of people that a person can be expected to effectively supervise. This number ranges from three to seven personnel, depending on the type of work being performed. Five personnel are considered the ideal number to be supervised by one person. The formula can be adapted to accommodate the supervisory requirements on an emergency scene. The operations section officer can effectively supervise three to seven branches. The branch officer can effectively supervise three to seven groups. The span of control is maintained because each officer is only communicating with three to seven subordinate officers.

Division of Labor: The emergency services organizational structure clearly defines a division of labor. That division is the line and staff functions of the organization. This is more evident in large departments than in smaller organizations that may require personnel to perform line and staff functions. Although they have different responsibilities, they are all interdependent of each other. Division of labor is based on the goals and objectives of the organization. The goals and objectives of the line section in based on emergency response. This would be under the operations section of the organizational chart. Line operations include nonemergency functions, such as cleaning and maintaining apparatus and living facilities, in-service training, and any other activities that would enhance the line personnel's ability to respond to an incident. The goals and objectives of the staff section are to provide for the needs of the line section. Logistics, planning, and finance are divisions of labor under the staff function.

Delegation of Authority: To successfully accomplish an objective, a supervisor may be required to delegate authority to a lower ranking individual. A fire chief cannot effectively supervise all of the day-to-day operations of a fire department and must rely on subordinate chief officers to accomplish the organizational objectives. The power to delegate authority allows the fire chief to step back and maintain an overview of the entire operation by communicating with those who have been delegated decision making authority. An incident commander (IC) has the ultimate responsibility of managing an incident and accomplishing all of the objectives needed to achieve a positive conclusion of the event. But he or she cannot complete the objectives alone. For example, during a disaster, rescue and treatment of victims must be accomplished. Fire suppression operations must be undertaken and resource management has to be organized. The IC would divide the operations section of the organizational chart into branches. Each branch would have an area of responsibility, such as EMS, fire, and rescue. An officer would be placed in command of each branch and would be given the authority to make operational decisions and manage his or her assigned areas. These branches could be divided into groups. Each group would have assigned duties, such as patient triage and transportation, search and rescue, or fire suppression. Group officers would be assigned to supervise the separate job junctions and would have decision-making authority. Each group officer reports to his or her branch officer. The branch officers report to the operations section officer and the operation section officer reports to the IC. The outcomes resulting from the decisions made by all of the officers will ultimately reflect on the IC.[6]

The History of The Incident Command Systems

An incident command system (ICS) is a kind of blueprint for organizing the response to an emergency. It takes into consideration all aspects of the response and attempts to integrate them into one smoothly operating, efficient, and effective whole. Facilities, equipment, personnel, communications, and standard operating procedures are all part of this well-oiled machine. Over the years, ICS development has evolved until it has become a field of study in its own right.

We will look first at the events that led to the development of the ICS. Then we will consider the later stages of development, which culminated in the **National Incident Management System (NIMS)**, a federally mandated program that is crucial to our nation's homeland security strategy. We will also consider aspects of ICS theory, its development, and its effect on firefighting incident management.

National Incident Management System (NIMS) ■ A reconstruction of NIIMS to accommodate additional needs arising from the nature of the new threat—terrorism—that the nation now faces.

FIRESCOPE

The concept of an ICS was born in the early 1970s in the wake of a disastrous wildfire in California that resulted in the loss of 16 lives and 700 structures. The major problems identified by postfire analysis were ineffective communications, lack of a common command structure, lack of accountability, and inability to coordinate available resources. Clearly, better and more centralized leadership was needed during large-scale wildfire operations.

Firefighting Resources of California Organized for Potential Emergencies (FIRESCOPE) was designed to answer this need. Under the **FIRESCOPE** model, a single IC would direct and coordinate multiple agency responses during wildland firefighting operations. FIRESCOPE pinpointed four requirements for the success of such a plan:

FIRESCOPE ■ Firefighting Resources of California Organized for Potential Emergencies.

- The organizational structure must be flexible to adapt to any size or type of incident.
- It must be able to be used for routine events as well as major incidents.
- The ICS must be standardized to allow the addition of personnel from different agencies and geographic locations into the operation without disruption to the organizational structure.
- The system must be economically feasible.[7]

This system was soon found effective in an all-hazard environment. FIRESCOPE's flexibility allowed it to be utilized in any emergency, regardless of the size and complexity of the operation or hazardous situation. FIRESCOPE established common terminology and procedures, defined job responsibilities, and developed an organizational structure. Although the system was created for wildland firefighting, it was applied by many fire departments throughout southern California for structure fires and other emergencies.

In 1982, the FIRESCOPE incident command system was revised and renamed the National Interagency Incident Management System (NIIMS) and adopted by the NFA for its incident command training programs.

One of the most enduring features of the FIRESCOPE ICS was its division of incident management into five areas of responsibility: command, operations, planning, logistics, and finance/administration. With only one addition (information and intelligence), these remain the basic command functions in today's ICSs. We will look at this command structure later in this chapter.

FIRE GROUND COMMAND SYSTEM

During the same time that FIRESCOPE was being developed, Chief Alan Brunacini of the Phoenix Fire Department was working on the **fire ground command (FGC) system**. FGC focused on urban fire issues: structural firefighting, hazardous materials emergencies, and mass casualty incidents. It was a simpler organizational model than FIRESCOPE, but it had the same central goal: allowing one incident commander to maintain control of fire ground operations. FGC was adopted by the NFPA as its model for fire ground organization and control.

FIRESCOPE VERSUS THE FGC

FIRESCOPE, adopted by the NFA, and the FGC, adopted by the NFPA, had much in common, but there also were differences. The most critical of these concerned the command structure. The FGC model was less formal and less rigid than that of FIRESCOPE. These differences created problems when jurisdictions working together on an incident were using different models. A common ICS was needed that would be accepted—and used—by all emergency responders. The pros and cons of the two systems were debated throughout the 1980s. In 1982, the Incident Management System was introduced for national acceptance and was widely adopted.

The Incident Management System

The **Incident Management System (IMS)** merged FIRESCOPE'S organizational design and command structure with the tactical and procedural components of the FGC. A set of operational protocols was established for structural firefighting, emergency medical incidents, structural collapses, hazardous materials emergencies, wildland interface fires (wildland fires that spread into urban areas), high-rise fires, and highway incidents. Central to the effectiveness of the system were three basic features:

- It established a common vocabulary to ensure effective lines of communication.
- It established clear lines of authority and span-of-control standards that would keep supervisors from being overwhelmed. A line officer could be in charge of no more than three to seven subordinates, and a line supervisor could command three to seven companies.
- As mentioned previously, the IMS adopted FIRESCOPE's command structure, clearly outlining five functional areas. This command structure is the bedrock on which all fire and rescue incident management systems rest.[8]

FUNCTIONAL AREAS OF THE IMS

The IMS protocol for managing any incident calls for responsibility to be divided into five functional areas: command, operations, planning, logistics, and administration/finance.

Command

Command consists of the IC and his or her command staff. The commander manages the incident at the strategic level and has the authority to control resources, establish functional areas, and ensure the safety of all responding personnel. Command is an executive function and should remain an executive function, no matter how small the incident may be. The commander is responsible for organizing the incident response to ensure that the mission is accomplished successfully. This includes establishing incident control objectives and work priorities, approving action plan development and resource allocation, coordinating with interagency and public officials, and approving all information given to the media.[9]

The command staff consists of several positions. A public information officer (PIO) is responsible for providing accurate information to the news media, outside agencies, and individuals who have been affected by the incident. The PIO is also responsible for providing information updates to the incident commander, incoming crews, and other organizations that may be involved in the operation.[10] A safety officer is responsible for the safety of all responding personnel. The safety officer has the authority to stop operations if responders are exposed to an immediate danger. The liaison officer maintains communication and contact among responding agencies.[11] The liaison officer also meets and greets any individual or agency representative reporting to the command post; gives directions to individuals and agencies; forwards messages from the commander; and assigns outside agencies to duties that will best meet the needs of the situation.[12]

Operations

The purpose of operations is to ensure that tactical assignments are carried out and to accomplish the strategic goals established by the incident commander. This functional area can become very complex, especially if the incident is large or multifaceted. The operations officer may be responsible for supervising simultaneous branch sections, assigned to operations during the incident, EMS, hazardous materials, search and rescue, recovery, and so on. Each branch officer reports directly to the operations officer, who reports directly to the IC. The operations officer usually comes from the agency or organization with the most jurisdictional involvement in the incident.[13]

Planning

Planning is responsible for evaluating incident information and developing the incident action plan. Its function is to provide the incident commander with up-to-date status reports on resource requirements and allocation, situation development, incident control, responder injuries, intelligence, and suggestions for stabilizing the incident.[14] It consists of four units: resources, situation, demobilization, and documentation. Technical specialists are utilized to interpret data, evaluate situational development, and forecast resource needs. Table 10-1 defines typical planning responsibilities.

TABLE 10-1	Planning Responsibilities
Incident Planning	• Works with command and operations to update the strategic plan
	• Assesses previous actions and strategies that have been used
	• Makes adjustments to current and future plans and makes recommendations to command and operations
	• Predicts the outcomes of the plans
	• Estimates and evaluates possible resource needs with operations
	• Reviews projected tasks established by command and operations
Resource Assessment	• Is responsible for personnel
	• Is accountable for responding units and assignments
	• Keeps track of special equipment
Situation Status	• Continually monitors events to maintain an up-to-date status of incident situations
Documentation	• Maintains a master sequence of events timeline and resource requests
	• Monitors operations to maintain span of control
Demobilization	• Develops a plan to dismantle the organizational structure and return all apparatus, equipment, personnel, and other resources to preincident status after the event is over[15]

Logistics

Logistics provides all of the outside resources and materials required for incident operations. This includes facilities (mass care, rehabilitation, family assistance centers, warehouses, and so on), transportation, supplies, equipment maintenance, fuel, food, communications and technological support, and EMS support services. The logistics officer usually assigns these requisition duties to a support and service section, collecting data from them, and then providing information to the incident commander.[16] The specific responsibilities of the logistics officer are defined in the following box.

RESPONSIBILITIES OF LOGISTICS OFFICER

- Locates and provides supplies and equipment
- Inventories all arriving equipment and notifies command of available resources
- Provides all support activities to all units in operations
- Provides critical incident stress debriefings
- Provides adequate and continual communications throughout incident operations[17]

Federal Emergency Management Agency (FEMA) ■ The federal agency responsible for supporting state and local municipalities during a disaster.

compensation and claims unit ■ Part of the finance branch of ICS that documents property loss by citizens during a disaster.

procurement unit ■ Part of the finance branch that establishes contracts for equipment that may be needed during an emergency.

cost unit ■ Part of the finance unit that tracks the cost of equipment and supplies.

time unit ■ Department that tracks all time sheets for every individual working on the emergency scene throughout the duration of the incident.

ADMINISTRATION AND FINANCE

The expense of managing a large, long-term incident can be staggering. For example, the cost of the 1970 California wildfire that led to the creation of the FIRESCOPE project averaged $18 million per day. Today, a large-scale operation can carry an even higher price tag. The finance and administration officer is responsible for documenting the cost of materials and personnel. This is very important, especially if the incident is declared a national disaster. If a national disaster declaration is issued, the **Federal Emergency Management Agency (FEMA)** will reimburse up to 75% of the incident's operational cost to the community. This reimbursement will occur only if all of the operational expenses are properly documented and verified.

Finance and administration includes four subunits. The **compensation and claims unit** ensures compensation documentation is completed correctly, maintains records on worker injuries and illnesses, and investigates damaged property claims. The **procurement unit** contracts for equipment that may be needed during the operation. This is usually accomplished during the development of the strategic plan prior to the incident. The **cost unit** keeps a record of expenses for equipment and supplies used during the incident. It provides cost estimates and suggests cost saving measures. The **time unit** keeps time sheets for all workers (including volunteers) employed throughout the operation (see Figure 10-2).[18]

FIGURE 10-2 The basic command structure is divided into five areas of responsibility. *Courtesy of the U.S. Department of Homeland Security.*

STOP AND THINK 10-1

You are assigned as officer for finance and administration during a major disaster in your community. The IC has informed you that there will be a thorough audit after the incident to ensure accountability.

1. What do you do to ensure that all expenditures are properly documented?
2. Why do these records have to be maintained so precisely?
3. What repercussions would result if the financial records were not kept?

NATIONAL INTERAGENCY INCIDENT MANAGEMENT SYSTEM

The **National Interagency Incident Management System (NIIMS)**, introduced in 1982, expanded on the IMS concept. It was a total systems approach to risk management that integrated all levels of national emergency incident planning—federal, state, and local. The NIIMS established five subsectors, each of which would tackle an area where standardization was vital. The goal was to develop and implement national standards in these five areas:

- The incident command system
- Training
- Qualifications and certification systems
- Publications
- Support technologies (satellite imagery, communications, geographic information systems [GIS], and so on)[19]

National Interagency Incident Management System (NIIMS) ■ A total systems approach to risk management that integrates all levels of national emergency incident planning—federal, state, and local.

INCIDENT COMMAND SYSTEMS—A REPORT CARD

In February 2000, Assistant Chief Dana Cole of the California Department of Forestry and Fire Protection submitted a research paper to the NFA. In his paper, entitled "The Incident Command System: A 25-Year Evaluation by California Practitioners," Chief Cole listed 16 attributes of the ICS. Former and present command and general staff officers of 17 state and federal major incident management teams in California were asked to rate the attributes based on their experience of how effective they were in actual practice. Based on these ratings, Chief Cole divided the attributes into three tiers, as follows:

Tier One.   These are the most effective attributes, which is not surprising because they are predetermined. The participants know the system and understand the rules that must be followed. Everyone has received training on how the system works. These are as follows:

- Predefined hierarchy, including chain of command and delineated responsibilities for every position
- Uniform terminology for identifying resources and organizational functions
- Modular organizational structure that is expanded and contracted as needed
- Incident action plans that are updated for each operational period
- Manageable span of control

Tier Two.   The next six attributes were rated lower. They contribute to the effectiveness of ICS but create some complications as well. These are as follows:

- Standardized forms used for all incidents
- Ample flexibility and authority given to staff for accomplishing objectives

- Cross-jurisdictional and cross-functional working relationships when ICS is used
- Communications plan that is coordinated among responding agencies
- Clear decision-making process
- Process for transitioning command authority from one level of government to another as incident complexity changes

Tier Three. The last five attributes were rated lowest by respondents. Although still perceived positively, they were also viewed as potential problem areas. These are as follows:

- Resource mobilization effectiveness
- Effectiveness of integrating nonfire government agencies into the ICS structure
- Consistency of implementation among various agencies
- Effectiveness of integrating nongovernment organizations into the ICS structure
- Agreement among agencies about who has authority to modify the ICS "rules of the game"[20]

Chief Cole emphasized that strengthening these last five attributes should be the focus of efforts to develop national incident command systems further. The effectiveness of ICS when implemented during large-scale operations could not be denied, but work was still to be done.

THE NATIONAL INCIDENT MANAGEMENT SYSTEM

If there was ever any question about the need for a nationwide IMS, the disastrous events of September 11, 2001, removed all doubt. Emergency operations demand multiagency or multijurisdictional responses, common terminology, and common communications. A system needed to be developed and implemented with the full weight of law to back it. This task was given to the newly created Department of Homeland Security.

Fortunately, as we have seen, much of the framework for such a system was already in place. NIIMS was restructured to accommodate additional needs arising from the nature of the new threat—terrorism—that the nation now faced. The new system was christened the National Incident Management System (NIMS). On February 28, 2003, President George W. Bush issued Homeland Security Presidential Directive (HSPD) 5, "Management of Domestic Incidents." This directive established NIMS as the official and universal "language" to be utilized by all local, state, tribal, and federal governments for emergency response.[21] On March 1, 2004, Secretary of Homeland Security Tom Ridge mandated that all agencies must adopt NIMS in order to receive federal preparedness funds.

The organizational structure of NIMS is comprised of five components[22]:

- *Preparedness:* This is an ongoing activity that is done prior to any emergency. It combines assessment, planning, procedures and protocols, training and exercise, personnel qualification, licensure, and certification, equipment certification, and evaluation and revision.
- *Communications and Information Management:* This component emphasizes the necessity for a standardized communication framework and common operating procedure based on the following:

Interoperability—ability to communicate within and across agencies and jurisdictions
Reliability—able to function in any kind of emergency
Scalability—functional use in both small and large incidents
Portability—standardized technologies, protocols, and frequencies
Resiliency—can perform when infrastructure is damaged or lost
Redundancy—have an alternative (back-up) system if primary system malfunctions

- *Resource Management:* Establishes the process for not only identifying resource requirements, but also identifies the process to order and acquire, mobilize, track, report, recover, demobilize, reimburse, and inventory resources.
- *Command and Management:* Provides a flexible, standardized incident management structure based on the ICS, multiagency coordination system, and public information.
- *Management and Maintenance:* HSPD-5 requires the establishment of a mechanism to ensure the ongoing management and maintenance of NIMS. The National Integration Center (NIC) manages administration, implementation, and compliance of NIMS. It is also responsible for standards and credentialing for emergency personnel, training and exercise support, and publication management.

PRINCIPLES OF NIMS

As we review the principles behind NIMS, it will become clear how much this system owes to earlier incident command systems.

Management by Objectives

The IC establishes specific, measurable objectives for each functional area (section). The section officers communicate progress reports to the IC as the objectives are met. All measurable performance is documented to correct deficiencies.

Reliance on an Incident Action Plan

The incident action plan (strategic action plan) guides the objectives for both operational and support activities. It is the foundation that supports tactical decisions and operations. The incident action plan is not written in stone. It allows changes to be implemented as the situation changes on the incident scene.

Manageable Span of Control

An incident management officer can supervise three to seven subordinates effectively. It has been suggested that five subordinates are the ideal number to maintain a manageable span of control.

Predesignated Incident Mobilization Center Locations and Facilities

Certain types of incidents require specialized support facilities. The requirements of these facilities should be designated in the incident action plan. The IC is responsible for the identification and location of these facilities, based on the situation. Examples of predesignated facilities are the incident command post, rehabilitation area, apparatus staging area, hazardous materials decontamination area, and so on.

Comprehensive Resource Management

Accurate and continual accountability must be maintained for all personnel, equipment, and resources. Resource management is responsible for organizing, requesting, delivering, tracking, and recovering all resources used during the emergency. This should include paid and volunteer personnel, equipment that has been contracted for use, and volunteer organizations.[23]

Integrated Communication

Interoperable communications and a common communications plan are an absolute necessity to maintain operational continuity during multiagency or multijurisdiction response incidents. This is accomplished by developing an emergency operations plan, with cooperation from all of the agencies that may participate during an emergency. Field exercises focusing on interagency communications can be conducted to determine weaknesses in the plan.

Establishment and Transfer of Command

Development of the incident command structure during strategic planning must include an established procedure for the transfer of command. If command is transferred, the new commander must be thoroughly briefed on past and present actions for the command process to continue safely and effectively.

Chain of Command and Unity of Command

The purpose of the incident command structure is to establish a line of authority to ensure accurate and efficient communications and information transfer. All participants must have designated supervisors to whom they report. These supervisors, in turn, are to report to their supervisor or directly to the IC.

Unified Command

Incidents involving multijurisdictions or multiagency response may require implementation of a unified command. This type of command allows different agencies to work together without compromising individual agency authority, responsibility, or accountability. The unified command structure is dependent on the incident's location and type. For example, if a military aircraft crashes into a civilian residential subdivision while attempting to land at a local air force base, the military fire department and the civilian fire department would respond to the scene. The military IC would assume command for all military operations—aircraft fire suppression, ordnance and possible secret technology security, and so on. The civilian fire department IC would assume command of all civilian activities—fire suppression, rescue, and so on. The two commanders would work together in a unified command structure to coordinate the two operations and successfully accomplish the mission.

Accountability of Resources and Personnel Deployment

Each level of the incident command structure requires implementation of an accountability system. All responding personnel must report in and be given a specific work assignment. Once assigned, they will report to only one supervisor (**unity of command**), and that supervisor will be assigned a limited number of subordinates to supervise (**span of control**). The arrival and distribution of resources must be recorded, and updated status reports must be provided to the IC.

unity of command ■ The concept that every member of the organization answers to just one supervisor.

span of control ■ The number of units or personnel that a supervisor can effectively manage.

Information and Intelligence Management

The effective and efficient management of an incident requires the gathering of accurate and timely information and intelligence. When the information is received, it must be shared with the appropriate agencies or authority. Depending on the size and complexity of the incident, information and intelligence can be added as a functional area of ICS, or it can be part of the command/staff function. The purpose of information and intelligence is to provide the IC with any information that may influence the outcome of the incident. This would include, but not limited to, national security, classified information, weather, environmental changes due to hazardous materials release, and risk assessment. Information and intelligence analyzes and then shares the information with others on a need-to-know basis (see Figure 10-3).[24]

All emergency personnel should have a thorough understanding of ICS and NIMS, regardless of rank. NIMS and ICS independent study courses are offered by the Emergency Management Institute and the NFA at the National Emergency Training Center in Emmitsburg, Maryland. The courses are offered online and free of charge. All emergency personnel should be required to complete these basic NIMS and ICS courses:

ICS 100 Introduction to ICS for Operational First Responders
ICS 200 Basic ICS for Operational First Responders
ICS 300 Intermediate All-Hazard NIMS ICS Review for Expanding Incidents
ICS 400 Fundamental Review for Command and General Staff
IS 700.a National Incident Management System (NIMS), an Introduction
IS 800.b National Response Framework, an Introduction

Knowledge is strength. The more knowledge that members of an organization possess, the stronger the organization becomes.

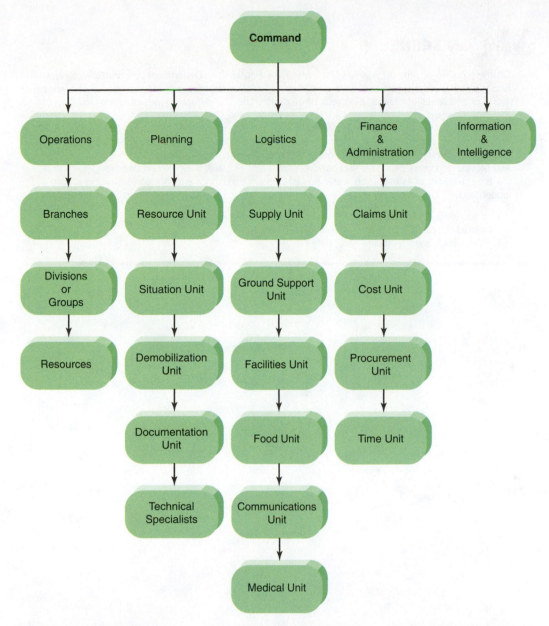

FIGURE 10-3 The basic NIMS command structure is divided into six areas of responsibility. *Courtesy of the U.S. Department of Homeland Security.*

STOP AND THINK 10-2

Consider the worst-case scenario in your area of response.

1. If you were responsible for developing the command structure, what part of NIMS would you consider using?
2. How will NIMS help you in this scenario?

ON SCENE

On the morning of June 24, 1995, companies of the Houston Fire Department responded to a warehouse complex on the east side of the city. The 175,000-square-foot burning warehouse contained 1,000-pound boxes of plastics, hundreds of oil drums, and 250 different types of chemicals. One of the exposed warehouses contained hydrazine, pesticides, and organic peroxides. A fully loaded liquid petroleum gas tank car, in close proximity to the burning warehouse, was parked on the railroad track separating a residential subdivision from the warehouse complex. The only access road to the complex and the initial water supply were inadequate. Additional alarms were requested to relay water supplies and provide additional personnel and handlines. By 12:30 PM, seven alarms had been requested. During the operation, oil-laden runoff water flowed across a parking lot and ignited, destroying an aerial ladder truck.

1. If you were incident commander, what parts of NIMS would you utilize?
2. Would this be an interagency operation?
3. If so, which agencies would participate?

Summary

The need for an incident command system (ICS) was inspired by forward-thinking emergency response officers who had witnessed the tragic consequences of chaotic large-scale operations. The evolution of ICS was slow and difficult. FIRESCOPE's ICS was developed for the management of large wildland fires; FGC was a less formal ICS that focused on structure fires, hazardous materials, and mass casualty incidents. Each concept was supported by different national organizations: FIRESCOPE by the National Fire Academy, and FGC by the National Fire Protection Association. It soon became evident that a national incident management system could not be developed without the full cooperation of all agencies responsible for public protection. After years of negotiations, all of the participants were able to resolve their differences and create the incident management system (IMS) that became the foundation for NIIMS. After September 11, 2001, components of NIIMS were modified to create NIMS.

NIMS has now been mandated by the Department of Homeland Security to be used for all incidents by all agencies receiving federal grant funding from the Department of Homeland Security. Of course, funding should not be the only reason for using NIMS. NIMS provides for unity of command, span of control, effective communications, and a whole host of other benefits. It is the best possible approach for emergency workers to bring their mission to a successful completion in a timely, cost-effective manner. For the IC, the chief value of NIMS is not the time or money savings, however. It is the flexibility and control that NIMS provides to him or her to protect the lives of emergency responders and civilians.

Review Questions

1. Create a simple fire department organizational chart with at least five levels of rank represented.
2. Give an example of division of labor in the structure of the fire department.
3. Why was the incident command system developed?
4. What is the main difference among FIRESCOPE, ICS, and FGC?
5. What are the main components of ICS?
6. What are the areas of responsibility of the ICS components?
7. Why is a national incident management system needed?

Suggested Reading

Cole, Dana. (2007). The incident command system: A 25-year evaluation by California practitioners. National Fire Academy. Retrieved April 23, 2006, from http://www.usfa.fema.gov/pdf/efop/efo31023.pdf.

Coleman, John. (1997). *Incident management for the street-smart fire officer*. Saddle Brook, NJ: PennWell Publishing Company.

Federal Emergency Management Agency (FEMA). (n.d.). Crosswalk/comparison of NIIMS/NIMS. NIMS. Retrieved April 23, 2006, from http://www.fema.gov/txt/nims/NIIMS-NIMSComparison.txt.

Federal Emergency Management Agency (FEMA). (2004). *NIMS and the incident command system*. NIMS. Retrieved February 11, 2006, from http://www.fema.gov/txt/nims/nims_ics_position_paper.txt.

Wurtz, Thomas. (2004). *Firefighter's handbook: Essentials of firefighting and emergency response*. (2nd ed.). Clifton Park, NY: Delmar Publishing.

Endnotes

1. Darter, Michael J., Jr. and Stein, Gary P. (2012, October). *U.S. Fire Department profile*. Retrieved January 19, 2013, from National Fire Protection Association website: http://www.nfpa.org/itemDetail.asp?categoryID=2486&itemID=55953&URL=Research/Statistical%20reports/Fire%20service%20statistics/.

2. Ibid.

3. Powell, Pam. (2011). Fire and emergency services as a career. In *Fire and Emergency Services Orientation and Terminology* (5th ed.). Stillwater, OK: International Fire Service Training Association, Fire Protection Publications, Oklahoma State University.

4. Firefly's Fire Service Information. (2011). *A little fire service history*. Retrieved April 10, 2013, from http://www.fireserviceinfo.com/history.html.

5. Morden, Tony. (2003). Principles *of management* (2nd ed.). Farnham, Surrey, UK: Ashgate Publishing.

6. Emergency Management Institute. (2005). *ICS-100 ICS overview*. Retrieved April 11, 2013, from FEMA website: http://training.fema.gov/EMIweb/IS/ICS100CR/ICS100SM/02ICS100ICSOverSM_NMSept05.pdf.

7. Coleman, John. (1997). *Incident management for the street-smart fire officer*. Saddle Brook, NJ: PennWell Publishing Company.

8. Emergency Management Institute. (n.d.). Lesson 3. *IS-100.b., Introduction to the Incident Command System*. Retrieved April 11, 2013, from FEMA website: http://emilms.fema.gov/IS100b/index.htm.

9. Coleman, op cit.

10. Wurtz, Thomas. (2004). *Firefighter's handbook: Essentials of firefighting and emergency response* (2nd ed.). Clifton Park, NY: Delmar Publishing.

11. Coleman, op cit.

12. Ibid.

13. Federal Emergency Management Agency. (2008). *NIMS* (p. 52). Retrieved April 11, 2013, from http://www.fema.gov/pdf/emergency/nims/NIMS_core.pdf.

14. Wurtz, op cit.

15. Coleman, op cit.

16. Wurtz, op cit.

17. Coleman, op cit.

18. Cole, Dana. (2007). The incident command system: A 25-year evaluation by California practitioners. National Fire Academy. Retrieved April 23, 2006, from http://www.usfa.fema.gov/pdf/efop/efo31023.pdf.

19. Coleman, op cit.

20. Federal Emergency Management Agency. (n.d.). Crosswalk/comparison of NIIMS/NIMS. NIMS. Retrieved April 23, 2006, from http://www.fema.gov/txt/nims/NIIMS-NIMSComparison.txt.

21. Emergency Management Institute. (2013). *Independent Study Course 700a*. Retrieved January 20, 2013, from FEMA website: http://emilms.fema.gov/IS700aNEW/NIMS01summary.htm.

22. Ibid.

23. FEMA, (n.d.), op cit.

24. Ibid.

Courtesy of James D. Richardson.

KEY TERMS

aspect, *p. 236*

barriers, *p. 236*

elevation, *p. 236*

external exposures, *p. 236*

internal exposures, *p. 236*

occupancy, *p. 228*

preincident planning, *p. 228*

shape of the area, *p. 236*

slope, *p. 236*

strategy, *p. 240*

tactics, *p. 240*

ventilation, *p. 242*

OBJECTIVES

After reading this chapter, you should be able to:

- Explain the difference between strategy and tactics.
- Describe the purpose for strategic planning.
- Identify the resources that must be considered in developing a strategic plan.

Introduction

To the untrained eye, firefighting operations may seem chaotic. This perception may be perfectly accurate if firefighters have arrived on the scene with no preplanning. It is a grave and sometimes fatal mistake to neglect the work that must be done before an emergency arises. Strategy and tactics are not developed on the fire ground; these should have been established and documented in writing, weeks, months, or even years prior to the event. This does not mean that strategy and tactics is written in stone. They are based on standard operating procedures (SOPs) developed from information gathered during preincident planning and designed to be flexible to allow the incident commander (IC) freedom to alter the incident action plan to meet the demands presented during the incident.

The terms *strategy* and *tactics* are often used interchangeably; however, they have quite different meanings. Strategy outlines what needs to be done to achieve a positive outcome in a given set of circumstances. Tactics have to do with how a task is accomplished, how personnel are deployed, and how resources are brought to the scene. Although they are very dependent on each other, strategy and tactics are separate functions, and they should be distinguished. The purpose of this chapter is to explain the purpose for preincident planning and introduce the basic principles of fire ground strategy and tactics.

Preincident Planning

preincident planning ■ Assessing risks and prioritizing potential hazards, then developing a strategic plan to mitigate the hazards.

The most accurate way to determine the amount and availability of resources is through **preincident planning**. Preincident planning usually involves an inspection. Identification of fire dangers is not only an important tool in the fire prevention program; it also provides valuable information on resource needs and availability that can be used to develop a strategic action plan. This information enables the first arriving officer to make sound strategic and tactical decisions and accomplish the mission effectively and safely. National Fire Protection Agency (NFPA) 1620 *Recommended Practice for Pre-Incident Planning*, provides preincident planning guidelines for fire departments. It includes the topics deemed important in the development and implementation of a preincident plan. NFPA 1620 requires the plan to provide the following information:

- Type of building construction
- Special elements that the site may have
- The number and type of occupancies
- Available water supply for firefighting
- Extra resources that may be needed for certain types of occupancies (child day care, assisted living, and so on)
- Develop operating instructions to contend with unusual hazards
- Develop procedures to test the plan and make corrections
- Develop standard operating procedures for the preparation, presentation, communication, and maintenance of information that would be of assistance to responding personnel
- Create a process for documenting and disseminating the data collected from the pre-plan inspection[1]

occupancy ■ General term for how a building structure or home will be used and what it will contain.

An effective preincident planning initiative is a sign of a strong departmental public relations program. Fire company personnel go into their areas of response, meet with local business owners, and ask permission to walk through their property. The company inspection helps build a feeling of trust between the fire department and the business community, and allows the company officer and personnel to gain first-hand knowledge of the target area or **occupancy**. The term *occupancy* refers to building use, such as residential, mercantile, and so on. By relating this new information to their past experiences and training, firefighters will have a foundation on which to build a strategic plan and to begin anticipating tactics. Company officers will also be able to make informed decisions about the kinds of training programs that will best prepare their companies for the challenges ahead (see Figure 11-1). The best preincident planning program will fail if it is not shared with all firefighters. Each fire company must provide the information gathered to other fire companies who may respond to that location. In the past, this required the development of hand-drawn floor, plot, and elevation plans and written documents detailing all features of the occupancy and occupant contact information. Today, new

UNIVERSAL CITY FIRE DEPARTMENT INSPECTION REPORT
DATE OF INSPECTION _____
PRE C OF O( ) TEMP-C OF O( ) FINAL ( ) REGULAR ( ) REINSPECTION ( )

BUSINESS NAME: PHONE NUMBER:
ADDRESS:

NOTED DISCREPANCIES ( )   NO DISCREPANCIES NOTED ( )

BUSINESS CONDITIONS
1. Building address not visible from the street. Size needs to be 4" or larger.
2. Outside needs cleaning-High grass, weeds, accumulation of trash. (Trees must have an overhead clearance of 13'6" over roadway).
3. Obstruction access to utilities. Electrical breakers Inside ( ) Outside ( ) Gas shutoffs ( ).

STORAGE AREAS
4. Storage closer than 24' from ceiling.
5. Provide metal containers with metal covers for oily rags.
6. Accumulation of waste or other unusable materials.
7. Improper storage of flammable or combustible chemicals.
8. Water heater closet or equipment room use for storage. No storage in these areas.
9. Compressed gas cylinder not secured. (Must be chained to wall and/or on a stand to prevent falling). Full or empty.
10. Storage closer than 18" from a sprinkler head.

FIRE EXTINGUISHERS (Due inspection once a year)
11. Fire Extinguishers need service. Inspection ( ) Recharge ( ) Last Inspection _____ Due Inspection _____.
12. Fire Extinguisher(s) Missing ( ) Not accessible ( ) Not visible ( ) Inspection Tag Missing.

BUILDING FIRE SPRINKLER SYSTEM (Due inspection once a year)
13. Sprinkler head impaired or obstructed ( ) Sprinkler control value obstructed or closed ( ).
14. Fire Department Connection (FDC) Blocked ( ) Caps Missing ( ) Obstructed with debris ( ).
15. Date of Last Inspection _____. Inspection due Date _____.
16. Hydro new building fire sprinkler system _____. Final fire sprinkler system _____.

FIRE SUPPRESSION SYSTEMS (Due inspection every six months)
17. Fire Suppression system for grills, fryers, etc. Last Inspection _____. Due Inspection _____.
18. Air test new fire suppression systems _____.

FIRE ALARM (Due inspection once a year)
19. Date of last inspection _____. Due inspection _____. Needs Service ( ).

PAINT BOOTH FIRE PROTECTION SYSTEMS (Due inspection every six months)
20. Date of last inspection _____. Due Inspection _____. Air test new paint booth _____.

FIRE LANES
21. Needs painted ( ) Signs missing ( ) Signs need replaced ( ) Lettering needs repainted ( ) Fire Hydrant obstructed ( ).

EXITS
22. Locked ( ) Blocked ( ) Double keyed lock needs changed to a thumb turn inside.
23. Exit not property marked ( ) Missing lighted exit sign ( ) Exit light needs repaired or not working ( ).
24. Need emergency lights ( ) Emergency lights need repaired or not working ( ).

HEATING AND AIR CONDITIONING
25. Heater not property vented ( ) Soot showing in air chamber or vent ( ) Filters not cleaned regularly ( ).
26. Combustibles kept to close to heater ( ) System not serviced regularly ( ).

ELECTRICAL
27. Improper use of extension cords ( ) Cords in poor condition ( ) Overload of outlet ( ).
28 Electrical outlet, junction boxes or panels missing covers. ( ) Broken covers ( ) Breakers not labeled ( ).

KITCHEN
29. Accumulation of grease on hood, ducts, filters, equipment ( ) Hood needs cleaning ( ) Vent Fan not working ( ).
30. Combustibles to close to cooking unit. ( ) Equipment not under hood ( ).

KNOX BOX
31. Need a Knox Box as required by Fire Code ( ) Need new key (lock changed) ( ).

This inspection is intended for your safety and your neighbor's safety. You are required by City Ordinance (Fire Code) to correct the above discrepancies by _____. _____ A re-inspection will be done on or about this date.

_____   _____
Signature of Owner/Manager/Employee   **Fire Official**
                                                                 (210) 659-0333   Ext. 788

U. C. Form No. 405

FIGURE 11-1 The inspection form eliminates oversights during the fire inspection. *Courtesy of the Universal City Fire Department, Universal City, Texas.*

technologies allow this information to be created on software, enabling each member of the department to view occupancy plans and potential hazards on electronic devices, such as laptop computers or i-Pads. Dashboard-mounted computers allow company officers to view an occupancy's characteristics while en route to the location.

In his book *Fireground Strategies*, Deputy Chief Anthony Avillo, who serves with the North Hudson (NJ) Regional Fire and Rescue, identified 13 factors that should be considered during the preincident planning process.[2]

- Construction
- Occupancy
- Apparatus and manpower
- Life hazards
- Water supply
- Auxiliary appliances
- Street conditions
- Weather
- Exposures
- Area and height
- Location and extent
- Time
- Hazardous materials

Two more items should be added to this list.

- Communications
- Command structure

BUILDING CONSTRUCTION

The importance of building construction cannot be overemphasized when developing a strategic plan. Strategic planners must take into account the building's size and configuration, its construction materials, and its interior finishes. They must also consider environmental influences, such as weather and topography, on fire behavior in different types of structures. The terrain on which the structure is built will determine personnel and apparatus needs. Seasonal changes may bring about drastic environmental conditions that could seriously affect firefighting strategies. Response times may be altered, or apparatus and personnel requirements may need revision. Strategic planners must include all of these variables.

Consider this incident: On March 9, 1995, San Francisco firefighters responded to a multistory residential structure fire. The residence was constructed on the side of a steep slope to take advantage of the panoramic view of the landscape. Entry to the structure was limited to the garage area. A fire crew made entry through the garage, which gave access to a living area on the second level of the structure. They were unaware that the main seat of the fire was on the first level below them and that it was being fed by 70-mile-per-hour winds. The intense heat caused the automatic garage door opener to activate and close the door, blocking the firefighters' escape route. One firefighter died, and two were seriously injured.[3] Weather, topography, and building construction added up to a deadly combination.

OCCUPANCY

The occupancy of a structure has a great influence on strategic planning. The term *occupancy* refers to the building's use, not to the number of people in a structure. Occupancy affects decisions involving life-safety and fire extinguishment methods.

Regularly scheduled preincident planning should be conducted to maintain up-to-date information on occupancy changes in a protected area, especially if the area is experiencing demographic changes, such as urban renewal or population growth. An occupancy change

FIGURE 11-2 Changes in occupancy may increase the fire load and hazards during fire operations. *Courtesy of James D. Richardson.*

may undermine the safety of a building by changing load impositions or by increasing fire loads to levels the structure was not designed to withstand (see Figure 11-2).

STOP AND THINK 11-1

In Houston, Fire Station Number One, a four-story structure, was sold and converted into a nightclub, restaurant, and aquarium. How does this change of occupancy change the load impositions on this structure and the occupancy load?

APPARATUS AND PERSONNEL

In 1954, Royer and Nelson, of the Fire Service Institute at Iowa State University, developed a method for determining personnel and apparatus needs based on required flow rates. In a study of 250 large fires during the late 1970s, it was determined that total personnel needs averaged one firefighter for every 50 GPM of required fire flow and one pumping apparatus for every 500 GPM of required flow. If a warehouse had a required flow rate of 3,000 GPM, personnel and equipment requirements would be calculated as follows:

$$\frac{3,000}{50} = 60 \text{ firefighters}$$

$$\frac{3,000}{500} = 6 \text{ pumping apparatus}$$

These figures are based on 100% involvement and would be reduced with less percentage of involvement. For example, if only 30% of the structure was involved, 20 firefighters and 2 pumping apparatus would be needed. For safety reasons, the number of firefighting personnel on the fire ground should not be less that recommended by the NFPA.

In 2010, the NFPA published NFPA 1710 *Standard for Organization and Deployment of Fire Suppression Operations, Emergency Medical Operations, and Special Operations to the Public by Career Fire Departments*. This standard recommends the following for minimum staffing of fire apparatus for an initial full alarm assignment for any structure fire:

1. The first arriving company must have a minimum of 4 personnel and arrive within 4 minutes of the 911 call.
2. A minimum of 15–17 personnel should arrive on the scene within 8 minutes of the 911 call.

This recommendation is for initial response. Additional personnel and apparatus may be needed once the estimated fire flow has been determined.[4] The NFPA 1710 standards are not based on the amount of fire flow but on safety and efficiency. The International City Managers Association (ICMA) conducted a study of fire ground operations. The results of the study indicated the following:

1. A minimum of 4 and often 8 or more firefighters are needed for simultaneous operations of raising ladders, ventilating, searching, and rescuing. Each operation must be under officer supervision.
2. Injuries, increased fire spread, and high dollar loss occurs if 16 firefighters are not on scene within the critical time period.
3. Five-person companies were 100% effective, 4-person companies were 65% effective, and 3-person companies were 38% effective.[5]

Each fire department has its own budgetary, equipment, and staffing problems. Many are not able to meet all of the recommendations suggested by the NFPA and other studies. It is very important that these departments reflect apparatus and personnel resources in their preincident planning.

LIFE-SAFETY HAZARDS

In most cases, preincident planning will identify the life-safety hazards of an occupancy. However, this may not always be the case. The 2010 census reported that 13.0% (40,267,984) of the population of the United States were 65 years of age or older.[6] With escalating demand for elder-care facilities, unlicensed nursing homes and care facilities are becoming more common. Unlicensed nursing homes and elder day-care facilities are often located in residential neighborhoods. When viewed from the street, they appear to be single-family dwellings. In residential areas, preincident planning activities are conducted on request from the homeowner or occupant. Therefore, these illegal businesses can go unnoticed for months or years. Many of the residents of these facilities have limited or no mobility. This creates a major problem for the incident commander and responders, especially when they are not aware of the facility use until an emergency occurs.

WATER SUPPLY

Water is the primary extinguishing agent in the majority of firefighting operations. Most strategic plans begin with water supply and water requirement considerations. This is a concern whether the plan is being developed for an urban area with an established water distribution system or for a rural, wildland area where there are no water distribution systems. In urban areas that have established water distribution systems, the main concern in strategic planning will be to determine if the water system can provide adequate fire flow.

The National Fire Academy, at Emmitsburg, Maryland, developed a formula for estimating the needed fire flow for a structure. It is based on studies of fire flows that successfully controlled a number of fires nationwide. The formula is a "rule-of-thumb" approximation and is based on 100% involvement of the structure.

$$\text{Fire flow} = \frac{\text{Length} \times \text{Width}}{3 \times 1}$$

EXAMPLE

If a single-story structure is 75 feet long and 45 feet wide, the needed fire flow would be

$$\frac{75 \times 45}{3} = 1125 \text{ GPM}$$

Most structure fires only involve one or two rooms, not the entire structure. In these circumstances, the fire flow must be adjusted to a lower GPM flow. For example, if only 50% of the structure was involved in fire, the fire flow would be calculated at 562.5 GPM.

If 25% of the structure were involved, the fire flow would be calculated at 281.25 GPM.

For multistory structures, multiply the results by the number of floors in the structure if the floors are involved in fire. If the floors were not involved, they would be considered interior exposures. Add 25% for each exposure.

For a two-story structure with the same dimensions with both floors involved in fire, the formula would be

$$\frac{75 \times 45}{3} = \frac{3375}{3} = 1125 \times 2 = 2250 \text{ GPM}$$

If the second floor were not involved in fire, the formula would be

$$1125 \times .25 = 281.25$$
$$1125 + 281.25 = 1406.25 \text{ GPM}$$

The formula is designed for offensive operations and becomes less accurate if there is over 50% involvement of the structure. Because these formulas are only approximations, they may require higher flows when defensive operations are initiated.[7]

Fire flow requirements can then be used to determine personnel and equipment needs on the fire ground. Another strategic problem faced by the fire service is domestic development in rural areas. These areas do not have readily available water distribution systems. Water is usually supplied to residences by private wells drawing from a groundwater supply. These wells rarely have adequate flow capacity to support the needs of firefighting apparatus. Therefore, the fire department must rely heavily on tank water carried in by the fire apparatus or on surface water supplies from lakes or ponds (see Figure 11-3). A preincident plan for rural firefighting must include provisions for transporting, relaying, or drafting water operations.

WARNING: It is critical to keep strategic preincident plans up-to-date. If demographic

FIGURE 11-3 Many rural areas rely on water tenders (mobile water supplies) for their fire operations. *Courtesy of James D. Richardson.*

changes have affected fire flow requirements in an area—urban or rural—and the preincident plan has not been altered to accommodate these changes, any future firefighting operations based on the outdated plan could be doomed to failure.

Rural and wildland areas pose a problem for preincident planning. As mentioned earlier, rural and wildland firefighting strategy should be based not only on the availability and amount of water but also on the method of its delivery. Water supplies may be provided by tanker trucks (tenders), relay operations from static water sources, direct drafting operations, or aircraft tanker operations. Moreover, wildland fires cover vast areas, and they produce extremely large amounts of heat. In the confined area of a structure fire, the conversion of water into steam displaces oxygen and aids in fire extinguishment. This phenomenon does not occur in open, wildland fires. Consequently, high volumes of water are needed to absorb the heat and cool the fuel for extinguishment. GPM flow rate computations will not be as important as total volume flow. The personnel and equipment requirements for wildland fires are also difficult to calculate because they vary widely depending on variables such as the size of the geographic area involved, the type of terrain, the amount of fuel, the water supply, and weather conditions.[8]

AUXILIARY APPLIANCES

Saving lives is the first priority in any emergency operation. Preincident inspections will identify the life-safety devices that are in already in place, such as sprinklers, standpipes, or special extinguishing systems. Inspections will also expose the need for additional appliances that could increase the safety of occupants in the event of a fire. Identifying, in advance, the location of these appliances in the building is vital. If a company knows where a standpipe is located, it can plan strategies for a rapid interior attack to attain environment tenability and possibly save lives.

Knowing which types of appliances are in use is also critical. For example, some total flooding and deluge systems use extinguishing agents, such as halon, halon replacement agents, and CO_2, that create toxic or hazardous atmospheres in the protected areas. Being forewarned of the presence of these systems allows firefighters to don the proper protective gear when responding to an emergency.

FIGURE 11-4 Knowledge of area streets and roads reduces response time.
Courtesy of James D. Richardson.

STREET CONDITIONS

Emergency responders should be very familiar with the streets in their area of responsibility. This is called "knowing your territory." However, knowing where the location is and getting to it are different matters. Road construction, special events, traffic, or adverse weather can make streets impassable for emergency apparatus. All personnel must be able to identify and navigate alternative routes quickly to any location in their response area (see Figure 11-4). A fire company may be sent to a location outside of its normal response area, or diverted to a new location while responding to a call. When this occurs, the officer must possess good map-reading skills to find the new location and direct the engineer/operator to the shortest route. Some fire departments have Global Positioning Systems (GPS) and Geographic Imaging Systems (GIS). GPS uses satellite coordinates to determine the geographical location of an apparatus or firefighter with accuracy within a few feet. GIS mapping software provides information on road location and topography. This new technology greatly enhances a fire company's ability to determine where they are in reference to any location that they may be dispatched to. GIS and GPS also allow dispatchers to determine the location of fire apparatus via digital display on a computer screen.

WEATHER

Weather is a major consideration when developing a preincident plan. Although certain seasonal weather patterns can be assumed, day-to-day weather conditions such as wind, temperature, and humidity are harder to predict. They must, however, be taken into account.

Wind

Wind is one of the most influential forces affecting the behavior of fire. It is common knowledge that wind intensifies burning by supplying additional oxygen to the fire. One pound of fuel uses 200 cubic feet of air during the combustion process. By increasing available oxygen, wind increases the rate of fuel consumption.

Wind also brings additional problems to the fire ground strategist because it greatly influences the direction of fire travel. This is no surprise to those who are experienced in wildland and rural firefighting. However, studies have shown that air currents inside buildings move in the same direction as the wind outside the building. Even light winds can overcome the air pressure created by the combustion process and affect the direction of the fire.[9] On April 16, 2007, a Virginia firefighter was fatally trapped during fire operations in a single-family structure. Wind conditions were high when the firefighter and his lieutenant made entry through the front door. They advanced up the staircase and into the master bedroom. Fed by a 25 mile per hour wind that gusted up to 48 miles per hour, the fire suddenly and dramatically increased. The firefighter was unable to escape and succumbed to thermal and inhalation injuries. Following this tragic event, the National Institute of Standards and Technology (NIST) conducted experiments on the effects of wind on the fire ground. NIST also investigated firefighter injuries and deaths resulting from wind-driven fires. Information about these studies can be retrieved from the NIST website at www.nist.gov/fire/wdf.cfm.[10] Wind can cause hot embers to be dispersed to unburned areas and move heated fire gases into contact with dry fuels. It is a very dangerous variable in any fire situation. In conjunction with high temperature and low relative humidity, wind can create a fire scenario that will tax any fire organization's resources.

Temperature

Preincident plans should consider both seasonal changes in temperature and daily weather fluctuations. The temperature will affect the type of firefighting operation planned. During structural firefighting operations, temperature extremes greatly affect personnel. Any preincident plan must provide for personnel rehabilitation and relief. In wildland situations, other considerations come into play. Wildland fuels heated by the sun ignite more readily than fuel in areas of dense shade, and temperature (as well as wind and relative humidity) greatly influence the speed and intensity of a wildland fire.

Relative Humidity

Relative humidity is the relationship between the current amount of moisture in the air and the total amount of moisture the air can hold at a given temperature. Air picks up moisture when it moves across surfaces. If the relative humidity drops below 30%, the air can absorb more moisture from such fuels as light vegetation. This lowers the ignition temperature of the fuel and increases the danger of rapid fire spread.[11]

Exposures

Protecting exposures are the next firefighting priority after life safety. An *exposure* is any structure or area of structure that is not yet on fire but could have possible fire damage. It can also refer to persons trapped in those areas. Exposures can be external or internal.

internal exposures ■ Areas or persons in close proximity that can be threatened by a fire within the involved structure.

external exposures ■ External areas or persons in close proximity that can be threatened by a fire.

Internal exposures include entrapped occupants, hazardous materials, or an uninvolved room. **External exposures** can be anything in close proximity to the fire, such as buildings, prairie grass, a forest, or any outdoor area. It is important that exposures be identified during preincident planning.

Many factors are considered when determining the priority of exposure protection. Naturally, life safety will come first, but building construction, private fire protections systems, hazardous materials, and access to the area are also important. In some circumstances the strategic plan makes exposure protection a priority over extinguishing the fire in an already involved structure.[12] For example, if a structure is totally involved with fire and rescue operations are not possible, it is better to focus on protecting exposed occupants and structures to prevent the fire from spreading.

AREA AND HEIGHT (TOPOGRAPHY)

Topography refers to the characteristics of terrain that include elevation, slope, and orientation. Because these features are fairly stable, topography tends to be the most predictable environmental element of a strategic plan.

Topography has always been considered an important factor when planning wildland firefighting operations. Some strategic planners, however, fail to consider its influence on structural firefighting. Urban preincident planning should identify conditions that could slow response times, prevent effective apparatus placement, or limit access to the interior of a structure. These may include bridges, road construction, and utility excavations. In fact, the urban landscape is actually less predictable than rural topography. Through an active preincident planning program, the strategic planner can stay abreast of changes that may affect firefighting response and operations. Cooperation and notification procedures should be established with other city departments and the state department of transportation so that responding companies will be aware of street or road closures because of construction or utility work.

Based on information from the preincident plan, the wildland fire strategist must take several topographic factors into consideration when developing an action plan. Each of these factors will affect ease of ignition and fire movement.

slope ■ The angle of a hillside.

aspect ■ The slope of land that is exposed to the sun.

shape of the area ■ Topographic shapes, such as canyons, ridges, and mountains.

elevation ■ The height of the land above sea level.

barriers ■ Natural or artificial firebreaks for controlling fire spread.

■ *Slope.* The angle of a hillside (its slope) affects the speed of fire traveling up the slope. The steeper the slope, the faster the fire will travel from the bottom to the top.

■ *Aspect.* This refers to the side of a slope of land that is exposed to the sun. In the Northern Hemisphere, slopes have a southern aspect. Slopes that are exposed to the sun usually have lighter and drier fuels. The ambient temperature is higher, and the relative humidity is lower. These are areas where fire ignition is more likely and spread tends to be more rapid.

■ *Shape of the Area.* Topographic shapes, such as canyons, ridges, mountains, and ravines, greatly influence fire behavior. Their effects on wind speed and direction influence the direction of fire travel, the intensity of burning, and the speed at which a fire will spread.

■ *Elevation.* This affects the condition of the fuel and the amount of fuel available. Because temperatures are normally higher at lower elevations, fuels at lower elevations become drier earlier in the summer than fuels at higher elevations.

As a rule, fuels are also less abundant at higher elevations; at extremely higher elevations, there may be no fuel at all. Precipitation and wind are also influenced by elevation.

■ *Barriers.* Lakes, rivers, roads, highways, and constructed firebreaks are effective tools for controlling fire spread. These natural and artificial barriers should be major considerations in strategic plan development.

Whether the emergency response is in an urban or rural area, topography should be considered in the strategic plan. Contingency plans must be developed in both settings. It makes no difference whether a road is blocked by a mudslide or by a water main repair project, the road is still blocked. A good preincident plan will take this into consideration and establish alternative routes for emergency responders (see Figure 11-5).[13]

STOP AND THINK 11-2

1. How does topography influence firefighting operations in your area of responsibility?
2. What can you do to obtain information on residential structures?
3. How does weather affect your strategic and tactical planning?

LOCATION AND EXTENT

The actual location and extent of a fire of course can be determined only after firefighters' arrival on the scene. Information gained from preincident planning and inspection reports, however, will give the incident commander clues to the probable location of the fire by identifying areas where a fire would most likely occur. Preincident plan information about a building's construction and design may also allow the IC to predict where the fire is most likely to travel.

The location and extent of a fire are primary factors in developing a strategic action plan and in deciding on tactics at the fire ground. These factors set the stage for all other activities, such as determining how and where the attack takes place, what exposures are to be protected, and which rescue operations may be necessary.[14]

TIME OF DAY

The time of day is a major factor in fire ground strategy. The preincident plan will indicate when a building would be at or near normal occupancy. For example, most businesses operate between the hours of 8:00 AM and 5:00 PM. Firefighters may assume that an office

FIGURE 11-6 The incident commander is responsible for protecting emergency responders from exposure to, or contamination from, hazardous materials during an operation. *Courtesy of James D. Richardson.*

building will be at normal occupancy levels during that time. This does not mean, of course, that the building will be unoccupied after 5:00 PM. It means only that the life-safety hazard will be expected to be lower because there are fewer people in the building at night.

Remember: Incident commanders should always expect the unexpected. That "ordinary" house fire to which you respond in the middle of a workday may turn out to be an unlicensed child- or elder-care center.

HAZARDOUS MATERIALS

Hazardous materials have become one of the primary concerns in firefighting response and operations. The location and quantity of these materials must be recorded during preincident planning and inspections. However, firefighters must be aware that almost every response will involve some type of hazardous material.

Fire personnel expect to encounter hazardous materials when responding to transport vehicle accidents but may not expect to encounter these during "routine" house or automobile fires. However, firefighters receive the heaviest exposure to hazardous materials during routine fires because they do not expect the exposure and are not properly protected. For example, when the insulation that protects an automobile's electronic components burns, the byproducts are carcinogenic. Many firefighters will not wear respiratory protection when combating an automobile fire. Firefighters also have a tendency to remove their self-contained breathing apparatus (SCBA) when conducting overhaul operations at structure fires. Even though fire has been extinguished, the IC should not allow firefighters to enter an enclosed area without SCBA until the atmosphere is tested and declared safe (see Figure 11-6).

STOP AND THINK 11-3

1. What type of hazardous materials might fire personnel find in a rural area?
2. Would the hazardous materials found in a rural area be different from those found in an urban area?

The basic strategic plan for responding to any hazardous material incident is simple—it should be put into effect whenever the presence of hazardous materials is suspected. The steps to follow for hazardous material fires are as follows:

1. Approach the location from the upwind and uphill side.
2. Cordon off the area to isolate the incident and prevent entry.
3. Identify the product.
4. Seek help from qualified experts.[15]

FIGURE 11-7 A common form of communication among the emergency responders is critical if the emergency operation is to succeed. *Courtesy of James D. Richardson.*

COMMUNICATIONS

An incident command structure cannot be effective without an efficient communication network. Without a command structure, even the best-laid strategic plan is built on quicksand. The complexity of a communication system depends on the size of the emergency service organization and the area being served. A written standard operating procedure for the system's use must include rules that increase efficiency and reduce unnecessary radio traffic.

With the advent of the terrorist threats facing the United States, a multiagency response to certain emergency incidents has become standard. These agencies must have access to a common communication system and be trained in the use of common terminology to communicate efficiently (see Figure 11-7). This is especially essential in smaller communities that rely heavily on mutual aid response from other municipalities. A multijurisdictional or multiagency response requires the use of an integrated communication system. A communications plan should contain the following elements[16]:

- *Systems.* Some jurisdictions are sharing the cost of setting up and maintaining a common communication system that can be used by all agencies responding to a multijurisdictional incident.
- *Frequencies and Resource Use.* When a common system is used, each agency can be assigned a frequency to use to communicate with its resource personnel and additional tactical channels to be used during emergency responses to communicate with other agencies.
- *Information Transfer Procedures.* A communications plan must allow for the transfer of information not only within an organization but also between different agencies. All involved entities should work together and agree on a set of written procedures for transferring information. Commonly agreed-on procedures will help to keep misinformation at a minimum during an emergency.[17]

Communications information should be included in every preincident plan.

COMMAND STRUCTURE

Developing an incident command structure during an incident may be construed as a tactical operation. The actual implementation of the system is tactical, but the basis on which the commander develops that command structure is the strategic plan and a standard operating procedure for command development. Fire ground operations must be strictly controlled and coordinated to ensure safety and effectiveness. All responders must know their assignments and know to whom they report. When an operation is well coordinated, the participants can focus on their individual assignments and accomplish their mission. The command structure establishes this well-defined line of authority.

Identifying all of these topics in a preincident plan and incorporating that information into the incident action plan will allow the IC to establish and prioritize strategic goals when an incident occurs. Without preincident plan information, the IC is working blind. For example, on December 29, 2009, a Wisconsin volunteer fire department responded to a dumpster fire at a local foundry. On arrival, they observed bluish green flames emitting from the container. Applying water and foam failed to extinguish the fire, and 12 minutes into the operation, the dumpster exploded, killing one firefighter and injuring eight more. After conducting an investigation of the incident, the National Institute of Occupational Safety and Health (NIOSH) identified four contributing factors:

- Lack of hazardous material awareness training
- No documented site pre-plan
- Insufficient scene size-up and risk assessment
- Inadequate disposal/storage of materials

The report listed eight key recommendations. The number one recommendation was to ensure that high-risk sites are pre-planned by walk-through inspections by all responding fire departments and update the plan annually. A thorough preincident plan would have identified those factors that contributed to the incident.[18]

Strategy

strategy ■ Planning and directing the maneuvering of personnel and equipment into the most advantageous positions prior to actual operations.

tactics ■ The actual maneuvering of personnel and equipment in an operation to achieve short-term objectives.

Strategy is a broad-based set of goals or objectives developed for controlling an incident and achieving a positive outcome. It is based on three priorities: life safety (firefighter and civilian), incident stabilization, and property preservation. The strategic plan, often referred to as the *action plan*, identifies the tasks to be accomplished, identifies the material resources that will be needed, and specifies personnel requirements.[19] Strategy defines the goals from which the fire ground **tactics** are implemented.

Several factors need to be considered when developing a fire ground strategy. The main concerns are available resources and environmental conditions. A strategy is well planned if there are enough resources to implement the plan and the plan is flexible enough to allow for unexpected or environmental changes. If there are not enough resources to execute the plan and it does not allow for flexibility, it does not matter how good it was in theory; it is virtually useless as a strategy. This is why preincident planning is so important. Knowledge of resource availability prior to an incident will allow the IC to identify and prioritize strategic goals.

In his book, *Strategic and Tactical Considerations on the Fireground*, Deputy Chief James P. Smith of the Philadelphia Fire Department identified seven basic strategies that must be accomplished during firefighting operations.[20] Although some text may recommend that these strategies be completed in the order listed below, objectives are often accomplished simultaneously. For example, ventilation and confinement may be used in conjunction with rescue. Ventilation and salvage operations can be performed at any time during the operation. All operations will be focused on the primary objective—life safety.

Rescue

The primary objective in firefighting operations is saving lives. The safety of the operations personnel must be the first consideration when initiating a rescue operation. If a number of victims must be rescued, it is recommended that the sequence of priority be as follows: The first victims to consider for rescue are those closest to the fire. The largest number of victims who are threatened by the fire should be considered next. The third priority should be any other victims in the fire area. The last priority is victims who are in danger of being exposed to the fire.

STOP AND THINK 11-4

You arrive on the scene of a fire in an apartment complex. One building is heavily involved and has trapped victims. The fire is moving in a northerly direction. There is a report that five people are in the building on the north side of the involved structure. In a building on the south side of the fire, 20 people are on balconies yelling for help.

What is your priority of rescue in this situation?

Exposures

As we have explained, an exposure can be anything that is in the path of a fire: a person, a room in a building, another building, a field of grass or other vegetation, and so on. Exposures can be protected either by removing them from the path of the fire or by keeping them cool with water spray. Protecting exposures can serve the dual purpose of confinement if the fire is surrounded and prevented from spreading to other areas.

Confinement

Rapid confinement deals a death blow to a fire. Conversely, if a fire cannot be quickly confined, heavy fire losses can be expected.

Concealed spaces in buildings, such as elevator and utility shafts, attics, or voids between trusses, allow fires to move rapidly from one area to another. Weather, wind, and topography affect wildland fire extension. The tactical officer should visualize a fire as having four sides, a top, and a bottom. Attention must be paid to all six dimensions if fire confinement is to be accomplished.

Extinguishment

The strategic aspect of fire extinguishment is a judgment call made by the first arriving fire officer. On arrival on the scene, the officer will estimate the type and amount of extinguishing agent that will be needed based on preincident plan information and observations made during scene evaluation.[21] An attack on the fire should not be attempted until an adequate amount of the proper kind of extinguishing agent is available at the incident scene. If on-hand supplies are inadequate, it may be prudent to use what is available to protect exposures and to contain the fire until enough agent is available for an all-out attempt at extinguishing the fire.

Overhaul

Overhaul is the process of ensuring that the fire is completely extinguished. There is nothing more embarrassing for a fire company than having to return to a fire scene because of a rekindle.

Overhaul operations are more complicated and considerably more dangerous than may be imagined. It is not simply a matter of just flushing everything out the back door and going back to the station. For example, care must be taken to prevent disturbance of possible crime scene evidence. Coordination between overhaul and fire scene investigation operations is an absolute necessity.

In many departments, overhaul operations are performed by the same fire companies that were assigned to the fire attack teams. These personnel may be suffering from fatigue and may lack the concentration necessary to identify hazardous conditions. The operations officer should consider bringing in fresh crews to perform the overhaul. Reducing exposure of overhaul operations personnel to toxic atmospheres should be a second major consideration for the operations officer. During firefighting operations, the firefighters wear respiratory protection. However, after the fire is extinguished, most firefighters remove their protective clothing. At this point, they are exposed to the highest concentrations of carbon monoxide and other products of combustion. The operations officer should ensure that all overhaul operations personnel continue to wear their breathing apparatus or appropriate respiratory protection until the atmosphere has been tested and deemed safe.[22]

Ventilation

ventilation ■ The systematic removal of pressure, gases, and heat from a structure or confined area.

Ventilation is the process of removing the byproducts of combustion (smoke, toxic gases, heat, and particulates of unburned fuel) from a confined area. The purpose of ventilation operations is to create an atmosphere that is as safe as possible, not only for firefighters entering the area but also for trapped victims. Ventilation is utilized to direct the movement of the fire, reduce the heat, increase visibility for firefighters, and increase the oxygen content of the atmosphere. The IC must place ventilation high on the priority list. It must be timed to coincide with attack crews to prevent fire spread.

The choice of method depends on the tactical requirements of the situation. Ventilation operations should be continued until the fire is completely extinguished and overhaul operations are completed.[23]

Salvage

Although salvage is the last operation in the prescribed tactical sequence, it can actually be initiated at any stage of the operation. The purpose of salvage is to protect property and unaffected areas from the damaging effects of the products of combustion and from water damage. In structure fires, this may involve covering furniture and personal property with salvage covers to protect them from water and smoke damage, removing water from the structure during or after the fire, or recovering valuables from fire debris.

Salvage operations during wildland firefighting operations may be incorporated into the confinement stage of the operation and may involve removing debris to inhibit fire spread and to protect structures exposed by the fire. Another form of wildland salvage operation is planting new vegetation in burned-out areas to prevent erosion or mudslides.

Deciding when salvage operations should take place and what they entail is the duty of the tactical operations officer.[24]

Tactics

Tactics involves the deployment of personnel and other resources in a manner that effectively and efficiently achieves strategic goals. Tactical decisions should be based on the incident action plan (IAP). The IAP identifies what should be accomplished, what resources are available to execute the plan, and who is responsible for which tasks. The tactical plan specifies how tasks are to be performed based on information from the strategic plan. Tactical planning is achieved at the operations level and is based on the availability of resources in relation to the situation at hand.[25]

TACTICAL DECISION MAKING

The tactical plan is heavily influenced by the decisions of the first arriving officer at a scene. He or she must note several factors in order to assess the situation. First, the location of the fire should be ascertained. Regardless of the type of incident (structure, wildland, and so on), location is the most important factor. Determining the present extent of the fire and what life hazards may be involved depends on locating the fire.

The next step is to judge the probability that the fire will spread. This is affected by many factors, such as the amount of heat buildup, the type of fuel, the building's construction, weather conditions, and topography.

Finally, the first arriving officer must immediately decide whether enough resources are available to start tactical operations. If the fire is larger than the available resources can handle, the officer must request more resources. OSHA 1910.134 and NFPA 1500 stipulate that two firefighters are required to initiate an interior attack on a fire. Two additional firefighters should be fully equipped and ready outside the structure to monitor the interior attack team and serve as a rapid intervention team (RIT) should the attack team need help. This is referred to as the *two-in/two-out rule*. However, situations arise in which enough personnel may not initially be on the scene to initiate the two-in/two-out rule, and the building's occupants are in imminent danger. Remember, life safety is the number one priority, and life safety of firefighters comes first. However, if it appears that occupants may be saved, the fire officer in charge may determine that immediate rescue operations must be conducted. A combination of fire attack and rescue operation is then initiated, and the two-in/two-out rule is waived.

To put it simply, the first arriving officer must determine what is burning, in which direction the fire is likely to travel, who or what is in its path of destruction, and whether additional help is needed at the scene. Before a tactical fire officer can answer these questions, he or she must be familiar with the strategic plan, understand fire behavior, and know the priority of tactical objectives.[26] Tactical objectives are met by implementation of an IAP. The IAP is a fluid document that can be amended as fire ground situations change. Once the plan is implemented, it must be analyzed for effectiveness. The tactical operations must be evaluated to ensure that the objectives are being met. If they are not, the plan must be altered to correct any deficiencies. The altered plan is then implemented. This cycle continues throughout the operation until all tactical objectives are completed.

ON SCENE

Midday on July 31, 1979, the Houston Fire Department responded to an apartment fire. When firefighters got to the scene, the first arriving companies found a structure housing several apartment units heavily involved in fire. The building was one of several on a 55-acre development. There were a total of 1,083 residential units in the development. The buildings were of wood-frame construction. The roofs and exterior walls were covered with cedar shingles. As the fire grew in intensity, a steady south wind sent burning embers downwind and ignited the shingle roofs of adjoining buildings. Before the fire was under control, seven alarms had been requested. A total of 35 fire apparatus and over 100 firefighters responded. An area over one-quarter of a mile long containing 437 apartment units was totally destroyed. Flying embers were reported in subdivisions 12 miles north of the fire.

1. What would you have done if you were the first arriving officer?
2. What do you think were the important factors to be considered on arrival at the scene?
3. What strategy and tactics would you have implemented?

Summary

A strategic (or action) plan is a broad-based protocol that establishes the goals for incident management. Based on information gathered from preincident surveys, the strategic plan identifies resources and establishes standard operating procedures to be used as guidelines during tactical operations. The plan must be flexible and capable of modification should unexpected events arise during the incident.

Tactical plans describe how strategic plans will be put into effect. The operations officer is responsible for implementing the tactical plan in the best way according to the actual situation. The operations officer must also stay abreast of changes in the situation and be able to react to unexpected developments. This does not mean the officer abandons the plan; it does mean that the tactical officer must have the skills and ability to adjust the plan to meet all of the tactical objectives. Tactical objectives must be accomplished in the proper sequence to ensure a positive incident outcome.

Constant communication and coordination between the operations officer and the incident commander are absolute necessities. A well-defined command structure gives responders confidence and improves morale.

Finally, whether your goal is developing a strategic plan or a tactical plan, always remember Murphy's famous law: "Anything that can go wrong, will go wrong." Plan for the unexpected, and plan relentlessly.

Review Questions

1. Why is it incorrect to assume that tactics and strategy are synonymous?
2. What is the purpose of a strategic plan?
3. Why is the prefire plan so important in developing a strategic plan?
4. What factors must be considered when developing the strategic plan?
5. What regulations and standards govern the two-in/two-out rule?

Suggested Reading

Avillo, Anthony. (2008). *Fireground strategies* (2nd ed.). Tulsa, OK: PennWell Corporation.

Bush, Loren, and McLaughlin, James. (1970). *Introduction to fire science*. Beverly Hills, CA: Glencoe Press.

Charlottesville Fire Department. (2006, February 19). *Emergency operations standard operating procedures*. Charlottesville, VA: Charlottesville Fire Department.

Clark, William. (1981). *Fire fighting principles and practices*. New York, NY: Technical Publishing.

Coleman, John. (1997). *Incident management for the street-smart fire officer*. Saddle Brook, NJ: Fire Engineering.

Demographics of the United States. (2006, May 1). *Wikipedia, the free encyclopedia*. Wikimedia Foundation.

Fuhrman, Paul. (1995, December). *Fatal house fire San Francisco, CA*. St. Louis, MO: Working Fire Subscription Films. Retrieved February 19, 2006, from Video Archive. http://www.workingfire.com.

Monkman, Chris. (n.d.). Murphy's laws origin. *Murphy's laws origin*. Retrieved January 15, 2006, from Murphy's Laws website: http://www.murphys-laws.com/murphy/murphy-true.html.

Page, James. (1973). *Effective company command*. Alhambra, CA: Borden Publishing Co.

Schneider, Rachel, and Breedlove, Deborah. (n.d.). Fire management study unit. *National weather service program*. Retrieved December 2, 2005, from http://www.fs.fed.us/conf/ee/fire_mgt.pdf.

Smith, James P. (2012). *Strategic and tactical considerations on the fireground* (3rd ed.). Upper Saddle River, NJ: Pearson Education.

Wurtz, Thomas. (2004). *Firefighter's handbook: Essentials of firefighting and emergency response* (2nd ed.). Clifton Park, NY: Delmar Publishing.

Endnotes

1. Unknown. (2003). *NFPA 1620 Recommended Practice for Pre-Incident Planning.* (2003). Retrieved January 29, 2013, from International Association of Fire Fighters website: http://www.iaff.org/et/jobaid/Intro1710/NFPA_1620.htm.

2. Avillo, Anthony. (2008). *Fireground strategies* (2nd ed.). Tulsa, OK: Fire Engineering, PennWell Corporation.

3. Fuhrman, Paul. (1995, December). *Fatal house fire San Francisco, CA.* St. Louis, MO: Working Fire Subscription Films. Retrieved February 19, 2006, from Video Archive. http://www.workingfire.com/

4. Royer, Keith, and Nelson, Bill, (n.d.). Water for fire fighting rate-of-flow formula. *Iowa State University Bulletin # 18,* Fire Service Extension, Iowa State University of Science Technology.

5. National Fire Protection Association. (2010). NFPA 1710 *Standard for Organization and Deployment of Fire Suppression Operations, Emergency Medical Operations, and Special Operations to the Public by Career Fire Departments.* Quincy MA: Author.

6. IAFF Local 3284. (2012, December). *NFPA 1710 Fact Sheet for the Public.* Osceola Fire Department, Osceola, FL. Retrieved January 25, 2013, from http://www.iafflocal3284.org/mobile/index.cfm?highlightId=5.

7. U.S. Department of Commerce Economics and Statistics Administration, United States Census Bureau. (2011, November). The older population. *2010 Census Briefs.* Retrieved April 12, 2013 from U.S. Census website: http://www.census.gov/prod/cen2010/briefs/c2010br-09.pdf.

8. Smith, James P. (2012). *Strategic and tactical considerations on the fireground Third Ed.* Upper Saddle River, NJ: Pearson Education.

9. Smalley James C. (2008). Wildland fire management. In *Fire protection handbook* (20th ed.,Volume II). Quincy, MA: National Fire Protection Association.

10. Avillo, op cit.

11. National Institute of Occupational Safety and Health. (2008). *Death in the line of duty, fire fighter fatality investigation Report F2007-12.* Retrieved from Centers for Disease Control website: http://www.cdc.gov/niosh/fire/reports/face200712.html

12. Wade, D. D., and Lundsford, J. (n.d.). Fire as a forest management tool: Prescribed burning in the Southern United States. Retrieved January 25, 2013 from http://www.woodlandstewardseries.com/landowner-information-for-managing-woodland/documents/Fireasaforestrytool.pdf.

13. Avillo, op cit.

14. Burgdorf. (2000). Wildland Fire Fighting Course S 190. *Introduction to wildland fire behavior.* Retrieved January 25, 2013, from http://www.wildlandfire.com/ppt/s190-1.ppt.

15. Smith, op cit.

16. Avillo, op cit.

17. Wurtz, Thomas. (2004). *Firefighter's handbook: Essentials of firefighting and emergency response* (2nd ed.). Clifton Park, NY: Delmar Publishing.

18. Ibid.

19. Fire Fighter Fatality Investigation and Prevention Program, National Institute of Occupational Safety and Health. (2010, July 6). Death in the line of duty: One fire fighter killed and eight fire fighters injured in a dumpster explosion at a foundry—Wisconsin. Retrieved January 29, 2013, from Centers for Disease Control website: http://www.cdc.gov/niosh/fire/pdfs/face200931.pdf.

20. Wurtz, op cit.

21. Smith, op cit.

22. Ibid.

23. Ibid.

24. Ibid.

25. Ibid.

26. Wurtz, op cit.

12

Emergency Medical Services and Emergency Management

Courtesy of TEEX-Texas A&M System.

KEY TERMS

DOT National Standard Curriculum,
p. 247

emergency support functions (ESF),
p. 253

Integrated Emergency Planning
System, p. 251

Local Emergency Management
Committee (LEPC), p. 252

National Response Framework (NRF),
p. 253

prehospital care, p. 247

OBJECTIVES

After reading this chapter you should be able to:

■ Describe the evolution of EMS
■ Describe the level of care provided by EMS
■ Explain the importance of higher education in EMS
■ Define emergency management
■ Identify the four areas of responsibility of emergency management
■ Explain the function of the LEPC

Evolution of EMS

This chapter will not go into great depth into the various levels of training and certifications for each kind of emergency medical technician (EMT) because whole books and courses have been dedicated to that endeavor. A brief summary of each will be described. *Because states vary on their rules and regulations, for more in-depth information about EMTs, please check with your particular state or local jurisdiction.* However, this chapter will demonstrate how the fire service and EMTs go hand in hand.

FIGURE 12-1 A modern air evac helicopter. *Courtesy of James D. Richardson.*

prehospital care ■ A term used for any emergency care bestowed to a person that is not provided in the hospital setting. This care can be administered by a firefighter, EMT, police officer or lay person. The standards for prehospital emergency care and the individuals who provide it are governed by the laws in each state and are typically regulated by the state office of EMS.

DOT National Standard Curriculum ■ The Department of Transportation first developed this standard in the 1980s. This curriculum is designed to instruct a student to the level of Emergency Medical Technician-Basic. This curriculum is critical because it is recognized that the majority of prehospital emergency medical care will be provided by the EMT-Basic who serves as a vital link in the chain of the health care team. In 2011, this curriculum was again reviewed and updated to better serve the field of emergency medical care in the respective fields in terms of scope of practice and practice guidelines.

Whether the individual is at the first responder basic level or a licensed paramedic, the health care professional provides **prehospital care** and should always be prepared for worst-case scenarios. In the civilian world, the first ambulances were motorized hearses operated by community funeral homes; in the military world, there were field medics who placed wounded soldiers on stretchers and transported them with carts or some kind of wagon. Later, rapid helicopter evacuations were used during wartime. Although helicopters have changed through the years, they are still used today for the military and private industry to utilize rapid transportation (see Figure 12-1). Many operations ran on the "scoop and run" protocol and many personnel had only cardiopulmonary resuscitation (CPR) training, if any. But in 1966, the National Academy of Sciences published a report entitled "Accidental Death and Disability: The Neglected Disease of Modern Society." This report described the magnitude of traffic-related death and disability while vividly describing the deficiencies in prehospital care in the United States. Furthermore, in 1966, the National Highway Traffic Safety Act (NHTSA) established the U.S. Department of Transportation (DOT) and awarded that agency the authority and responsibility to improve EMS education, including the development and implementation of training standards. States were encouraged to develop state EMS offices with part of the costs paid by the Highway Safety Programs. This was the beginning of change for emergency services.

In the early 1970s, the DOT developed the first **National Standard Curriculum** to serve as the guideline for the training of EMTs. Also in the 1970s, the National Registry of Emergency Medical Technicians (NREMT) was established. The NREMT allows reciprocity into different states, but it is important to note that EMS is regulated by the state in which you are certified. By the 1980s, EMS had become established and added the component of advanced life support (ALS) care into its standards. Housing automatic external defibrillators (AED) in ambulances was a dramatic change in the way EMTs administer prehospital care. In 2010, the American Heart Association set new guidelines for CPR and emergency cardiovascular care, and NREMT follows those standards. The recent transition of having AEDs in the ambulance and the new standards now allow lower level EMTs to perform a number of advanced skills in the field that were formerly reserved only for paramedics. Local and state regulations must be reviewed and adhered to since each are different. A new technology used in the field helps the medics in performing lifesaving CPR. Automated CPR devices are sometimes equipped in ambulances. (See Figure 12-2.) According to the Bureau of Labor Statistics (BLS), employment of EMTs and paramedics is expected to grow by 33% from 2010 to 2020, much faster than the average for all occupations. According to the BLS, more than 201,000 paramedics and EMTs are employed throughout the country. Many are employed by EMS and private ambulance companies, local government agencies, or fire departments.[1]

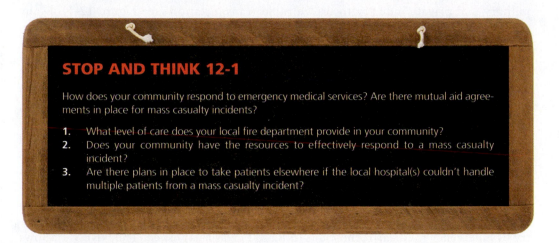

STOP AND THINK 12-1

How does your community respond to emergency medical services? Are there mutual aid agreements in place for mass casualty incidents?

1. What level of care does your local fire department provide in your community?
2. Does your community have the resources to effectively respond to a mass casualty incident?
3. Are there plans in place to take patients elsewhere if the local hospital(s) couldn't handle multiple patients from a mass casualty incident?

Level of Emergency Medical Care

Emergency medical technician is a term broadly used to refer to personnel in prehospital care; however, there are four basic levels of training that will be briefly discussed. Entire books have been written on each level of training, so this description is not all-inclusive of each job duty.

■ **Emergency Medical Responder** (EMR—previously known as First Responder). This is an entry-level certification for first responders who can administer basic first aid such as CPR and other basic life support functions. It prepares individuals for a variety of prehospital settings.

- **Emergency Medical Technician (EMT**—previously known as EMT-Basic). This certification teaches the student how to initially assess a patient's condition and provide basic care such as airway management and management of fractures. It also allows the student to practice these skills. At this level, most providers do not administer anything intravenously.
- **Advanced EMT (AEMT**—previously known as EMT-Intermediate). This certification builds off the basic training of EMT. This level can perform more invasive procedures than an EMT-B can perform. For example, in most states an AEMT provider can insert an IV into a patient. At the AEMT level, some states or jurisdictions allow them to administer some drugs to patients. This level goes beyond the national standard of 1999. The certification requires state-approved Advanced Medical Technician course.
- **Paramedic** (previously known as EMT-Paramedic). This certification is the highest an EMT can receive. The training for this certification focuses on advanced science courses such as pharmacology, which teaches the paramedic how different drugs interact with each other, along with courses in cardiology and anatomy.

It should be noted that each state or jurisdiction is different and a person interested in these fields of study should check with his or her local rules and regulations.

Fire and Emergency Services Higher Education Guidelines

Fire and Emergency Services Higher Education (FESHE) was first discussed in Chapter 4 but has an impact in emergency medical services (EMS) as well. A professional development model has been developed to reflect the sequential evolution of this field. These models are not promotional ones; rather, they are competency-based professional development paths supported by training, higher education, and certification elements. Each model has prescribed recommendations for how each competency can be addressed by each element. (See Figure 12-3.)

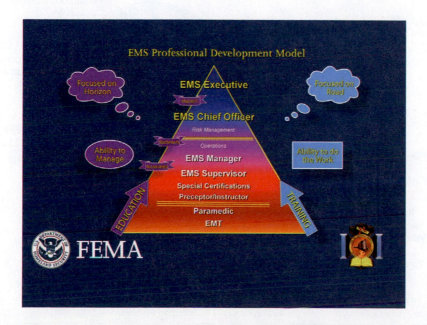

FIGURE 12-3 EMS professional development model as noted by the United States Fire Administration/National Fire Academy for Higher Education.

Postsecondary Education

As with the fire service, postsecondary education is just as important and crucial for the EMT. Long hours of training and continuing education are just a small part of what these providers have endured just to gain a position in the career they have chosen. Continuing education (CE) hours are required to maintain most certifications and licensures through the state level and national level.

Many higher education institutions offer associate of applied science degrees for the paramedic and even bachelor's degree in para-medicine, health care management, or health care administration. It is of the utmost importance that a student interested in higher education should look into the regional and national accreditation as described in an earlier chapter of the institution of his or her choice. The EMS degree plan has many variations in its core courses depending on the emphasis taken and the institution involved.

Firefighters and EMTs

Due to the nature of firefighting and EMT's role in the community, it is no wonder that these two jobs sometime become entangled. The skills and training needed to respond to emergency calls for both are very similar and the personalities required for these types of jobs are equivalent. Because almost all firefighting jobs involve some level of emergency medical care training, most firefighters continue to keep their medical certification current. Thus most firefighters are also EMTs because of the nature of the emergencies that they respond to. Due to the shift schedule in both careers, it enables the individual to work a second job and many chose a firefighting/EMS career. The National Fire Protection Association (NFPA) has published a fact sheet that depicts this relationship between fire and EMS personnel.[2] (See Figures 12-4A and B.)

The fire service does more than fight fires.
- Almost two-thirds of fire department responses were medical aid calls in 2010.
- Only 5% of fire department calls in 2010 were due to actual fires.

Fire Department Provision of Emergency Medical Service
2008–2010 Annual Averages

FIGURE 12-4 The United States Fire Service Fact Sheet. National Fire Protection Association. *Reprinted with permission from* Fire Loss in the United States During 2011 *by Michael J. Karter, Jr., Copyright© 2012, National Fire Protection Association.*

Emergency Management and Homeland Security

Emergency management has become a very important career field in both the public and private sectors. The BLS lists this profession as one of the best 50 jobs in 2011 and expects employment to grow 22% between 2008 and 2018.[3] What is emergency management? It is a discipline that encompasses many professions. In the public sector, the person assigned as emergency manager may be a civilian, firefighter, law enforcement officer, or an emergency medical professional. An individual who serves as an emergency manager must have the following characteristics:

1. Exceptional oral and written communication skills
2. Organizational skills
3. The ability to coordinate multidepartment and multiagency operations

The process of acquiring a position as an emergency manager (EM) varies from state to state. This position is usually assigned to the head of an emergency service department—fire, law enforcement, or EMS. However, some municipalities will hire nonemergency service individuals for this position. Currently most public sector EM positions do not require a college degree but do require work experience and certificates of training from nationally accepted programs, such as the Professional Development Series, the Advance Professional Development Series, Incident Command Series, and the National Incident Management Systems Series from the Emergency Management Institute in Emmitsburg, Maryland. However, a bachelor's degree is becoming a much more desirable job qualification. The International Association of Emergency Managers (IAEMS) website has a jobs link that lists available positions and qualifications.

History of FEMA

Emergency management is not a new concept. The federal government has always seen a need to aid communities after disasters have struck. In 1803, Portsmouth, New Hampshire, was devastated by fire. Congress passed the Congressional Act of 1803 authorizing federal funding for disaster relief. This was the first time that federal funds were used in disaster recovery. During the Great Depression of the 1930s, President Franklin D. Roosevelt authorized the use of federal loans to local governments to repair public facilities, roads, and bridges damaged during natural disasters. The Flood Control Act of 1934 authorized the U.S. Army Corps of Engineers to develop and implement flood control projects. The major drawback of these efforts was the lack of a central agency to coordinate disaster relief. This weakness in disaster relief became self-evident during the 1960s and 1970s when a series of hurricanes and earthquakes brought destruction and financial devastation to areas of the United States from Florida to Alaska. Again, laws were passed to help fund relief efforts but there was no coordination to centralize the relief operations.

In 1979, President Jimmy Carter signed Executive Order 12127 that authorized the creation of the Federal Emergency Management Agency (FEMA). The Federal Insurance Administration, National Fire Prevention and Control Administration, National Weather Service Community Preparedness Program, Federal Preparedness Agency of the General Services Administration, Federal Disaster Assistance Administration, and Civil Defense responsibilities were placed under the FEMA banner. Under FEMA, an "all hazards" approach to disasters was developed. There were four areas of focus: preparedness, response, recovery, and mitigation. Under a new management system, known as the **Integrated Emergency Management System**, local, state, and federal agencies could now coordinate their efforts under one agency.[4]

FEMA is not a "take control of the situation" agency. Its primary duty is to offer financial assistance to communities struck by disaster. To qualify for FEMA assistance,

Integrated Emergency Management System ■ A management philosophy that builds awareness and skills that is designed for personnel who fill specific emergency support positions within their community. This system implements policies, plans, and procedures to protect life and property through applications of sound emergency management principles in all phases of emergency management.

certain requirements must be met by the community. Some of the requirements are as follows:

1. The community must have a disaster plan.
2. The disaster plan must be tested through exercises.
3. The National Incident Management System must be adopted and implemented.

If the community meets all FEMA requirements, FEMA will refund 75% of the community's cost for infrastructure repair and provide small business loans to noninsured individuals for personal property loss. If the community does not meet FEMA requirements, federal aid to the community may be rejected.

Focus of Emergency Management

There are two types of disasters—natural and human-made. Neither can be prevented. The only thing that can be done is to prepare for any possibility. Emergency management begins at the local level. Each community must have an Emergency Manager (EM) who is responsible for coordinating disaster preparedness, response, recovery, and mitigation efforts. This is not a one-person function. It involves every agency and department in the community, as well as the private sector. For example; the community's **Local Emergency Planning Committee (LEPC)** is comprised of representatives from all of the emergency services, local medical community, public works, the news media, volunteer relief organizations (Red Cross, Baptist's Men's Club, and so on), and private business (heavy equipment, retail stores, and so on), to name a few. The LEPC coordinates with the EM to develop a disaster response plan. Each member of the LEPC is assigned an area of responsibility. Each member develops a list of resources that are required to accomplish their respective task. Written agreements are made with private companies to supply additional resources when needed. This could be anything from heavy equipment to drinking water. When the plan is completed, it is sent to the state for approval. The community must complete this process to qualify for federal grants and funding.

The National Fire Protection Association standard 1600 (NFPA 1600, *Standard on Disaster/Emergency Management and Business Continuity Programs*) addresses prevention, mitigation, preparedness, response, continuity, and recovery. Some of these areas will now be discussed. The standard recommends that emergency management programs "shall identify hazards, monitor those hazards, the likelihood of their occurrence, and the vulnerability of people, property, the environment, and the entity itself to those hazards."[5] Emergency management is truly a community function.

In 2007, a group of emergency management practitioners and academics was brought together at the Emergency Management Institute in Emmitsburg, Maryland. Working with Dr. Wayne Blanchard of the Emergency Management Higher Education Project, this group was assigned the task of identifying and defining the principles of emergency management. The following is a summary of their findings.

Definition of emergency management:

"Emergency Management is the managerial function charged with creating the framework within which communities reduce vulnerability to hazards and cope with disasters."[6]

"Emergency management must be:

1. *Comprehensive*—emergency managers consider and take into account all hazards, all phases, all stakeholders and all impacts relevant to disasters.
2. *Progressive*—emergency managers anticipate future disasters and take preventive measures to build disaster-resistant and disaster-resilient communities.

Local Emergency Planning Committee (LEPC) ■ An LEPC is composed of representatives from all of the emergency services, local medical community, public works, news media, volunteer relief organizations (Red Cross, Baptist's Men's Club, and so on), and private business (heavy equipment, retail stores, and so on), to name a few. The LEPC coordinates with the EM to develop a disaster response plan.

3. *Risk-driven*—emergency managers use sound risk management principles (hazard identification, risk analysis, and impact analysis) in assigning priorities and resources.
4. *Integrated*—emergency managers ensure unity of effort among all levels of government and all elements of a community.
5. *Collaborative*—emergency managers create and sustain broad and sincere relationships among individuals and organizations to encourage trust, advocate a team atmosphere, build consensus, and facilitate communication.
6. *Coordinated*—emergency managers synchronize the activities of all relevant stakeholders to achieve a common purpose.
7. *Flexible*—emergency managers use creative and innovative approaches in solving disaster challenges.
8. *Professional*—emergency managers value a science and knowledge-based approach based on education, training, experience, ethical practice, public stewardship and continuous improvement."[7]

PREPAREDNESS

Preparedness is the cornerstone of emergency management. It begins with identifying all hazards in the community. After the hazards are identified, they must be prioritized based on their potential threat. This process is referred to as risk management, determining the probability and possibility of a critical event occurring. After the hazards have been prioritized, planning must be implemented. Planning identifies the needs of the community and the agencies that will respond to those needs. Local plans are based on the guidelines set forth in the National Response Framework. All emergency responders must understand how the NRF is related to their own department. The NRF identifies fifteen **emergency support functions (ESF)** that are to be included in all emergency plans. ESFs provide coordination of federal interagency support when federal agencies respond to an incident. They provide a structure for grouping functions that are most frequently used to support federal aid to states. These ESF are as follows:

ESF 1. Transportation
ESF 2. Communications
ESF 3. Public Works and Engineering
ESF 4. Firefighting
ESF 5. Emergency Management
ESF 6. Mass Care, Emergency Assistance, Housing, and Human Services
ESF 7. Logistics Management and Resource Support
ESF 8. Public Health and Medical Services
ESF 9. Search and Rescue
ESF 10. Oil and Hazardous Materials Response
ESF 11. Agriculture and natural Resources
ESF 12. Energy
ESF 13. Public Safety and Security
ESF 14. Long-Term Community Recovery
ESF 15. External Affairs

The **National Response Framework (NRF)** and the support annexes are available for download at the government website: www.fema.gov/emergency/nrf/.[8]

The NRF is a national emergency management guideline for federal, local, state, and tribal governments. It establishes an all-hazards approach to emergency management.

The local plan determines the feasibility of the disaster response and recovery actions. The plan must be written to clarify how response and recovery organizations will function and identify the resources that they will need to accomplish their tasks. Finally, it must be constantly updated and maintained through training and exercises.[9]

emergency support functions (ESF) ■
The ESF provide coordination of federal interagency support when federal agencies respond to an incident. They provide a structure for grouping functions that are most frequently used to support federal aid to states.

National Response Framework (NRF) ■
The NRF is a national emergency management guideline for federal, local, state, and tribal governments. It establishes an all-hazards approach to emergency management.

The preparedness function establishes the guidelines for response, recovery, and mitigation. It is an evolving process that identifies hazards and threats, determines the resources needed to protect the community, and analyze the community's critical infrastructure vulnerability.[10]

STOP AND THINK 12-2

Is your community's Emergency Response Plan up to date? Has it been tested?

1. Why is it important to maintain a current Emergency Response Plan?
2. In your community, who should be involved in developing the plan?
3. Does your community have mutual aid agreements with other communities and the private sector?

RESPONSE

Response is the most visible and shortest phase. During this phase all the activities required during an emergency are placed into action. This is the "boots on the ground" phase of emergency management. It involves the activation of the Emergency Response Plan and ESF from the NRF. Response begins with the preincident deployment and staging of resources and ends when the recovery efforts become the priority. Although there are many elements to response, three basic goals must be accomplished. The population must be protected, the damage from the primary impact (tornado, hurricane, and so on) must be limited, and the damage from secondary impacts (fires, hazardous materials release, and so on) must be minimized.[11] We are all familiar with the operational aspects, firefighting, search and rescue, medical response, fatality identification, maintaining civil order, and so on. There are, however, other duties that must be performed during the response phase. An important element of response is maintaining an efficient and effective command structure and emergency operations center. (See Figure 12-5.) One person in charge of field operations and a well-coordinated emergency operations center, to provide the logistics, are absolutely necessary to ensure efficient operations during the response

FIGURE 12-5 Incident management command center. *Courtesy of TEEX-Texas A&M System.*

phase. Accurate records must be maintained throughout the emergency, especially during response. The tracking, use, and cost of resources must be recorded. This includes person-hours accrued by paid and volunteer personnel and equipment and supplies provided from the private sector through mutual aid agreements. Proper records must be maintained to qualify for FEMA reimbursement.[12]

STOP AND THINK 12-3

How would your community respond to a major disaster? How would the mutual aid agreements aid in the response?

1. Does your community have the resources to effectively activate the Emergency Response Plan?
2. How would the resources be tracked as they are used?

RECOVERY

The process of recovery begins during the preparation stage. Like response, recovery must be recognized as a vital part of emergency management and incorporated into the disaster plan. The recovery process often begins during the response stage because the decisions made during the response will affect how the recovery phase will progress. Its purpose is to restore the community to the same status as it was before the disaster. It is estimated that thirty percent of the response is still in operation when recovery begins.[13]

There are two types of recovery, short-term and long-term. Short-term recovery involves those activities needed to restore the vital infrastructure required to sustain the health and welfare of the community, such as mass care, feeding, and sheltering victims; temporary housing; return of basic utilities; damage assessments; public information; restoring government and economic activities; and debris management. Short-term recovery may only take a few day or weeks to complete. Long-term recovery begins when the community starts to rebuild the critical infrastructure—roads, bridges, highways, public buildings—and the local economic base. This would include supporting the local business community by buying supplies from those businesses still in operation and helping destroyed businesses acquire small business loans to help them reopen. According to FEMA, more than 40% of businesses never reopen after a natural disaster. This number increases to 70% if the disaster is a result of a terrorist act. Recovery also includes unemployment assistance, crisis counseling, and donations management.[14] Long-term recovery may take months or years to complete. Three years after hurricane Ike, Galveston County, in Texas, submitted applications for funding for the ongoing recovery. The Galveston County Commissioners Court authorized applications for $31,558,883 for infrastructure and an additional $115,725,091 for the demolition and reconstruction or repair of single and multifamily homes.[15]

STOP AND THINK 12-4

How long would it take for your community to recovery from a major disaster?

MITIGATION

Mitigation is a phase of emergency management that is conducted throughout the other three phases. Mitigation may be proactive or reactive. Therefore, it is most prominent during the preparedness and recovery phases. As a proactive function, the overall intent of mitigation is to prevent a disaster from occurring. Natural disasters cannot be prevented. Some may think that human-made disasters can. However, as stated earlier in this section, disasters cannot be prevented, even human-made ones. The only thing that a community can do is reinforce the protection that it currently provides. Proactive mitigation begins with hazard and risk assessment during the preparedness phase. The type of hazards and the amount of risk the hazards will present will determine where additional resources will be utilized and the amount of resources that will be needed. These resources can be utilized to "harden the target" so as to eliminate or reduce the amount of damage that the disaster may incur. Proactive mitigation efforts also include, but are not limited to, activities involving building and fire code modification, floodplain management, and structural design.[16] Testing the Emergency Response Framework brings all of the community's stakeholders together in a mock, realistic exercise that will place demands on all departments and resources. This allows the EM to identify and mitigate weaknesses in the plan and allocate additional resource support before a real disaster occurs. The Disaster Mitigation Act of 2000 provides funding to state, local, and tribal governments to assist in mitigation planning and projects.[17]

Reactive mitigation begins at the end of the response phase and throughout the recovery phase. There is no such thing as a perfect Emergency Response Plan. There are many aspects to a disaster. Not all can be identified. For example, after the 1993 bombing of the World Trade Center in New York, mitigation procedures were placed into effect to help prevent or reduce the effects of a future bombing. However, no one anticipated someone using large commercial aircraft to attack the buildings. Reactive mitigation is relatively easy. The weaknesses are exposed and the consequences of the failure are evident. The questions to be asked are

1. How did this failure occur?
2. What can be done to prevent or reduce the impact of future disasters?
   a. Control of the hazard.
   b. Control of the human use of the hazard
   c. Control of both.[18]

The cost of reactive mitigation is much higher. Damaged infrastructure may need to be dismantled and reconstructed with new improvements. The damage to buildings may require new building codes with higher standards to withstand the forces faced during the disaster. This increases the cost of reconstruction. However, proactive mitigation is harder to "sell" to the public. Long periods of time between disasters may cause a community to become apathetic. It is estimated that approximately 20,000 residents refused to evacuate Galveston, Texas, before hurricane Ike struck on September 13, 2008. The reported fatalities from this storm range from 50 to 200. A final fatality count has never been reported.[19] Over 11,000 building permits were issued by the City of Galveston after Hurricane Ike made landfall. Most of the destruction was in the older residential areas of Galveston where houses were built under less restrictive building codes of the past. Convincing the taxpayers and the private sector to invest money for mitigation that will protect them in the future is very difficult.

Postsecondary Education for Emergency Management

As with the EMS degree plan, to strengthen a person's leadership skills there is an option to obtain higher education in the field of emergency management. You can now find accredited emergency management programs across the United States. Just to name a few examples, one could earn an Associate of Applied Science in Emergency Management Administration and Homeland Security, Bachelor of Arts in Emergency and Disaster Management, or a Master of Arts in Emergency and Disaster Management. Obtaining this type of higher education typically leads to jobs such as emergency program manager or director, emergency preparedness instructor, emergency operations center chief, director of security, risk management expert, hospital coordinator, or technical training supervisor. Furthermore, the recent state of natural disasters and potential of terroristic attacks demonstrates the need for individuals with skills and qualifications in emergency management.

ON SCENE

On October 29, 2012, Hurricane Sandy struck the northeast coast of the United States near Atlantic City, New Jersey. The storm measured 2,000 miles across and covered most of the northeastern United States. Its 80-mile-per-hour winds extended 175 miles from it center. The wind, however, was not the most damaging part of the storm. The 6- to 11-feet storm surge flooded many areas of New York and New Jersey. In some areas of New York City, the storm surge topped 13.88 feet. Several fires erupted in the New York metropolitan area. One fire in Queens, New York, destroyed 80 homes. It is estimated that Hurricane Sandy caused $20 billion in property damage and $10 to $30 billion in lost business.[20]

Summary

Fire and emergency services go hand in hand. One cannot go into either of these professions without considering some aspect of the other. As shown in this chapter, a majority of the calls that fire department personnel will encounter will be medically related. There are various levels of training required for EMTs that will enhance their careers and personal satisfaction in their job. With each level of medical training comes added responsibility and guidance from their state or local jurisdiction. As with firefighters, EMTs must always check their state and local rules and regulations that govern them.

Emergency management is a growing profession. It requires individuals with oral and written communication skills. The EM must also be able to organize and coordinate multidepartment and multiagency functions in a common Emergency Response Plan. He or she must also understand the functions of emergency management, preparedness, response, recover, and mitigation. Working with other departments and agencies, the EMT must identify and prioritize hazards through risk management. Resources must be allocated to allow the community to respond adequately to the disaster. Plans must be developed that will provide assets for community recovery activities Finally, based on information gathered during the preparedness phase or after the event has occurred, mitigation activities should be initiated to strengthen potential weaknesses against future events.

Review Questions

1. What are the required characteristics for an emergency manager?
2. What are the four phases of emergency management?
3. How many emergency support functions are in the National Response Framework?
4. Which phase involves the activation of the Emergency Response Plan and emergency support functions from the National Response Framework?
5. Describe the processes for short-term and long-term recovery.
6. Describe the two types of mitigation.

Suggested Reading

American Academy of Orthopaedic Surgeons (AAOS). (2010). *Emergency care and transportation of the sick and injured* (10th ed.). Burlington, MA: Jones & Bartlett.

Goss, Kay C. (2007). *Emergency management handbook.* Stillwater, OK: International Fire Service Training Association, Oklahoma State University.

Limmer, Daniel, and O'Keefe, Michael F. (2011). *Emergency care* (12th ed.). Upper Saddle River, NJ: Pearson Education.

Lindell, Michael K., Prater, Carla, and Perry Ronald W. (2007). *Introduction to emergency management.* Hoboken, NJ: Wiley.

Milete, Dennis S. (2004). *Disasters by design.* Washington, DC: Joseph Henry Press.

Endnotes

1. Bureau of Labor Statistics, U.S. Department of Labor. (2012, March 29). EMTs and paramedics. In *Occupational Outlook Handbook, 2012–13 Edition.* Retrieved January 29, 2013, from Bureau of Labor Statistics website: http://www.bls.gov/ooh/healthcare/emts-and-paramedics.htm.

2. National Fire Protection Association. (2012, March). The United States Fire Service Fact Sheet. Retrieved

October 6, 2012, from http://www.nfpa.org/assets/files/
PDF/Research/FireServiceFactSheet.pdf.

3. Webster, Henry Clay. (2010, December 6). Best careers: Emergency management specialist. Retrieved from *US News* website: Retrieved November 10, 2012, from http://money.usnews.com/money/careers/articles/2010/12/06/best-jobs-2011-emergency-management-specialist.

4. Federal Emergency Management Agency. (2013, April 9). About the agency. Retrieved from http://www.fema.gov/about/history.shtm.

5. National Fire Protection Association. (2007). *NFPA 1600 Standard on Disaster/Emergency Management and Business Continuity Programs* (Section 5.3). Quincy, MA: Author.

6. Blanchard, Wayne. (2007, September 7). *Federal Emergency Management Higher Education Project, North Dakota State University*. Retrieved April 14, 2013, North Dakota State University website: http://www.ndsu.edu/fileadmin/emgt/PrinciplesofEmergencyManagement.pdf.

7. Ibid.

8. Federal Emergency Management Agency. (2012, August 12). *National Response Network*. Retrieved October 5, 2012, from http://www.fema.gov/emergency/nrf.

9. Lindell, Michael K., Parter, Carla, and Perry, Ronald W. (2007). *Introduction to emergency management*. Hoboken, NJ: Wiley.

10. Goss, Kay C. (2007). *Emergency management handbook*. Stillwater, OK: International Fire Service Training Association, Oklahoma State University.

11. Lindell et al., op cit.

12. Ibid.

13. Becker, Christine. (2009, March). Disaster recovery: A local government responsibility. *PM Magazine*, *91*(2). Retrieved October 21, 2012, from http://www.ibts.org/news/press-releases/2009/disaster-recovery-a-local-government-responsibility.html.

14. Goss, op cit.

15. Colley, Jack. (2008, December 8). Texas Division of Emergency Management. *Hurricane Ike impact report*. Retrieved October 25, 2012, from FEMA website: http://www.fema.gov/pdf/hazard/hurricane/2008/ike/impact_report.pdf.

16. Goss, op cit.

17. Ibid.

18. Lindell et al., op cit.

19. Were Ike's missing washed away? (2008, September 18). Associated Press. Retrieved October 17, 2012, from NBC News website: http://www.nbcnews.com/id/26765035/ns/weather-hurricane_ike/t/some-ikes-missing-may-have-washed-away/.

20. Sharp, Tim. (2012, November). Superstorm Sandy: Facts about the Frankenstorm. *LiveScience Reference*. Retrieved February 9, 2013, from LiveScience website: http://www.livescience.com/24380-hurricane-sandy-status-data.html.

On Scene Suggested Answers

Chapter 1

1. Organizations such as the International Association of Fire Chiefs (IAFC) have programs, grants, and scholarships that can help firefighters and departments move forward while embracing fire service history and traditions.

2. The United States Fire Administration (USFA) and other entities listed in the text. Facts and figures from these agencies can be utilized to formulate a plan to start to remedy these issues.

Chapter 2

1. Hurricanes Katrina and Rita of 2005 and other weather-related disasters. Volunteers have been trained to respond to these types of calls, and if they feel a sense of patriotic duty to respond to the call, they should not be penalized in their job. Job protection can enhance a person's loyalty to his or her employer and can encourage more volunteerism.

Chapter 3

1. The fire service has always attracted different types of individuals. Reasons range from the excitement of the job to the respect and honor they receive from the community they protect. Most firefighters will tell you that they feel a great deal of accomplishment and believe that they are making a difference in the community they serve.

2. Prepare before you ever submit the application of employment. You might not realize it, but the interview process starts once the application is submitted when there's a position for hire. Research and practice can make all the difference between success and failure in the application and interview process.

Chapter 4

1. The National Professional Development Model. This was developed by the U.S. Fire Administration/National Fire Academy's Fire and Emergency Services Higher Education network.

2. The fire and emergency services field is taking on more responsibility in the industry. For this reason, fire departments are setting training and education standards for initial hiring because of this occupation's complexity and increased interest in it. Employers need individuals who can read, write, fill out reports correctly, follow directions, and understand policies and procedures pertaining to the job.

Chapter 5

1. The chief should work to build a positive relationship with the volunteer organization. Future plans for building the paid department should include active participation of the volunteers.

2. The chief should approach the city administration to negotiate the purchase of the assets of the volunteer department. A combination department should be formed as a transition department. This would include allowing the volunteers who wish to become full-time paid firefighters to receive the state training required to achieve certification. These firefighters can be absorbed into the paid department as needed. This may reduce some of the opposition to change. As the paid fire department grows, the volunteer department can be reduced.

Chapter 6

1. The incident commander must prioritize the course of action. The first priority is the safety of the firefighters. A defensive action may be needed to prevent exposure to and contamination by the chemicals and pesticides.

2. a. Chemical exposure and contamination of personnel and equipment.
   b. Exposure of the residents in the surrounding area to toxic or hazardous airborne pollutants.
   c. Contaminated water runoff.
   d. Possible explosive hazard if fire expands to fertilizer storage area.

3. Water will be an effective extinguishing agent, but it must be used sparingly, and runoff must be controlled. Using water will cause the chemicals and pesticides to combine, forming a toxic and hazardous mixture that will contaminate personnel and equipment. Some fertilizers become explosive when wet and then exposed to heat.

4. In this situation, it would be better to let the fire burn to destroy the chemicals and pesticides. Limit firefighter exposure by implementing a defensive strategy and ensuring that all firefighters wear full personal protective equipment and SCBA. An evacuation of areas downwind of the fire would be advisable. The area should be diked to prevent runoff water from entering storm drains or waterways. The EPA should be notified. A hazardous materials response team would have the expertise and equipment to acquire air and water samples to determine the level of contamination. All equipment and personnel must be decontaminated before leaving the area. All contaminated personal protective equipment must be decontaminated and secured so that it will not be used by any personnel. No equipment will be placed back into service until thorough decontamination operations have been conducted. All firefighters who responded to the fire would be required to complete a medical exposure report for documentation in case of future complications.

Chapter 7

1. The implementation of a neighborhood fire inspection program may be a tool to allow firefighters an opportunity to get a visual of construction, deterioration, and other changes in the community. Although home inspections are only voluntary, just driving around the community will help firefighters get a better idea of possible problems that may arise during a fire.

2. The two-story addition to the residence might have been detected if a community preincident plan had been developed.

Chapter 8

1. The officer must be familiar with the state laws concerning warrantless entry. It would probably be wise to request law enforcement response to the scene.

2. This depends on the state law. Some states defend the rights of landowners if the fire poses no threat to other property.

1. The U.S. Supreme Court has ruled that warrantless entry is allowed in the event of an emergency. It also has ruled that a fire is an emergency.

   Warrantless entry is allowed if an inspector observes a life-safety violation that can be viewed from a public way.

Chapter 9

1. The main issue is determining if the present water supply is adequate to supply this additional demand. If it isn't, a secondary water supply must be determined.

2. The chief should demand that storage facilities be constructed to ensure adequate water supply during drought. The size of the storage tanks will be based on flow requirements for the new development based on residential, industrial, and firefighting needs.

3. The chief should ask if there are plans to expand the development and what is the timeline to do so. He or she should ask the city administration if there are plans to develop industrial or business activities in the area. This will help determine the size of the infrastructure for initial construction.

Chapter 10

1. Command, Operations, Logistics, Planning, Finance and Administration, Information and Intelligence.

2. Yes.

3. The Environmental Protection Agency, the Department of Transportation, and the Occupational Safety and Health Agency.

Chapter 11

1. Request additional alarms and send apparatus downwind to protect exposures. Prepare for a strong defensive operation.

2. Life safety, evacuation of the buildings in close proximity to the fire, and downwind from the fire. Crowd and traffic control. Water supply.

3. Ensure all buildings in the complex are evacuated, set up hose lines to protect exposures downwind, set up heavy stream devices for a defensive attack.

Chapter 12

1. Public information is very important. Areas that are subject to severe weather should always be prepared to respond when the event occurs. Public apathy occurs when there is a long time span between severe weather events. Communities should exercise the emergency response plan often, to keep the public aware of the potential dangers. Public information notices should be scheduled periodically, through the various media outlets, to emphasize the importance for the public to be prepared to be self-sufficient for at least 72 hours until additional resources can be provided.

Answers to Review Questions

Chapter 1

1. Ancient Egypt.
2. 1648 at New Amsterdam, New York.
3. A group of men who carried wooden rattles that they would twirl to sound a fire alarm.
4. The apparatus Aldini was one of the earliest breathing apparatus. It was tested by the French in 1825.
5. Maltese cross.
6. The Fire and Emergency Services Higher Education organization was founded in 2000 by the National Fire Academy and the U.S. Fire Administration. Its purpose is to provide strategic approaches to professional development by moving the fire and emergency services from a technical occupation to a profession through higher education.

Chapter 2

1. The firefighter recruit is a position that has been created to help a person develop the skills to gain employment as a firefighter. The recruit trains as a firefighter, and a department or organization typically guarantees the recruit a job afterward, whereas a firefighter who attends an academy has to search for a job on graduation.
2. NFPA, FEMA, state and local agencies, the Forestry Service, police departments, medical directors, and so on.
3. The position requires advanced study and learning skills. The training required to become a paramedic is more extensive, and responsibilities at an emergency scene are dual. In many areas, paramedic training takes an additional six months to a year to complete.
4. Contract firefighters overseas, disaster assistance, hazardous materials specialists with a specialty in bioterrorism, and emergency medical specialty teams.
5. A fire prevention specialist teaches citizens about fire prevention and builds awareness to fire safety, whereas a training specialist is a person who provides firefighters or emergency personnel with programs to develop and maintain skilled firefighting performance.
6. Smokey Bear.
7. Dispatcher, fire prevention inspector, safety specialist, salesperson, insurance adjuster, and arson investigator.
8. Fire inspector, industrial fire brigade member, emergency medical technician, or paramedic.

Chapter 3

1. A valid driver's license and a GED or high school diploma.
2. Usually securing and filling out an application.
3. a. Subscribe to a notification service.
   b. Check newspapers.
   c. City human relations department.

4. Libraries and bookstores.

5. a. Engage in general physical training.
   b. Visit the fire station and use its equipment to prepare.
   c. Become a reserve/cadet or volunteer firefighter, which provides the training and practice at manipulating fire-related equipment.

6. a. Have firefighters give you practice with a mock oral interview.
   b. Consider possible questions and practice your delivery on tape or in front of a mirror.
   c. Learn all that you can about the fire chief and his or her department before the interview.

7. To let the employer assess your performance, both mentally and physically.

Chapter 4

1. Primarily technical.

2. Firefighter I, Firefighter II, Hazardous Materials Operation, Fire Instructor I, Fire Instructor II, Fire Investigator, Public Fire and Life Safety Educator I, Marine Firefighting for Land Based Firefighters—Awareness Level, Fire Officer I, Fire Officer II, Inspector I, Hazardous Materials Technician, and Airport Firefighter.

3. Associate of Applied Science.

4. National and regional.

5. Cadet/reserve programs and volunteer firefighting.

Chapter 5

1. Administrative personnel are close to day-to-day operations in the fire station. This close relationship tends to reduce resistance to policy and procedure changes because station personnel are better informed about the reasons for the changes.

2. Pumpers are designed for structural firefighting or for support operations that require sustained pumping activity. Initial attack pumpers are designed to combat small grass and vehicle fires. They may be used on structural fires but cannot be used for sustained operations.

3. Aerial platforms are designed to either operate with a telescoping or articulating box beam or truss beam boom. They are different from aerial ladder platforms, which consist of a heavily reinforced telescoping aerial ladder with a platform at the tip. Personnel use the ladder for rescue and to move equipment and tools to and from the platform. The ladders on the aerial platforms are designed only for emergency evacuation of the platform personnel.

4. Ambulance, hazardous materials apparatus, heavy rescue apparatus, aircraft rescue, and firefighting apparatus.

5. The Level B ensemble is similar to the Level A except that it does not give encapsulation protection from vapors. It offers splash protection only.

6. The PASS device is designed to emit a loud audible alarm when the wearer is motionless for more than 30 seconds. It may also be activated manually by the wearer if he or she becomes entrapped but not unconscious. Some PASS devices are also provided with a bright strobe light that serves as a location beacon.

Chapter 6

1. There are three components to an atom. The nucleus contains an equal number of protons and neutrons. The protons carry a positive electrical charge; the neutrons have no electrical charge. Negatively charged electrons circle the nucleus at different levels called shells.

2. When heat is transferred to matter, molecular movement increases. As the heat increases, molecules may collide with one another, and the matter may begin to decompose (pyrolyze), with some molecules escaping the surface of the matter as free radicals. Free radicals combine readily with oxygen in a chemical reaction known as oxidation, which is an exothermic or heat-releasing reaction. The newly created heat causes further pyrolysis in a chain reaction that keeps the fire burning.

3. Matter can exist in a solid, liquid, or gaseous state. When in a solid state, the molecules are packed closely together. They cannot move freely but only vibrate. The tighter the molecules are packed, the more dense the matter. The molecules in a liquid are more energetic and are able to move freely and slide past one another. This allows the liquid to flow and take the shape of the container in which it is placed. The molecules in a gas travel in a straight line at high speeds. Matter in the gaseous state occupies much more space than when it is in the solid or liquid state.

4. Heat energy can be classified as chemical, electrical, or mechanical. Chemical energy is released in oxidation reactions that produce heat. Chemical energy reactions can be in the form of spontaneous heating, heat of decomposition, heat of solution, or heat of reaction. Electrical heat energy is the passing of electrons from one atom to another. This movement is called current. There are six forms of heat energy: resistance, arcing, sparking, static, lightning, and induction. Mechanical energy comes in two forms: friction and compression. Friction heating occurs when two objects strike each other. Compression heating occurs when gaseous matter is compressed. As the gas is compressed, the molecules begin to collide with one another and with the walls of the container, and the temperature increases.

5. The four stages of fire are ignition, growth, fully developed, and decay. As soon as the ignition process is initiated, the growth stage begins. During the growth stage, heat from ignition is transferred to the molecules in the fuel, which begin to move more rapidly, colliding with one another and escaping from the surface of the fuel as vapor. Ignition of the vapor produces more heat, which radiates back to the fuel and initiates more molecular movement. If an adequate oxygen supply is available, this process will continue into the fully developed stage. The fully developed stage culminates with a phenomenon known as flashover. During this stage, burning fire gases travel laterally along the ceiling of the room and radiate heat down to the room's flammable contents. The contents begin to pyrolyze and give off vapors. The ignition of these vapors creates total area involvement. The decay stage occurs when adequate oxygen supply allows sufficient oxidation to deplete the fuel supply to the point that the fire will self-extinguish.

6. Fires are classified by the type of fuel that is burning. Class A fires are fueled by ordinary combustibles, such as wood, paper, and plastic. These products pyrolyze and can be extinguished by cooling with water. Class B fires involve flammable liquids and gases. These fires are extinguished by removing the oxygen from the fuel. Water-based foams create a film on the surface of the liquid, and the foam separates the vapors from the liquid. Class C fires are energized electrical fires. When the electricity is removed, a Class C fire becomes a Class A or B fire. Class D fires are combustible metal fires. Most of these fires require special extinguishing agents. Water may be used on some combustible metal fires but only if a very large quantity is available and used. Some metal fires burn at such high temperatures that water is decomposed into its constituent elements of hydrogen and oxygen and burns away, increasing instead of reducing the fire's intensity.

Chapter 7

1. a. NFPA 220
   Type I fire-resistive
   Type II noncombustible

Type III ordinary construction

Type IV heavy timber construction

Type V wood-frame construction

b. NFPA 5000®

Types I and II noncombustible

Types III, IV, and V combustible

2. a. First digit represents the fire-resistive rating in hours for the exterior load-bearing walls.

   b. Second digit represents the fire-resistive rating in hours for the interior load-bearing walls, beams, girders, trusses, arches, and columns.

   c. Third digit represents the fire-resistive rating in hours for the floor and roof assemblies.

3. Type II construction is noncombustible but unprotected. The unprotected steel has no resistance to fire exposure and can fail when heated to 1,000°F.

4. Columns and foundations are designed to resist compression forces.

5. The floors of old-style Type III structures are designed to collapse without destabilizing the exterior walls. Modern tactics and strategy provide for aggressive interior fire attack, thus exposing today's firefighters to the hazard of structural collapse.

6. a. Prestressed concrete components are constructed at a factory and transported to the worksite. Tension is placed on the cable inside the form before the concrete is poured into the form. The concrete bonds with the cable.

   b. Poststressed concrete components are constructed on the worksite. Tension is placed on the cable after the concrete has cured. The cable is not bonded with the concrete.

Chapter 8

1. Building codes are designed to address fire safety through building construction and design. Fire prevention codes are enforced by the fire department and focus on fire safety as it relates to building maintenance and use, fire protection systems, and hazardous materials and processes.

2. In 1866, Iowa State Supreme Court Justice John Dillon ruled that local jurisdictions have only the legislative powers that are granted to them by the state because the Tenth Amendment to the Constitution of the United States gives power only to the states, not to local governments. The Cooley Doctrine claims that the Tenth Amendment gives power to the states "or to the people," interpreting "to the people" to include local jurisdictions. The Cooley Doctrine is based on the principle of home rule, which was embodied in the English Carta Civibus Londonarium (1100 AD) and in the Magna Carta (1215 AD). Dillon's Rule was upheld by the U.S. Supreme Court in 1903 and again in 1927.

3. The Fourth Amendment to the U.S. Constitution protects the rights of citizens from unlawful search and seizure. If a fire official is denied entry for the purpose of code compliance inspection, entry can be acquired only through the issuance of a warrant. Warrantless entry can occur only if a life-safety issue exists or if the area being inspected is visible from a public way.

4. Minimum codes are adopted by the state but allow local jurisdictions to implement more stringent code requirements as needed. Mini-maxi codes are statewide codes and prevent local jurisdictions from modifying or adopting any other regulations.

5. Model codes can be adopted by reference or by transcription. If they are adopted by reference, the adopting ordinance simply refers to the edition of the model code that is to be used. If they are adopted by transcription, the entire model code must be published in the adopting ordinance.

6. Prescriptive codes establish specific requirements for building and fire safety. The mandates are based on past fire incidents that resulted in a large loss of life and property. These codes are usually not changed unless there is another tragic incident. They do not identify the fire safety objectives. If the building is constructed using the mandated requirements, fire safety will be achieved. Performance codes specifically identify fire safety goals and objectives and mandate the use of building materials and assemblies that meet those objectives.

Chapter 9

1. A water source, a method to deliver the water to its final destination, a method to treat the water, and a method to distribute the water to the individual customers.

2. A gravity system, a direct pumping system, or a combination system.

3. Water distribution systems are categorized as either separate or dual systems.

4. The base-valve dry-barrel hydrant is constructed of cast iron and designed to be freestanding. A valve operating system is located on top of the hydrant for easy access. When the valve is closed, water drains from the hydrant barrel. This hydrant is the most common hydrant used in the fire service. It is found in areas where temperatures are susceptible to dropping below freezing. The barrel of the hydrant is dry because the base-valve is installed below the frost line. Water does not enter the barrel until the base-valve is opened.

   Wet-barrel hydrants do not require a base valve. Individual valves control each outlet on the hydrant. This simple design allows additional hose lines to be attached to the hydrant without disruption of the water flow. Wet-barrel hydrants have fewer moving parts and are easier to maintain than base-valve dry-barrel hydrants. They are located in areas that are not subject to freezing temperatures.

5. 500 GPM from the first standpipe and 250 GPM from each additional standpipe.

6. NFPA 13.

7. Outside screw and yoke (OS&Y) valve, post indicator valve (PIV), wall post indicator valve (WPIV), and post indicator valve assembly (PIVA).

8. The post indicator valve assembly.

9. Fixed-temperature, rate-of-rise, or a combination fixed-temperature/rate-of-rise.

10. Light-obscuration or light-scattering.

11. When Halon 1301 is decomposed by heat, the halogens react with hydrogen formed by the combustion process to create toxic byproducts: hydrogen fluoride, hydrogen bromide, and bromine. Decomposition by-products of Halon 1211 include all of these compounds, plus hydrogen chloride and chlorine. Halon 1211 is the more toxic of the two agents. Total flooding CO_2 systems are designed to create a minimum CO_2 concentration of 34%. Life cannot be sustained in atmospheres with CO_2 concentrations above 9%. Asphyxiation is the primary danger posed by CO_2 systems. Dry chemical agents are nontoxic and nonconductive, but they can irritate the eyes, mucous membranes, and respiratory system. Some agents may cause chemical burns on moist skin. Personnel must always wear full protective equipment, including SCBA, when entering areas that have been extinguished with dry chemical agents.

Chapter 10

1. a. Unity of command means everyone answers to only one person.
   b. Chain of command means all communication and decision making travels in an orderly fashion up and down an interlinked chain.
   c. Span of control means the organization is set up so that people do not supervise more personnel than they can manage effectively, usually three to seven.

d. Division of labor means the jobs to be done are divided into manageable workloads and allow for specialization where necessary.

e. Delegation of authority means that when personnel are assigned a task, they are empowered with the authority as well as the responsibility to see that it is completed.

f. Exception principle means the supervisor is advised of situations or occurrences that are out of the ordinary.

2. This chart should begin with the fire chief at the top and, by using various titles, proceed down through company officer to firefighter. The answer to this question can include any of the specialty positions of line and staff.

3. The incident management system was developed as a result of a disastrous wildfire in 1970 in California. There was no command and control, resources were not effectively used, and communication between agencies was nonexistent.

4. FIRESCOPE ICS was developed for large wildland fires and coordinating multiagency response. FGC was a simpler command system that focused on structure fires, hazardous materials, and mass casualty incidents.

5. ICS has five main components: command, planning, finance, logistics, and operations.

6. a. Command controls the incident at the strategic level and has authority to control resources, establish functional areas, and ensure the safety of all responding personnel.

b. Operations is responsible for the implementation of the tactical assignments, fire, EMS, hazardous materials operations, and so on.

c. Planning is responsible for evaluating incident information and developing the incident action plan. Its function is to provide the incident commander with up-to-date status reports on resource requirements and allocation, situation development, incident control, responder injuries, intelligence, and suggestions for stabilizing an incident.

d. Logistics provides all of the support requirements for the incident operations, including facilities, family assistance centers, transportation, supplies, and equipment, and so on.

e. Finance/administration is responsible for documenting the cost of materials and personnel.

7. With the advent of the terrorist threat toward the United States, it is evident that emergency operations will include multiagency or multijurisdictional responses. The NIMS gives these organizations a common command structure, common terminology, and common communications. NIMS establishes who is in command and allows for a unified command on very large and complex incidents.

Chapter 11

1. Strategy is a broad-based plan for controlling an incident based on standard operating procedures and focuses on life safety, incident stabilization, and property conservation. Tactics involves the actual deployment of resources to resolve the situation and focuses on rescue, exposures, confinement of the hazard, extinguishment, overhaul, ventilation, and salvage.

2. The strategic plan identifies the task to be accomplished, identifies the resources needed, and specifies personnel requirements.

3. It is the most effective way to identify the hazard and determine the amount and availability of resources needed should an emergency occur.

4. Water supply, apparatus and equipment, available qualified personnel, weather conditions, topography, building construction, communications, and command structure.

5. OSHA 1919.134 and NFPA 1500.

Chapter 12

1. The emergency manager must have exceptional oral and written communications skills, organizational skills, and the ability to coordinate multidepartment and multiagency operations.

2. The four phases of emergency management are preparedness, response, recovery, and mitigation.

3. There are 15 emergency support functions.

4. The activation of the Emergency Response Plan and emergency support functions is implemented during the response phase.

5. Short-term recovery involves those activities needed to restore the vital infrastructure needed to sustain the health and welfare of the community, such as mass care, feeding and sheltering victims, temporary housing, return of basic utilities, damage assessments, public information, restoring government and economic activities, and debris management. Short-term recovery may only take a few day or weeks to complete. Long-term recovery begins when the community starts to rebuild the critical infrastructure, roads, bridges, highways, public buildings, and the local economic base. This would include supporting the local business community by buying supplies from those businesses still in operation and helping destroyed businesses acquire small business loans to help them reopen.

6. Proactive mitigation begins with hazard and risk assessment during the preparedness phase. The type of hazards and the amount of risk the hazards will present will determine where additional resources will be utilized and the amount of resources that will be needed. These resources can be utilized to "harden the target" so as to eliminate or reduce the amount of damage that the disaster may incur.

   Reactive mitigation begins at the end of the response phase and throughout the recovery phase. There is no such thing as a perfect Emergency Response Plan. There are many aspects to a disaster. Not all can be identified.

Stop and Think Suggested Answers

Chapter 1

Box 1-1

1. Implement the 16 Firefighter Life-Safety Initiatives.
2. Collect data and research to identify hazard areas that need to be improved.

Chapter 2

Box 2-1

1. Industrial fire brigades and contract firefighting in the United States and overseas.
2. Fire protection engineer or fire and emergency services consultants.

Box 2-2

1. Budget constraints.
2. Cost-effectiveness.
3. Human resource benefits are at a minimum. Specialized training is not available to local host nationals. The U.S. military is cutting back on its personnel, so it needs all members of its team to do what they are specifically trained to do—fight wars and protect our boundaries. Furthermore, overseas jobs offer incentives that conventional jobs do not, such as extremely high pay for hazard duty, travel opportunities, and tax-free incentives.

Chapter 3

Box 3-1

It is important for fire departments to have a baseline physical for each firefighter in case he or she ever becomes ill for unknown reasons. Unfortunately, there are still several fire departments that do not require a medical physical for an applicant when hired. Because firefighting is a hazardous occupation, firefighters must be monitored throughout their careers in case they are ever exposed to a hazardous material. This must be documented and kept on file in case the firefighter later develops symptoms from the exposure. Fire departments also need to develop and implement medical screenings and physical fitness standards that are equally applicable to all firefighters to make sure prior to hiring that the individual can perform the job in extreme working conditions. Firefighters must be in good health and in good physical condition to perform at a moment's notice.

Box 3-2

1. The definition given for diversity in this chapter is ethnic, gender, racial, and socioeconomic variety. However, diversity in America is much more than a simple list of differences. Diversity is at the core of what our country is, and it is our respect for diversity that draws so many people from around the world to our shores.
2. It is important to have diversity in the fire service because the service should reflect the entire composition of the community. Small towns to large cities often have residents

from different ethnic backgrounds, and the fire service should reflect the workforce in these areas.

3. The benefit of having a diverse workforce in fire and emergency services is a better working relationship with the community. An individual who does not speak English will receive better patient care from a firefighter who speaks the same language. A female patient will feel more comfortable talking to a female firefighter/paramedic than to a male firefighter/paramedic. Diversity in the workforce will allow better customer service to the taxpayers who provide the funding for the service.

Chapter 4

Box 4-1

1. Rope rescue, surface water rescue, vehicle and machinery rescue, confined space rescue, structural collapse rescue, trench rescue, dive rescue, and wilderness rescue.

2. NFPA 1006, *Standard for Rescue Technician Professional Qualifications*.

3. Some NFPA standards require that an individual accrue a certain number of continuing education hours annually. Furthermore, most state fire regulatory agencies require that certified individuals must complete a certain number of continuing education hours annually to stay certified in a particular discipline.

Box 4-2

1. It is important to have a seamless transition between courses taken by fire and emergency services personnel to ensure that all necessary subjects are covered and to reduce duplication of subject matter between courses.

2. A national professional development model would assist local, state, and federal government agencies in deploying resources and fire and emergency services personnel in the event of a natural disaster or terrorist incident.

Chapter 5

Box 5-1

Rural communities may need specialized apparatus, such as wildland (brush truck) apparatus or mobile water supply apparatus, which are not needed in some urban areas.

Box 5-2

Entering a hazard zone with inadequate protective clothing is like entering a hazard zone with no protective equipment at all.

Chapter 6

Box 6-1

The rate of heat release and intensity of the combustion process depends on the form of the fuel. Solids, liquids, and gases react differently to various environmental conditions, depending on their chemical composition. The form of the fuel also determines the strategy and tactics used in mitigating the situation.

Box 6-2

Firefighters should familiarize themselves with the types of hazards that exist in their jurisdictions. This can be accomplished only by conducting thorough preincident inspections and developing a preincident plan.

Box 6-3

Firefighters should always be aware of potential electrical hazards. These hazards are not limited to downed power lines. Caution must be taken when using aerial apparatus around any power lines or during stormy weather. Polyester material, often part of the firefighter's uniform, has a tendency to develop a static charge as the wearer moves. Flammable atmospheres created by gas or vapor leaks can be ignited by static discharge.

Chapter 7

Box 7-1

When structural components are weakened by fire, they may shift from their normal position and direct loads in a way that the support structure was not designed to withstand. This may cause a domino effect in which structural components will fail in sequence as loads and forces shift, weakening structural integrity. This is also referred to as a *progressive collapse*.

Box 7-2

Even though Types I and II structures are noncombustible, they are limited in their fire resistance. Unprotected steel structures are susceptible to early collapse when exposed to intense heat. It is very important that firefighters obtain information about the contents of Types I and II structures in their respective areas of response. The types of contents determine the fire load and rate of heat release to which the structure will be exposed if the contents burn. This will help firefighters estimate the length of time that the building will maintain structural integrity under heavy fire conditions.

Chapter 8

Box 8-1

Unscheduled inspections may be performed in response to a written complaint concerning a code violation. In some jurisdictions, if the violation is witnessed from a public way, an unscheduled inspection may be requested.

Box 8-2

Failure to perform a follow-up inspection after documenting a code violation will result in the inspecting officer's being held legally liable for any damage, deaths, or injuries resulting from a fire that is related to the violation that was documented.

Chapter 9

Box 9-1

Many factors may cause a large drop in residual pressure. First, a review of the water main maps in the area is needed to determine the main size and age. If the main is equal to the size of the mains in areas tested and is determined to be adequate, a report of the pressure drop should be made to the water department because it may indicate a possible partial blockage of the main. This can greatly affect fire operations in the area. Low residual pressure will reduce the ability of an apparatus to provide an adequate fire flow. This will limit the number of handlines or master streams that the apparatus can supply.

Box 9-2

Attempting to use the extended hose line from a Class II system may prove to be a fatal error. The water pressure may be low or nonexistent. Even if the pressure is adequate, the GPM flow may not be, especially if this is a well-involved fire. Extending hose lines up a stairwell is a time-consuming, arduous task and may be the only solution. If the situation allows, an aerial ladder or platform can be raised to a window on the fire floor. If the apparatus is not equipped with handline outlets, the nozzle tip on the master stream device can be removed and replaced with a gated wye (a valve with reduced outlets to which handlines can be attached). This allows the master stream device on the aerial apparatus to be used as a Class I standpipe.

Box 9-3

Most newspaper printing rooms have total flooding or local application carbon dioxide extinguishing systems. When these systems activate, not only is the oxygen displaced in the room, but the condensed environment also forms a fog that limits visibility. It is imperative that all responding personnel wear full protective equipment, especially self-contained breathing apparatus.

Chapter 10

Box 10-1

1. The finance sector can be divided into smaller sectors, with each sector having an assigned sector officer. The compensation and claims sector records all operational damage claims by citizens. The procurement sector ensures that no leased or rented equipment will be used unless there is a written contract of agreement. Records of all costs are documented by the cost sector. The time sector ensures that no personnel are allowed to work unless they are logged in and out and all hours worked are documented on a time sheet.

2. In order to receive aid from FEMA, an accurate accounting of all expenditures must be maintained. FEMA will reimburse 75% of the community's cost of a disaster. Small business loans are available to citizens who have suffered property loss.

3. For a community to qualify for reimbursement, FEMA requires that accurate spending records be maintained throughout the emergency. Written contracts must accompany all expenditures for leased or rented equipment, and records of operational costs for the equipment must be made available. Time sheets must be kept on all personnel working during the operation, including volunteers. If these records are not maintained properly, FEMA can deny any claim and leave the community to absorb all of the costs for response and recovery.

Box 10-2

1. Emergency response personnel should plan for a worst-case scenario and develop the NIMS structure that would be utilized should the event occur. After the plan is developed, emergency response personnel should conduct an exercise to determine the strengths and weaknesses of the plan. The exercise will also identify the responsibilities of all participants and their function in the National Incident Management System.

2. Conducting an exercise will help the incident commander identify the part of NIMS that needs to be activated. NIMS will identify which agencies need to be involved.

Chapter 11

Box 11-1

There is a great increase in occupancy load and live load forces. The increased amount of equipment, furniture, people, and the weight of the water for the aquarium adds live and dead loads that were not considered in the original design of the structure. The structure had to be reinforced to withstand these additional loads.

Box 11-2

1. Topography influences fire operations regardless of the geographical area. In some areas of the country, topography may be advantageous to emergency response and operations. In other areas, the topography may be viewed as an obstacle to emergency response and operations. This can be determined only by establishing a preincident strategic plan and exercising that plan so that all participants understand their roles and responsibilities.

2. It is very difficult to gain permission to create a prefire plan for a residential structure. This can be done only by obtaining permission from the occupants of the residence or, under limited circumstances, by obtaining a warrant. The best way to obtain information on a private residence is during its construction. The building contractor must have a permit that allows for periodic inspections by the building inspectors. Although a fire company may not have legal authority to inspect the property, depending on the local fire and building codes, it may get permission from the building contractor to perform a walk-through to familiarize personnel with the type of construction and the basic floor plan.

3. Weather, in conjunction with topography, can create major problems for strategic and tactical planners. All areas of the country experience seasonal changes as well. Strategic planners should incorporate response and operational alternatives into the plan to handle possible extremes due to the time of year or to extreme weather conditions.

Box 11-3

1. Barn fires in a rural setting are very dangerous. These structures often contain such hazardous materials as flammable gases (acetylene and propane), flammable liquids (gasoline and diesel fuel), ammonium nitrate (fertilizer), and various types of pesticides and herbicides.

2. Many of the hazardous materials found in rural settings can also be found in urban areas. Flammable liquids and gases are common to both environments, as are pesticides and herbicides. Powdered chlorine is found more often in an urban setting, especially in residential areas that have private swimming pools. This can pose a severe problem to firefighters. Burning chlorine gives off vapors that cause hydrochloric acid to form wherever there is moisture. Standard structural firefighting protective clothing is not effective in this type of situation. If dry chlorine powder mixes with certain hydrocarbons, a violent exothermic reaction occurs, producing high heat and toxic fumes.

Box 11-4

The people in the involved structure should be rescued first. The 5 people in the exposed building on the north side of the fire should be rescued next, because they are the largest number immediately threatened by the fire. The 20 people on the south side of the fire should be rescued last, because they are the victims least exposed to danger. The fire is moving away from them.

Chapter 12

Box 12-1

1. *(This should be answered according to each individual community's response plan.)*

2. *(This should be answered according to each individual community's response plan.) Explain resources available such as specialized teams, written protocols, and so on.*

3. *(This should be answered according to each individual community's response plan.) Mutual aid agreements are critical to have in place prior to mass casualty incidents and practice incidents are vital for training.*

Box 12-2

1. The local plan determines the feasibility of the disaster response and recovery actions. The plan must be written to clarify how response and recovery organizations will function and identify the resources that they will need to accomplish their tasks. Finally, it must be constantly updated and maintained through training and exercises.

2. It involves every agency and department in the community, as well as the private sector.

3. *(This should be answered according to each individual community's response plan.)*

Box 12-3

1. *(This should be answered according to each individual community's response plan.)*

2. Accurate records must be maintained throughout the emergency, especially during response. The tracking, use, and cost of resources must be recorded. This includes human-hours accrued by paid and volunteer personnel and equipment and supplies provided from the private sector through mutual aid agreements. Proper records must be maintained to qualify for FEMA reimbursement.

Box 12-4

1. *(This should be answered according to each individual community's response plan.)*

Expand Your Learning Activities

Newark Fire Department

Standard Operating Procedure

OCCUPATIONAL SAFETY AND HEALTH			
Subject:	*INCIDENT RESPONSE*	SOP#	403.03
		IAFF	
		Notice:	06/08/05
Written By:	CAPT. SPURGEON	Initiated:	07/01/05
Approved:		Revised:	
Fire Chief:	JACK STICKRADT	Date:	05/19/05
Safety Director:	KATHLEEN BARCH	Date:	06/06/05

I. Scope
A. This standard applies to the driver of a vehicle owned or operated by the department while responding to an incident. It was developed to establish safety guidelines during *Priority One* and *Priority Two* responses.

B. This standard was developed to ensure that the department's response to both *Priority One* and *Priority Two* incidents are safe, appropriate, and efficient.

II. Categories of Response
A. *Priority One:* Those incidents that pose a significant risk to life or property. *Priority One* response requires the use of all audio (siren and air horns) and visual (lights) warning devices in accordance with ORC 4511.041. These devices must be in use during the entire duration of the response unless an incident commander downgrades the response to a *Priority Two*. Units responding on a *Priority One* status shall identify such by the radio transmission *"Priority One"* (i.e., "Rescue One is responding *Priority One* to 123 Your St"). The initial response to the following types of incidents shall be considered *Priority One* emergencies:

1. *A reported fire in a structure.* This will receive a full first-alarm assignment.
   a. A full first-alarm assignment shall normally consist of all stations with a minimum of 2 engines, 1 ladder company, and 1 medic unit.
      (1) A "company" is defined as a firefighting apparatus staffed by at least three personnel.
   b. If "nothing showing" is reported by the incident commander, all companies excluding the first due shall downgrade their response to *Priority Two*.

Newark Fire Department

Standard Operating Procedure

Subject:	*INCIDENT RESPONSE*	SOP#	403.03

2. *A reported fire outside of a structure that involves the potential destruction of property or poses a risk to human or animal life* (e.g., dumpster, refuse, brush, etc.). This will receive 1 engine or ladder company assignment.

3. *A reported car fire.* This will receive 1 engine or ladder company assignment.
4. *Carbon monoxide checks with a reported illness.* This will receive 1 engine or ladder company and 1 medic unit assignment.
5. *Automatic fire alarms* shall receive a full assignment but are emergencies for the first due engine company only. All other units shall initially respond *Priority Two* and can be upgraded if needed.
6. *Inside gas leaks* shall receive the same initial assignment as an automatic fire alarm.
7. *Outside gas leaks* shall receive 1 engine or ladder company as the initial assignment.
8. *Mutual aid requests for fire support* shall receive 1 engine or ladder company as the initial assignment.
9. *Life-threatening Priority One medical incidents* (reports of unresponsiveness or unconsciousness) shall receive 1 medic unit and 1 engine or ladder company as the initial assignment.
10. *On all other Priority One medical incidents where an assist is required* the first due unit shall respond on a Priority One status and the additional unit shall respond Priority Two.
11. *On all Priority One medical incidents where an assist is not required* the response shall consist of 1 medic unit.
12. *Violent scenes* (stabbings, shootings, or fights in which a weapon is involved) shall receive the same assignment as a report of unresponsive or unconscious with the addition of 1 command unit. No units shall respond on a *Priority One* basis unless law enforcement personnel have secured the scene.
    1. In the event that a unit arrives on an incident scene that has not been secured, those unit(s) shall stage.

B. *Priority Two:* Those incidents that do not pose a significant risk to life or property. Audio and visual warning devices shall not be used during *Priority Two* responses unless ordered by an incident commander to upgrade the response to *Priority One* status. Units responding to such incidents shall identify themselves only as responding (i.e., "Rescue 1 is responding 123 Your St"). The initial response to the following types of incidents shall be considered *Priority Two:*
1. Carbon monoxide checks without a report of illness.
2. Public service calls to assist the public when there is no immediate threat to life or property.

III. **Response Guidelines**
   A. Apparatus and vehicles engaged in *Priority Two* responses shall obey all applicable traffic safety rules and regulations and shall not exceed the posted speed limit.
   B. Any and all traffic preemptive systems shall be utilized before exiting a division building prior to a *Priority One* response.

Newark Fire Department

Standard Operating Procedure

Subject:	*INCIDENT RESPONSE*	SOP#	403.03

C. Apparatus and vehicles engaged in a *Priority One* response shall at all times govern their response by the traffic, weather, and road conditions present at the time of response.
   1. The maximum speed of travel shall *not* exceed posted limits by more than 10 mph.

D. During a *Priority One* response, drivers shall bring their vehicles to a complete stop for any of the following:
  1. When directed by a law enforcement officer.
  2. Stop signs.
  3. Red traffic signals.
  4. Blind intersections.
  5. When the driver cannot account for all lanes of traffic in an intersection.
  6. When other intersection hazards are present.
  7. When encountering a stopped school bus with flashing warning lights.
E. Drivers shall proceed through an intersection only when the driver can account for all lanes of traffic in the intersection.
F. Drivers shall bring their vehicles to a complete stop at all railroad grade crossings and shall not cross the tracks until instructed do so.

IV. Responsibilities
  A. Drivers shall be directly responsible for the safe and prudent operation of their vehicles in all situations.
  B. When a driver is under the direct supervision of an officer, the officer shall assume responsibility for the actions of the driver and shall be responsible for immediately correcting any unsafe condition.

GLOSSARY

ability/agility test   A demanding physical test that can include dry hose drag, charged hose drag, halyard raise, roof walk, attic crawl, ventilation exercise, victim removal, ladder removal and carry, stair climb with hose, ceiling breach and pull, crawling search, or stair climb with air bottles and hose hoist—each measuring physical strength and stamina.

absolute temperature   Temperature that is measured from absolute zero.

absolute zero   The temperature at which all molecular movement ceases. This is established at −459.67° Fahrenheit, 0° Rankine (based on the Fahrenheit scale), −273° Celsius, and 0 Kelvin (based on the Celsius scale).

advanced life support units (ALS)   Medical units that provide advanced medical treatment and transport to the severely injured or ill.

aerial ladder   A rotating, power-operated ladder mounted on a self-propelled automotive fire apparatus.

aerial ladder platform   A power-operated aerial device that combines an aerial ladder with a personnel-carrying platform supported at the end of the ladder.

aerial platform apparatus   A fire apparatus that carries a hydraulically operated elevating platform that can be utilized for rescue, fire, or observation operations.

andragogy   The term relevant to lifelong learning for adult education methodology. Distance education is particularly suited for andragogical concepts.

apparatus Aldini   One of the earliest recorded attempts to improve breathing difficulties in toxic environments was tested in 1825 in France. This was not the most desirable breathing apparatus for firefighters to wear, but scientific testing was conducted under actual fire conditions.

articulating boom   A boom constructed of several sections that are connected by a hinged joint that allows the boom to function much like the human elbow.

aspect   The slope of land that is exposed to the sun.

Associate of Applied Science   Degree program usually two years in length.

atom   The smallest particle that constitutes the matter of an element.

axial loads   Loads that are placed perpendicular to the central axis of the support structure and utilize the full support capabilities of that structural component.

baccalaureate programs   Degree programs usually four years in length. Students receive a BS or BA degree.

backdraft event   Instantaneous explosion or rapid burning of superheated gases that occurs when oxygen is introduced into an oxygen-depleted space.

background check   Verification that a candidate's experience or credentials are what they represent.

Bambi Bucket   A large-capacity collapsible container used to transport water, via helicopter, to wildland fire locations.

barriers   Natural or artificial firebreaks for controlling fire spread.

basic life support units (BLS)   Medical units that provide medical and transportation needs for patients who do not have life-threatening injuries or illness.

bioterrorism   Relating to or involving the use of toxic biological or biochemical substances as weapons of war.

branch lines   The smallest pipes of a sprinkler system on which the individual sprinkler heads are mounted.

British thermal unit (BTU)   The amount of heat required to raise the temperature of one pound of water at 60°F by 1°F at constant pressure.

building codes Laws adopted by states, counties, or cities or other government sources to regulate all aspects of building construction.

burn building Training tower used so that new recruits can practice their skills.

caloric (or calorific) value The amount of heat necessary to raise the temperature of one gram of water by 1°C.

ceiling breach and pull This event is designed to simulate the critical task of breaching and pulling down a ceiling to check for fire extension. It uses a mechanized device that measures overhead push-and-pull forces and a pike pole. The pike pole is a commonly used piece of equipment that consists of a six-foot-long pole with a hook and point attached to one end.

certification programs Shorter educational programs that provide a certificate verifying expertise in a specific trade or discipline.

chain reaction of self-sustained combustion When a fuel is heated to the point that it releases flammable vapors, which burn and generate more heat on the fuel, which releases more flammable vapors.

citation Same as a summons except it is issued by a law enforcement officer instead of a judge.

clean extinguishing agent An extinguishing agent that leaves no residue or collateral damage resulting from the discharge of the agent.

combination flame detectors Alarm devices that can detect abnormal conditions by several means.

combination system A water supply system that uses mechanical pumps and gravity to supply operating pressure for moving water.

compensation and claims unit Part of the finance branch of ICS that documents property loss by citizens during a disaster.

compressed air foam system (CAFS) A high-energy foam-generation system consisting of an air compressor (or other air source), a water pump, and foam solution that injects air into the foam solution before it enters a hoseline.

compressive force A load that causes structural elements to shorten.

compressive stress The stress applied to materials that results in their decrease in volume.

conflagration A fire that increases in size and spreads beyond human-made and natural barriers.

corbels Pieces of stone jutting out of a wall to carry weight.

cost unit Part of the finance unit that tracks the cost of equipment and supplies.

couvre feu A French name for a metal lid used to cover an open hearth. The English word *curfew* was derived from this name.

criminal conviction record check A search of an individual's history of any convictions under the law for felony, misdemeanor, and motor vehicle convictions.

crossmains Pipes that supply the branch lines from the main riser. They are slightly smaller than the riser but are larger than the branch lines.

dead-load The structure and any equipment or building components that are permanently attached to the building.

Department of Homeland Security Created in 2002 primarily from a conglomeration of existing federal agencies in response to the terrorist attacks of September 11, 2001. Its purpose is to protect the nation against threats to the homeland. Its efforts affect firefighters and emergency personnel.

direct pumping system A water supply system that relies on mechanical pumps to move water.

discretionary authority Allows the fire official to use discretion when enforcing the law.

distribution grid The system in which power travels from a power plant to individual customers.

distributor pipes Piping located inside a building.

diversity Ethnic, gender, racial, and socioeconomic variety in a situation, institution, or group.

DOT National Standard Curriculum The Department of Transportation first developed this standard in the 1980s. This curriculum is designed to instruct a student to the level of Emergency Medical Technician-Basic. This curriculum is critical because it is recognized that the majority of prehospital emergency medical care will be provided by the EMT-Basic who serves as a vital link in the chain of the health care team. In 2011, this curriculum was again reviewed and updated to serve the field of emergency medical care even better in the respective fields of scope of practice and practice guidelines.

drafting pit A poured-in-place concrete pit used to simulate drafting operations.

drill tower A tower and training facility for firefighters.

eccentric loads Loads that are placed perpendicular to the support structure but do not pass through the central axis, thus causing the support structure to bend.

electrons Negatively charged particles that orbit the nucleus of an atom.

elevation The height of the land above sea level.

Emergency Management Institute (EMI) Serves as the national focal point for the development and delivery of emergency management training to enhance the capabilities of federal, state, local, and tribal government officials; volunteer organizations; and the public and private sectors to minimize the impact of disasters on the American public.

Emergency support functions (ESF) Areas that are included in all emergency response plans.

EMT Emergency Medical Technician. A specified level of medical training that usually consists of around 100 hours of classroom and on-site training.

encrustation The buildup of chemical deposits and biological growth on the inside walls of a water pipe, which reduces the inside diameter of the pipe.

endothermic reactions Chemical reactions in which substances absorb heat energy.

energy The capacity to do work.

engineering corrections Corrections that enhance fire safety. These corrections may include the installation of fixed fire protection systems, such as automatic sprinklers.

enhanced 911 An emergency phone system that automatically gives a dispatcher the caller's location.

ergonomic The scientific principle that bases design around human needs, and the profession that applies science and data to this concept.

exothermic Chemical reaction between two or more materials that changes the materials and produces heat.

external exposures External areas or persons in close proximity that can be threatened by a fire.

FAA An agency in the U.S. Department of Transportation that oversees all aspects of civil aviation.

familia publica "Servants of the commonwealth." These men were strategically positioned near the city gates to protect the city from fire.

Federal Emergency Management Agency (FEMA) The federal agency responsible for supporting state and local municipalities during a disaster.

FESHE Fire and Emergency Services Higher Education. Working with coordinators of two- and four-year academic fire and emergency medical services (EMS) degree programs, the U. S. Fire Administration's National Fire Academy (NFA) has established the FESHE network of emergency services–related education and training providers. The FESHE mission is to establish an organization of postsecondary institutions to promote higher education and to enhance the recognition of the fire and emergency services as a profession to reduce loss of life and property from fire and other hazards.

fire ecology The study of the interrelationship of wildland fires and living and nonliving things in the environment.

fire flow Quantity of water available for firefighting in a given area.

fire ground command system (FGC) The incident management system developed for fire ground and emergency medical operations.

fire lines Boundaries around a fire area to prevent access except for emergency vehicles and relevant professionals.

fire load The amount of fuel in a compartment expressed in pounds per square foot obtained by dividing the amount of fuel present by the floor area.

fire prevention codes Body of laws designed to enforce fire prevention and safety.

fire science Study of the behavior, effects, and control of fire.

fire wardens Volunteers who patrolled the cities, inspected chimneys, and issued fines for noncompliance. Funds collected from fines were used to purchase new fire equipment.

firemark Metal marker that used to be produced by insurance companies for identifying their policyholders' properties.

fireplug A wooden plug inserted in logs that carried the town's water supply approximately every half block. These plugs could be removed to access the water supply for firefighting.

FIRESCOPE Firefighting Resources of California Organized for Potential Emergencies.

fire tetrahedron Identifies the four basic components required for the combustion process to take place.

flashover (rapid fire progress) Stage of a fire at which all surfaces and objects within a space have been heated to their ignition temperature and flame breaks out almost at once over the surface of all objects in the space.

flash point Minimum temperature at which a liquid gives off enough vapors to form an ignitable mixture with air near the liquid's surface.

flow pressure The pressure created by the forward velocity of a fluid at the discharge orifice.

friction loss Pressure loss caused by turbulence as water flows through a conduit.

GIS Provide access not only to street maps but also to geographical data such as water sources, hazardous materials storage sites, residences with disabled persons, and municipal hydrants.

gravity system A water supply system that uses only gravity to move water.

Great Fire of Boston Boston's largest urban fire and one of the largest fires in American history.

Great Fire of London A devastating fire in central London that was considered one of the major events in the history of England.

groundwater Water located beneath the ground surface.

gusset plates Metal plates used to unite multiple structural members of a truss.

hand pumper A fire engine that has long bars (also known as brakes or pumping arms) running parallel to the body that operate the pump. Firefighters physically use the pumping arms to make a full up and down motion to build up pressure, allowing water to spray out of a hose.

head pressure The height of the surface of the water supply above the discharge orifice is equal to 0.434 pound per square inch (psi) per foot of elevation.

heat of combustion Total amount of thermal energy that could be generated by the combustion reaction if a fuel were completely burned.

heat sink A material that has the ability to accumulate and store heat faster than it can be dissipated and to release the heat when the surrounding environment cools.

heavy rescue squad A team of firefighters who are highly skilled in various types of rescue operations that require the use of highly specialized equipment and training.

helitack Used during wildfires when the location of the fire(s) is inaccessible to firefighting crews.

higher education Obtaining knowledge for certifications, professional development, or college credit. The act of accumulating new facts to augment your personal and mental database.

high-pressure systems Carbon dioxide is stored at room temperature in standard Department of Transportation–approved cylinders at a pressure of approximately 850 pounds per square inch.

Homeland Security The department created after the events of September 11, 2001, that is responsible for assessing the nation's vulnerabilities.

hotshot Highly trained firefighters used primarily in handline construction.

hydraulically operated Equipment that operates by utilizing pressure created by compressing oil in a cylinder.

hydrophoric materials Substances that react with water.

ignition temperature or fire point The temperature at which the liquid will emit vapors at a rate that will support a sustained flame after the ignition source is removed.

incident command system (ICS) A blueprint for organizing the response to an emergency situation.

incident effectiveness The ability of the department's personnel to perform their functions safely and effectively during an emergency. This requires physical fitness, training, and equipment.

Incident Management System (IMS) A merger of FIRESCOPE'S organizational design and command structure with the tactical and procedural components of the Fire Ground Command system.

indicating valve A valve that can be visually inspected to determine whether it is open or closed.

infrared flame detectors Flame detectors that operate in extreme environment conditions.

initial attack pumper A fire pumping apparatus that is utilized for fast response and rapid attack on a fire.

injunction A legal document from a judge that orders a person or entity to cease or to perform certain activities.

Integrated Emergency Management System A management philosophy that builds awareness and skills that is designed for personnel who fill specific emergency support positions within their community. This system implements policies, plans, and procedures to protect life and property through applications of sound emergency management principles in all phases of emergency management.

internal exposures Areas or persons in close proximity that can be threatened by a fire within the involved structure.

International Fire Service Accreditation Congress (IFSAC) Focuses on ensuring that training and certification within its member jurisdictions meet strict National Fire Protection Association Professional Qualification standards. IFSAC is divided into two assemblies: the Certificate Assembly and the Degree Assembly.

ionization detectors Smoke detectors that use a finite amount of radioactive material to make the air within a sensing chamber conduct electricity.

latent heat The amount of heat energy absorbed or released by a substance when it changes physical states without a change in temperature.

latent heat of fusion The amount of heat absorbed by a substance when converting from a solid to a liquid at the same temperature.

latent heat of vaporization The amount of heat absorbed by a substance when converting from a liquid to a gas or vapor at the same temperature.

Level A Ensemble Highest level of skin, respiratory, and eye protection that can be afforded by personal protective equipment.

Level B Ensemble Personal protective equipment that provides the highest level of respiratory protection but a lesser level of skin protection.

Level C Ensemble Level B personal protective ensemble with filtered respirators for respiratory protection.

Level D Ensemble The standard work uniform worn at the fire station.

light-obscuration systems Smoke detection systems that are activated when smoke particles block a light beam from a photoelectric light-sensitive cell.

light-scattering systems Smoke detection systems that are activated when light beams are scattered by smoke particles and strike a light-sensitive photoelectric cell.

line operation   Those activities that involve the direct physical tasks necessary for the completion of the mission.

load-bearing   A component of a structure that is designed to bear the weight of the structure and loads and transmit this weight to the foundation.

local application systems   Permanent suppression systems that are required to cover a protected area with 2 feet of foam depth within two minutes of system activation.

Local Emergency Planning Committee (LEPC)   This committee comprised of representatives from all of the emergency services, local medical community, public works, the news media, volunteer relief organizations, and private business. The LEPC coordinates with the local emergency manager to develop a disaster response plan. Each member of the LEPC is assigned an area of responsibility.

low-pressure systems   Liquefied carbon dioxide at 0°F is contained in large refrigerated storage tanks pressurized to 300 pounds per square inch.

Maltese cross   Symbol representing the traditions and ideals of the fire service. The arms and the tree of the Maltese cross are equal in length. The arms and the tree widen as they extend from a central point. In most cases, the edges are flat or curved slightly outward.

mandatory authority   Makes the actions mandatory under the law.

manipulative training   Hands-on learning of operations, equipment, and tools. In the fire service, this type of learning exhibits safety and proper techniques.

matter   The substance or substances of which all physical things are composed. It occupies space, has weight, and can be measured. It exists as a solid, liquid, or gas.

mini-maxi code   Establishes minimum and maximum building code requirements.

minimum codes   Lowest acceptable standards.

mobile data terminal (MDT)   Mobile computer that communicates with other computers on a radio system.

model codes   Building codes developed and maintained by a standards organization separate from the governing body for building codes.

molecule   Groups of two or more atoms.

multipurpose dry chemical systems   Dry chemical systems that use extinguishing agents that are rated to be effective on more than one type of fuel.

National Board of Fire Underwriters (NBFU)   Organized in 1866, the NBFU developed standards for fire safety in building construction and control of fire hazards.

National Fire Academy (NFA)   Concerned with education—designing fire-control courses and programs and making them available to firefighters across the nation.

National Incident Management System (NIMS)   A reconstruction of NIIMS to accommodate additional needs arising from the nature of the new threat—terrorism—that the nation now faces.

National Interagency Incident Management System (NIIMS)   A total systems approach to risk management that integrates all levels of national emergency incident planning—federal, state, and local.

National Response Framework (NRF)   A national set of guidelines on which all local emergency plans are based.

negligence   The failure of a fire official to perform in the same manner a prudent person would under the same circumstances.

neutron   A particle in the nucleus of an atom that has no electrical charge.

NFPA   National Fire Protection Association.

NFPA 1001   Standard for firefighter professional qualifications.

NFPA 1041   A standard that requires instructors to be cognizant of the safety of their students to ensure that classroom and practical evolutions are conducted in a safe, controlled manner. Prerequisites and requirements are mandated.

NFPA 1403   A standard that provides the framework to improve safety during live fire evolutions and covers conditions such as safety, site preparation, water supply, training plan, fuel, and ventilation.

NFPA 1582 *Standard on Comprehensive Occupational Medical Program for Fire Departments.* Standard covers minimum medical requirements for firefighters, including full-time or part-time employees, and paid or unpaid volunteers.

NIMS National Incident Management System.

non–fire suppression activities Any fire department activity that is not directly involved in fire suppression.

nonindicating valve A valve that has to be physically operated to determine if it is open or closed.

non-load-bearing A partition or wall that supports only itself or interior finish components.

nonpressurized dry hydrant system A hydrant system that is used to draft water from a static water source.

normal atmospheric pressure The pressure of the atmosphere at sea level is 14.7 pounds per square inch.

NREMT National Registry of Emergency Medical Technicians.

occupancy General term for how a building structure or home will be used and what it will contain.

occupation classification Classification given to a structure by a model code used in that jurisdiction and pursuant to the given use of that structure.

ordinance A law that is established by a local jurisdiction, such as a county, city, or town.

ordinary dry chemical systems Dry chemical systems designed primarily to extinguish flammable liquid fires (Class B fires).

OSHA Occupational Safety and Health Administration.

OSHA Code of Federal Regulations 29 CFR 1910.120 The federal code that mandates the level of training for anyone working with hazardous materials.

oxidation Chemical process that occurs when a substance is combined with oxygen.

paramilitary organization A group of civilians organized in a military fashion.

permit model Businesses that have a high potential for loss of life, such as nightclubs, restaurants, and hazardous materials processing and storage facilities, require a permit to operate.

personal alert safety system (PASS) Electronic lack-of-motion sensor that sounds a loud tone when a firefighter becomes motionless.

photoelectric detectors Smoke detectors that use small light source to detect smoke.

pictometry An enhancement of GIS that generates a 3-D perspective of a given area.

prehospital care A term used for any emergency care bestowed to a person that is not provided in the hospital setting. This care can be administered by a firefighter, EMT, police officer, or layperson. The standards for prehospital emergency care and the individuals who provide it are governed by the laws in each state and are typically regulated by the state office of EMS.

preincident planning Assessing risks and prioritizing potential hazards, then developing a strategic plan to mitigate the hazards.

prescribed fire A fire that is purposely ignited by fire personnel or agencies under controlled conditions for specific management objectives.

pressurized dry hydrant system Piping grid that connects several dry hydrants. A fire department connection is placed at the base of the grid near a water source. During firefighting operations, a pumping apparatus connects to the water supply and supplies water to the dry hydrant system.

primary feeder mains Large-diameter pipes that form the outside perimeter of the water distribution grid and receive water from the water source.

Pro Board National Board on Fire Service Professional Qualifications Accrediting Fire Service Training Organizations.

probationary period Introductory period of employment that allows the employee and agency to determine if the employee is suited for the job.

procurement unit Part of the finance branch that establishes contracts for equipment that may be needed during an emergency.

proton A particle in the nucleus of an atom that carries a positive electrical charge.

public safety department A department that employs a person who is triple-certified as a police officer, emergency medical technician (Basic or Advanced level), and firefighter.

pump-and-roll operations The capability of an apparatus to pump water while it is in motion.

pyrolysis The chemical decomposition of a substance by heat.

pyrophoric material substances that can spontaneously ignite when exposed to air.

rate of heat release The speed at which a fuel releases its caloric value when burning.

regulations Mandated by law.

residual pressure The pressure that remains in the system when water is being discharged.

resistance The opposition of the flow of an electric current in a conductor component.

riser The largest pipe in a sprinkler system. It is connected directly to the main water supply line and contains the sprinkler alarm valve. It supplies water to the crossmain.

secondary feeder mains Piping that interconnects with the primary feeder main.

sedimentation The accumulation of mud, sand, and other debris inside a water pipe, reducing the inside diameter of the pipe.

self-contained breathing apparatus (SCBA) Wearable respirator supplying a breathable atmosphere that is carried or generated by the apparatus and is independent of the atmosphere.

shape of the area Topographic shapes, such as canyons, ridges, mountains, and so on.

shear forces Loads that cause the structural element to deform or fracture in a direction parallel to the force, by sliding one section of the element along others.

siphona A large syringe used to deliver a stream of water.

slope The angle of the hillside.

smoke jumpers Firefighters who parachute from airplanes to suppress forest fires in remote locations.

Smokey Bear Created in 1944. The longest running public service campaign in U.S. history for fire prevention.

sole plate The bottom-most component of a wall assembly. The bottom of the wall studs are connected to this component, which in turn is connected to the foundation of the structure.

spalling The process in which parts of a concrete structure break or chip off when the structure is exposed to extreme heat.

span of control The number of units or personnel that a supervisor can effectively manage.

Sparky the Fire Dog Created in 1951. Used for NFPA's Risk Watch and Learn Not to Burn programs and has a special appeal to children's programs.

specific gravity Weight of a substance compared with the weight of an equal volume of water at a given temperature.

specific heat The amount of heat required to raise the temperature of a specified quantity of material by 1°C.

staff operation The administrative activities necessary to accomplish day-to-day operations.

stair climb Applicant must lift a prepared hose bundle from the floor onto the shoulder and climb a specific number of the stairs in a specified time frame. Applicant must be wearing an air tank and harness (without valves, hose, or mask) during this evolution.

standard operating procedures (SOP) Rules by which an organization or fire department operates for its day-to-day functioning. Usually these procedures are documented in a handbook or related source.

standards A recommended course of action that is not mandated by law.

static electricity heating The heat generated by static electricity arcing.

static pressure Energy that is available to force water through pipes and fittings.

statute A law established by a state or federal legislative body.

steam engine   A heat engine that performs mechanical work using steam as its working fluid.

stoichiometric reaction   A reaction in which all reactants are present in fixed, definite proportions for the reaction to go to completion. The most violent of all combustion processes occurs when the fuel and oxygen proportions provide complete combustion, leaving no fuel or oxygen residue.

strategy   Planning and directing the maneuvering of personnel and equipment into the most advantageous positions prior to actual operations.

subsidence   A phenomenon in which the surface of the ground sinks as water is drawn from the underground aquifer faster that it can be replenished.

summons   A written document from a judge or a law enforcement officer ordering a code violator to appear in court on a certain date.

surface water   Water collected on the ground from a stream, lake, ocean, or wetland.

tactical training   Comprehensive training, hands-on learning, and analysis for emergency incidents.

tactics   The actual maneuvering of personnel and equipment in an operation to achieve short-term objectives.

target areas   Areas that are of high value or where there is a potential for a large loss of life.

telescoping boom   Aerial device raised and extended through sections that slide within each other.

tensile force   A force that causes the structural element to lengthen.

tensile stress   Stress in a structure that tends to pull it apart.

thermal imaging camera (TIC)   Gives firefighters the ability to see heat generated by an object or person.

time unit   Department that tracks all time sheets for every individual working on the emergency scene throughout the duration of the incident.

top plate   Horizontal part between where the roof and studs finish.

torsional loads   Loads that are placed only on the longitudinal axis of the structural component, causing it to twist.

total flooding systems   Extinguishing systems that are designed to completely fill the protected area with extinguishing agent when activated.

training   Obtaining knowledge for the maintenance of skills required for actual tactical tasks performed during an emergency response.

training bureau   A division of an organization or department whose mission is to provide training to firefighters so that they have the knowledge, skills, and abilities to mitigate emergency incidents safely while minimizing the risks to themselves, civilians, and the environment.

tuberculation   The buildup of rust on the inside surface of a metal water pipe, which reduces the inside diameter of the pipe.

Type I combustion   Direct oxidation. Pyrolysis or decomposition of the matter is not required to occur in order to sustain combustion.

Type II combustion   Sequential oxidation. The molecular structure of the substance must be broken down to produce combustible gases, which will react with oxygen in the air.

ultraviolet flame detectors   Fire detection devices that are activated when ultraviolet light waves are detected.

unity of command   The concept that every member of the organization answers to just one supervisor.

vapor density   Weight of a given volume of pure vapor gas compared with the weight of an equal volume of dry air at the same temperature and pressure.

vapor pressure   The pressure exerted on the walls of a closed container by a liquid when the number of molecules escaping from the liquid reaches an equilibrium with the number of molecules returning to the liquid.

ventilation   The systematic removal of pressure, gases, and heat from structure or confined area.

warrants   Orders from a judge allowing fire officials the right to search or seize private property or arrest an individual.

watersheds   Geographic areas around a body of water that have a topography that allows all rainfall in that area to flow into and replenish that body of water.

wet chemical   Potassium carbonate, potassium acetate, potassium citrate, or a combination these chemicals, and phenolphthalein or phosphoric acid, mixed with water. Used for extinguishing cooking oil fires.

wildland firefighting apparatus   A fire pumping apparatus with a high ground clearance and off-road capabilities.

INDEX

A

Ability/agility test, 38
Absolute temperature, 102
Absolute zero, 102
Accelerator, in sprinkler system, 190
Accrediting Commission of Career Schools and Colleges of Technology (ACCSCT), 61
Action plan. *See* Strategic action plan
Adjusters. *See* Insurance adjusters
Administration
    IMS, 218
    of model codes, 159–60
Administration buildings, 65–66
Adoption by reference/transcription, model code, 159
Advanced EMT (AEMT), 249
Advanced life support units (ALS), 77
Advancement, 51–54
Aerial apparatus, 73–76
Aerial ladder platform, 74
Aerial ladders, 68, 73–74
Aerial platform apparatus, 74
Agility test. *See* Ability/agility test
Aircraft, firefighting, 81–82
Aircraft Rescue and Firefighting Apparatus (ARFF), 80–81
Aircraft rescue firefighter (ARFF), 22–23
Air pack, 10
Alarm-indicating device circuits, signaling systems, 192
Alarm systems, local, 193
Alarm valve, 186
Aldini, apparatus, 10
Ambulances, 77
America Burning, 12, 50
America Burning—Revisited, 13
American Insurance Association (AIA), 153
American Water Works Association
    Standard C502–94 *(Standard for Dry Barrel Fire Hydrants),* 172
    Standard C503–88 *(Standard for Wet Barrel Fire Hydrants),* 172
Andragogy, 56
Apparatus
    early, 6–10
    fire (*See* Fire apparatus)
    and preincident planning, 231–32
    specialized (*See* Specialized apparatus)
Apparatus Aldini, 10
Apparatus bay, 68
Apparatus engineers, 23
Aqueous Film Forming Foam (AFFF) foam, 204
Aquifers, 164–65
Arch roof, 134, 135
Arcing heating, 108–9
Arson units. *See* Fire Investigation Units (FIU)
Articulating boom, 74, 75
Aspect, 236
Assistant chiefs, 212
Associate of Applied Science degree, 55
Atkins v. Kansas, 156
Atmospheric pressure, 99
    normal, 100
Atom, defined, 97

Augustus, Emperor, 2
Authority, delegation of, 214
Automatic detection systems, 195–98
    carbon monoxide detectors, 198
    flame detectors, 197
    heat detectors, 195–96
    smoke detectors, 197
Automatic dry system, 179
Automatic sprinkler system, 180–92
    components, 182–89
    deluge system, 191
    dry-pipe system, 190
    preaction system, 190–91
    residential, 192
    wet-pipe system, 189–90
Automatic Vehicle Location (AVL), 31
Automatic wet system, 179
Auxiliary appliances, and preincident planning, 234
Auxiliary signaling systems, 194
Avillo, Anthony, 230
Axial loads, 121

B

Baccalaureate programs, 56
Backdraft event, 114
Backflow preventer, 182–83
Background check, 41–42
Badges, 11
Balloon-frame construction, 146, 147
Baltimore, great fire, 154
Bambi Bucket , 81
Barn roof. *See* Gambrel roof
Barriers, 236–37
Base-valve dry-barrel hydrants, 171–72
Basic life support units (BLS), 77
Bathroom facilities, 68
Battalion, 211, 213
Bay, apparatus, 68
Beams, 124–25
    unrestrained steel, 137
Bilingual ability, 43
Bioterrorism, 32
Board for the Coordination of the Model Codes (BCMC), 136
Body language, 40–41
Boots & Coots International Well Control, Inc., 30
Boston
    Cocoanut Grove nightclub tragedy, 152
    fire engine, 7
    Great Fire, 7–8
Bracing, in walls, 126
Braithwaite, John, 7
Branch lines, 186–87
British thermal units (BTUs), 102
Brush trucks, 71–72
Bucket brigades, 4
Building codes, 153. *See also* Fire prevention codes
Building construction, 119–48
    loads/forces, 120–22
    overview, 120
    and preincident planning, 230

    structural components (*See* Structural components, in building construction)
    Type I, 136–40
    Type II, 140–41
    Type III, 141–44
    Type IV, 144–45
    types, 135–48
    Type V, 146–48
Building Officials and Code Administrators International (BOCA), 154
Burn building, 49
Burning process, 111
    decay stage, 114
    fully developed stage, 113
    growth stage, 112–13
    ignition stage, 112
Burns, Eddie, 60
Burns, Eddie W., Sr., 32
Butterfly roof, 134

C

Cadets. *See* Fire cadet/recruit
Caloric/calorific value, 104
Camara v. Municipal Court, 157
Candidate Physical Ability Test (CPAT), 38
Cantilevered walls, 126
Captain, 211
Carbon dioxide extinguishing systems, 200–201
Carbon monoxide (CO) detectors, 198
Career opportunities, 18–33. *See also* Selection process
    firefighting, 19–26
    future, 31–32
    nonfirefighting, 26–30
    overview, 18–19
    private sector, 30–31
    volunteer, 31
Carey, John, 180
Ceiling breach and pull test, 38
Ceiling joist, 130–31
Central station system, 195
Certificate Assembly, IFSAC, 59, 60
Certifications, 52–53
    programs, 55
    specialty, 52, 53
Chain of command, 214, 222
Chain reaction of self-sustained combustion, 99
Champion Forest Volunteer Fire Department, 115
Charge, electrical, 107
Charles II, 3
Charlotte (NC) Fire Department, 55
Check valve. *See* Backflow preventer
Chemical energy, 104–7
    heat of combustion, 104–5
    heat of decomposition, 106
    heat of reaction, 107
    heat of solution, 106–7
    spontaneous heating, 105–6
Chicago
    Iroquois theater tragedy, 152
Chicago Fire Department, 9
Chief's interview, 39–41

Cincinnati Fire Department, 8
Citations, 160
City of Clinton v. Cedar Rapids and Missouri Railroad Company, 156
Clapper, in sprinkler system, 189
Class A fires, 114
Class B fires, 114–15
Class C fires, 115
Class D fires, 115
Class III standpipe systems, 179
Class II standpipe systems, 178
Class I standpipe systems, 178
Class K fires, 115
Clean extinguishing agents, 199
Closed-circuit SCBA, 89
Clothing, 82–88
  proximity ensemble, 86–87
  station work uniforms, 82–83
  structural firefighting ensemble, 83–85
  wildland firefighting ensemble, 85–86
Cocoanut Grove nightclub tragedy, Boston, 152
Codes. *See also* Ordinances
  building, 153
  defined, 152
  enforcement, legal authority of, 156–58
  fire prevention (*See* Fire prevention codes)
  writing of, 152
Cole, Dana, 219
Columns, 125–26
Combination fire departments, 210
Combination flame detectors, 197
Combination heat detectors, 195, 196
Combination water transport systems, 166, 167
Combined-agent apparatus, 80
Combustion
  process, 110–11
  self-sustained, chain reaction of, 99
Command, principles of, 214–15
Command structure, and preincident planning, 239–40
Communications
  as NIMS principle, 221
  and preincident planning, 239
Communications operators, 26–27
Community Emergency Response Team (CERT), 31
Compensation and claims unit, 218
Compressed Air Foam (CAFS) extinguishing systems, 71, 204
Compressive stress, 122, 123
Concrete
  moisture content of, 140
  structural qualities, 139–40
Conduction, 102–3
Conductor, electrical, 107
Confinement, as strategy, 241
Conflagrations, 2
Congreve, William, 180
Construction. *See* Building construction
Continuous line detectors, 196
Contract firefighter opportunities, 30
Control units, signaling systems, 192
Control valves, 168
Convection, 103
Cooley Doctrine, 156
Cooling by extinguishing systems, 202
Corbels, 141, 142
Cost unit, 218
Council for Higher Education (CHEA), 61
Couvre feu, 3
Criminal conviction record check, 41
Crossmains, 186
Ctesibius of Alexandria, 6
Current, electrical, 107
Customary system, 102

D

Dalmatians, 12
Davila, Martin D., 60
Dead loads, 121, 129
Decay stage, burning process, 114
Decision making, tactical, 242–43
Degree Assembly, IFSAC, 59, 60
Degrees at a Distance Program (DDP), 57
Delegation of authority, 214
Deluge automatic sprinkler system, 191
Department of Defense (DoD)
  firefighting opportunities, 21
Department of Homeland Security, 32
Deputy chiefs, 212
Detection systems. *See* Automatic detection systems
Dillon's Rule, 156
Direct flame contact, 104
Direct pumping water transport systems, 166, 167
Disaster Mitigation Act of 2000, 256
Discretionary authority, 159
Dispatchers, 26–27
Distance education, 56–57
Distance Education and Training Council (DETC), 61
Distribution grids, 168
Distributor pipes, 168
District chief, 211
District organizational unit, 213
Diversity, 43–44
Division of labor, 214
Documentation of inspections, 160
Dormitories, 68, 69
Double-seated clapper valve, 190
Drafting pit, 49
Drill tower, 49
Driver, 211
Driving record, 43
Dry chemical extinguishing systems, 201–2
Dry hydrants, 174
Dry-pipe system, 190
Dual water distribution models, 170
  advantages/disadvantages of, 171
Dynamic load, 121

E

Eccentric loads, 122
Education. *See* Higher education
Electrical energy, 107–10
  resistance heating, 107–8
Electrons, defined, 97
Elevation, 236
Eligibility lists, 42
Emergency incident management systems. *See* Incident Management System (IMS)
Emergency management, 251
  defined, 252
  mitigation, 256
  postsecondary education for, 256–57
  preparedness, 253–54
  recovery, 255
  response, 254–55
Emergency manager (EM)
  characteristics of, 251
Emergency Medical Responder (EMR), 248–49
Emergency Medical Service (EMS), 14
  evolution of, 246–48
  professional development model, 249
Emergency Medical Technicians (EMT), 14, 246, 249
  firefighters and, 250
  postsecondary education for, 250

Emergency support functions, NRF, 253
Employment. *See* Career opportunities
EMT Basic status, 42
Encasement, structural steel protecting method, 138
Encrustation, 169
Endothermic reactions, 106
Energized electrical fires, 115
Energy
  chemical, 104–7
  defined, 101
  electrical, 107–10
  heat, 102–4
  mechanical, 110–14
Engineering corrections, 153
Enhanced 911 (E911), 194
Equipment, 91–92
Ergonomics, 27
Executive Fire Officer Program (EFOP), 50–51
Exhauster, in sprinkler system, 190
Exothermic reactions, 98, 106–7
Exposures
  defined, 235
  and preincident planning, 235–36
  as strategy, 241
Extent, fire
  and preincident planning, 237
External exposures, 236
Extinguishers, 8–10

F

Facilities, fire department, 65–69
  administration buildings, 65–66
  fire stations, 67–69
  training facility, 66–67
Factory Mutual Research Corporation (FMRC), 195
Fahrenheit (F) scale, 102
Familia publica, 2
Federal Aviation Administration (FAA)
  ARFF regulations, 22, 81
Federal Emergency Management Agency (FEMA), 50, 218
  CERT, 31
  history of, 251–52
Federal firefighters, 21
The Federal Fire Prevention and Control Act of 1974, 50
Field of study, 2
Finance, IMS, 218
Fire
  classification, 114–15 (*See also* specific classes)
  defined, 97
  dynamics (*See* Fire dynamics)
  maintenance and flow testing, 174–77
Fire academies, 20
Fire and emergency services, 13–14
  career opportunities (*See* Career opportunities)
Fire and Emergency Services and Higher Education (FESHE), 14, 57, 249
  guidelines, 58–59
Fire apparatus, 69–76
  aerial apparatus, 73–76
  mobile water supply, 72–73
  pumpers, 70–72
Fire apparatus engineer, 23
Fire apparatus operator (FAO), 211
Fireboats, 10, 81
Fire cadet/recruit, 19–20
Fire chief, 212
Fire company, 212
Fire Corps, 31

Fire-cut design, 141–42
Fire department connection (FDC), 174, 185
Fire Department Explorers, 49
Fire departments
    facilities, 65–69
    organizations, 212–13
    rank structure, 211–12
    resources (*See* Resources, fire department)
    types, 209–10
Fire department training specialists, 29
Fire dynamics
    atoms, 97
    chemical energy, 104–7
    electrical energy, 107–10
    heat energy, 102–4
    matter, 97–101
    mechanical energy, 110–14
    molecules, 97
Fire ecology, 22
Fire engines
    internal combustion, 9
    red, 11
Fire equipment salespersons, 27
Fire extinguisher service technicians, 28
Fire extinguishment, as strategy, 241
Fire facilities, 10–11
FireFighter 1 (FF1), NFPA level, 21, 42–43
FireFighter 2 (FF2), NFPA level, 21
Firefighter/EMT, 21
Firefighters, 211
    and EMTs, 250
    federal, 21
    hazardous materials technicians, 25–26
    higher education (*See* Higher education)
    hiring requirements for, 20–21
    industrial, 25
    responsibility of, 21
    selection process (*See* Selection process)
    training (*See* Training)
    wildland, 21–22
Firefighting aircraft, 81–82
Firefighting Resources of California Organized
        for Potential Emergencies (FIRESCOPE),
        215
    vs. FGC system, 216
Firefighting service
    evolution of, 2–6
    insignias and traditions, 11–12
Fire flow
    computation, 105
    defined, 105
    requirements, 233
Fire ground command (FGC) system, 216
    FIRESCOPE *vs.*, 216
Firehouses, 11
Fire hydrants, 171–77
    base-valve dry-barrel, 171–72
    classification and color-code
        recommendation, 177
    dry, 174
    gallon per minute flow of, 176
    wet-barrel, 172–73
Fire inspections
    goals of, 29–30
Fire inspectors, 29–30
Fire Investigation Units (FIU), 26
Fire investigators, 30
Fireline Handbook, NWCG, 71
Fire lines, 22
Fire load
    computation, 104
    defined, 104
Firemark, 3
Fire marshals, 28
Fireplugs, 5

Fire point, 99
Fire prevention codes. *See also* Ordinances
    adoption, 158–59
    defined, 153
    legal authority of code enforcement, 156–58
    model codes, history of, 153–55
    types, 158–59
Fire prevention specialists, 29
Fire Prevention Week, 29
Fire protection engineers (FPE), 30
Fire protection systems
    automatic detection systems, 195–98
    automatic sprinkler system (*See* Automatic
        sprinkler system)
    fire hydrant, 171–77
    overview, 164
    signaling system, 192–95
    special extinguishing systems, 198–204
    standpipe systems, 177–80
    water distribution models, 170–71
    water supplies, 164–70
Fire pump, steam-powered, 7
Fire-resistive coating, structural steel protecting
        method, 138–39
Fire science, 2
Fire science technology curriculum
    standards for, 55–56
FIRESCOPE (Firefighting Resources of
        California Organized for Potential
        Emergencies), 215
    vs. FGC system, 216
Fire stations, 67–69
    apparatus bay, 68
    bathroom facilities, 68
    kitchen area, 68–69
    living area, 69
    offices, 69
    sleeping accommodations, 68, 69
Fire tetrahedron, 111
Fire tragedies
    historic, 152
Fire triangle, 111
Fire wardens, 4
First Law of Thermodynamics, 101
Fixed foam extinguishing systems, 203
Fixed-temperature heat detectors, 195–96
Flame/light detectors, 197
Flammable liquids, 114–15
Flammable metals fires, 115
Flashover, 112
Flash point, 99
Flat-roof, 131
Flexibility, steel, 137
Floors, in Type III building construction, 141
Flow pressure, 175, 179
Flow rate
    calculations, 232–33
    standpipe systems requirements, 179–80
Flow ratio, pipes, 169
Foam and water extinguishing systems, 204
Foam extinguishing systems, 203–4
Forces, in building construction, 120–22. *See
        also* Loads, in building construction
    shear, 139
Forest rangers, 22
Foundations, 123–24
    pier and beam, 124
Fourth Amendment, U.S. Constitution, 160
FPE. *See* Fire protection engineers (FPE)
Franklin, Benjamin, 4
Frank v. Maryland, 157
Frequencies, communication, 239
Friction, defined, 110
Friction loss, 180
Full-time paid firefighter, 5

Fully developed stage, burning process, 113
Fully paid (career) fire departments, 210
Functional areas, IMS
    command, 216–17
    logistics, 218
    operations, 217
    planning, 217

G

Gable roof, 132
Gallon per minute flow of hydrants, 176
Gambrel roof, 132, 133
Gaseous matter, 100–101
Gate valve, 183
General Duty Clause, 79–80
General Services Administration Standard
        KKK-A-1822, 77
Geographic Information Systems (GIS), 31,
        234–35
Gibbs, 10
Global Positioning System (GPS), 31, 234–35
Graff, Frederick, 171
Graphite, 174–75
Gravity water transport system, 165, 166
Great Baltimore fire, 154
Great Fire of Boston, 7–8
Great Fire of London, 3, 6
Greenburg, Morris, 172
Grinnell, Frederick, 181
Groundwater, 164–65
Growth stage, burning process, 112–13
Gusset plates, 131

H

Halon extinguishing systems, 199–200
Hand-operated pumps, 6
Hand pumper, 5
Hazardous materials
    and preincident planning, 238
    protective ensembles, 87–88
Hazardous materials (HazMat) apparatus,
        79–80
Hazardous materials response team
        (HazMat), 26
Hazardous materials technicians, 25–26
Hazards, life-safety
    and preincident planning, 232
HAZWOPER (Hazardous Waste Operations
        and Emergency Response), 80
Head pressure, 163
Headquarters office, 65–66
Heat detectors
    combination, 195, 196
    fixed-temperature, 195–96
    rate-of-rise, 195, 196
Heat energy, 102–4
    conduction, 102–3
    convection, 103
    latent heat, 102
    radiation, 104
    specific heat, 102
Heating
    arcing, 108–9
    induction, 109–10
    resistance, 107–8
    sparking, 109
    spontaneous, 105–6
    static electricity, 109
Heat of combustion, 104
Heat of compression, 110
Heat of decomposition, 106
Heat of reaction, 107
Heat of solution, 106–7

Heat sink, 102
Heavy rescue apparatus, 79
Heavy rescue squad, 76
Helitacks, 22
Higher education. *See also* Training
    defined, 48
    distance and online education, 56–57
    fire science technology curriculum, 55–56
    overview, 47–48
    post secondary education, 54–57
High-expansion foam extinguishing systems, 203–4
High-pressure carbon dioxide extinguishing systems, 200
High School Pathway Program, 59
High-temperature cooking oils fires, 115
Hip roof, 132, 133
Homeland Security, 13, 251
Home rule, 156
Horse-drawn fire wagons, 7
Hotshot crews, 22
Houston Fire Department, 224, 243
Humidity, and preincident plans, 235
Hurricane Katrina, 23–24
Hurricane Sandy, 257
Hydrants. *See* Fire hydrants
Hydraulically operated ladders, 73
Hydrophoric materials, 107

I

I-codes, ICC, 154–55
Ignition stage, burning process, 112
Ignition temperature, 99
I-joist, 129–30
Immediately dangerous to life and health (IDLH), 89
Incident action plan (IAP), 221, 242, 243
Incident commander (IC), 214
Incident command system (ICS), 211
    FIRESCOPE, 215
    history of, 215–16
    ratings, 219–20
Incident effectiveness, 209
Incident Management System (IMS), 216–24
    administration, 218
    defined, 216
    features, 216
    finance, 218
    functional areas (*See* Functional areas, IMS)
    NIIMS, 219
    NIMS, 220–23
Indicating valve, 183
Induction, 109–10
Industrial firefighter, 25
Information management, as NIMS principle, 222
Information transfer procedures, 239
Infrared flame detectors, 197
Initial attack pumper, 70–71
Initiating device circuits, signaling systems, 192
Injunction, 160
Insignias, firefighting service, 11–12
Inspection model, 157
Inspections
    documentation of, 160
    warrantless entry, 156–57
Instructor, training
    codes/standards for, 54
Insulator, 107
Insurance adjusters, 29
The Insurance Office, 3
Insurance Service Office, Inc. (ISO), 6, 66
Integrated Emergency Management System, 251

Intelligence management, as NIMS principle, 222
Internal combustion fire engines, 9
Internal exposures, 236
International Association of Emergency Managers (IAEMS), 251
International Building Code (IBC), 135
International City Managers Association (ICMA), 232
International Code Council (ICC), 135–36, 154–55
International Conference of Building Officials (ICBO), 154
International Fire Service Accreditation Congress (IFSAC), 56, 58, 59–60
International Fire Service Training Association
    aerial apparatus categories, 73
Interview
    chief's, 39–41
    mock, 39
    sample questions, 40
Interviewing skills, 39–41
Inventors, 31
Ionization detectors, 197
Iroquois theater tragedy, Chicago, 152

J

Job performance requirements (JPRs), 54
Jobs. *See* Career opportunities
Joists
    ceiling, 130–31
    I-joist, 129–30

K

Katrina, Hurricane, 23–24
Kellogg Brown & Root (KBR), 30
K factor coefficient, 189
Kitchen area, fire station, 68–69
Knights of St. John, 11

L

Labor, division of, 214
Laboratory testing, 159
Ladders, 68
    aerial, 73–74
    hydraulically operated, 73
Language skills, 43
Lantern roof, 132, 133
Latent heat, 102
Latent heat of fusion, 102
Latent heat of vaporization, 102
Latter, Alexander, 8
Law of the Conservation of Mass and Energy, 101
Learn Not to Burn program, 29
Level A ensemble, 87
Level B ensemble, 88
Level C ensemble, 88
Level D ensemble, 88
Liaison officer, 217
Lieutenant, 211
Life-safety hazards, and preincident planning, 232
Light detectors. *See* Flame/light detectors
Lightning, as static electricity form, 109
Light-obscuration smoke detector systems, 197
Light rescue squads, 77
Light-scattering systems, 197, 198
Line operation, 211
Line organization, 212
Line-type rate-of-rise detector, 196
Liquid matter, 99–100
Live load, 121
Living area, 69

Load-bearing walls, 126
Loads, in building construction, 120–22
    axial, 121
    defined, 120
    distribution, 121
    eccentric, 122
    imposition, 121
    torsional, 122
    types of, 121
Local alarm systems, 193
Local application systems, 200, 201
Local Emergency Planning Committee (LEPC), 252
Location, fire
    and preincident planning, 237
Lockheed Martin, 30
Logistics, as IMS functional area, 218
London, Great Fire, 3, 6
Long-term recovery, 255
Looped grid, 168
Lote, Thomas, 7
Louis F. Garland Department of Defense Fire Academy, 21
Lower water seat clapper, 190
Low-pressure carbon dioxide extinguishing systems, 200

M

Major firefighting vehicles, 80
Maltese cross, 11–12
Manby, George, 8
Mandatory authority, 159
Manipulative training, 49, 55
Mansard roof slopes, 135
Manual dry system, 179
Manual wet system, 179
Mapp v. Ohio, 157
Master-coded alarm systems, 193
Mathews, Samuel R. C., 171
Matter, 97–101
    defined, 97
    gaseous, 100–101
    liquid, 99–100
    solid, 98–99
Mechanical energy, 110–14
    burning process, 111–14
    friction, 110
    heat of compression, 110
    process of combustion, 110–11
Medical exams, 41
Memorial Park, 6
Merrill, William Henry, 154
Metal fires, 115
Mill construction, 145
Mini-maxi code, minimum codes *vs.,* 158
Minimum codes *vs.* mini-maxi code, 158
Minorities, 44
Mitigation phase, emergency management, 256
Mobile data terminal (MDT), 92
Mobile water supply fire apparatus, 72–73
Mock interviews, 39
Model codes. *See also* Fire prevention codes
    administration/enforcement of, 159–60
    adoption of, 158–59
    history of, 153–55
Molecules, 97
Montreal Protocol, 199
Motorized era, 5
Multipurpose dry chemical systems, 201

N

National Academy of Sciences, 247
National Aeronautics and Space Administration (NASA), 10

National Board of Fire Underwriters (NBFU), 153
National Building Code, 135, 154
National Electric Code, NFPA, 155
National Fallen Firefighters Foundation, 6
National Fallen Firefighters Memorial Park, 6
National Fire Academy (NFA), 210
  Degrees at a Distance Program, 57
  fire flow estimation, 232–33
  training/education programs, 50–51
National Fire Incident Reporting System (NFIRS), 30
National Fire Protection Association (NFPA), 26, 65, 135, 155
  certifications, 52, 53
  codes/standards for training instructor, 54
  combustion process, defined, 111
  on hydrants flow-testing, 174–75
  levels, 21
  manufacturing standards, 65
  medical exam standard, 41
  pipe size requirements, 168–70
  sprinkler standards, 181–82
  standpipe systems, 177–80
National Highway Traffic Safety Act (NHTSA), 247
National Incident Management System (NIMS), 215, 220–23
  command structure, 223
  organizational structure, 220–21
  principles, 221–22
National Incident Management System (NIMT), 14
National Institute of Occupational Safety and Health (NIOSH), 240
National Interagency Incident Management System (NIIMS), 219
National Professional Development Model, 59
National Registry of Emergency Medical Technicians (NREMT), 14, 247
National Response Framework (NRF)
  emergency support functions, 253
National Standard Curriculum, 247
National Wildfire Coordinating Group (NWCG), 86
  Fireline Handbook, 71
Natural Resource Conservation Service (NRCS), 174
Neally's Smoke Excluding Mask, 10
Negligence, 159
Nelson, Bill, 231
Nero, Emperor, 2
Neutrons, defined, 97
New York City
  fire wardens, 4
  plug uglies, 5
NFPA 10 *(Standard for Portable Fire Extinguishers),* 115
NFPA 12 *(Standard on Carbon Dioxide Extinguishing Systems),* 200
NFPA 13 *(Standard for the Installation of Sprinkler System),* 182, 190
NFPA 14 *(Standard for the Installation of Standpipe and Hose Systems),* 177, 179
NFPA 17 *(Standard for Dry Chemical Extinguishing Systems),* 202
NFPA 25 *(Standard for the Inspection, Testing, and Maintenance of Water-Based Fire Protection Systems),* 177
NFPA 70 *(The National Electrical Code),* 193
NFPA 71 *(Standard for the Installation, Maintenance, and Use of Proprietary Central Station Signaling System),* 195
NFPA 72 *(The National Fire Alarm Code),* 193, 194

NFPA 101 *(Life-Safety Code),* 155, 181
NFPA 220 *(Standard on Types of Building Construction),* 136, 144, 145, 146
NFPA 251 *(Standard Methods of Tests of Fire Endurance of Building Construction and Materials),* 136
NFPA 291 *(Recommended Practice for Fire Flow Testing and Marking of Hydrants),* 177
NFPA 414 *(Standard for Aircraft Rescue and Fire-Fighting Vehicles),* 81
NFPA 720 *(Standard for the Installation of Carbon Monoxide (CO) Warning Equipment in Dwelling Units),* 198
NFPA 1001 *(Standard for Fire Fighter Professional Qualifications),* 21
NFPA 1035 *(Standard for Professional Qualifications for Public Fire and Life Safety Educator),* 29
NFPA 1041 *(Standard for Fire Service Instructor Professional Qualifications),* 54
NFPA 1403 *(Standard on Live Fire Training Evolutions),* 49
NFPA 1404 *(Standard for a Fire Department Self-Contained Breathing Apparatus Program),* 90
NFPA 1408 *(Standard for Thermal Imaging Training),* 92
NFPA 1500 *(Standard on Fire Department Occupational Safety and Health Program),* 82, 90, 243
NFPA 1582 *(Standard on Comprehensive Occupational Medical Program for Fire Departments),* 41
NFPA 1600 *(Standard on Disaster/Emergency Management and Business Continuity Programs),* 252
NFPA 1620 *(Recommended Practice for Pre-Incident Planning),* 228
NFPA 1710 *(Standard for Organization and Deployment of Fire Suppression Operations, Emergency Medical Operations, and Special Operations to the Public by Career Fire Departments),* 232
NFPA 1801 *(Standard on Thermal Imagers for the Fire Service),* 92
NFPA 1901 *(Standard for Automotive Fire Apparatus),* 65, 77
NFPA 1903 *(Standard for Mobile Water Supply Fire Apparatus),* 73
NFPA 1904 *(Standard for Aerial Ladder and Elevating Platform Fire Apparatus),* 76
NFPA 1906 *(Standard for Wildland Fire Apparatus),* 72
NFPA 1971 *(Standard on Protective Ensembles for Structural Fire Fighting and Proximity Fire Fighting),* 85
NFPA 1975 *(Standard on Station/Work Uniforms for Fire and Emergency Services),* 83, 88
NFPA 1976 *(Standard on Protective Ensemble for Proximity Fire Fighting),* 86
NFPA 1977 *(Standard on Protective Clothing and Equipment for Wildland Fire Fighting),* 86
NFPA 1981 *(Standard on Open-Circuit Self-Contained Breathing Apparatus for Fire Service),* 90
NFPA 1982 *(Standard on Personal Alert Safety Systems),* 90
NFPA 1991 *(Standard on Vapor-Protected Ensembles for Hazardous Materials Emergencies),* 87

NFPA 1992 *(Standard on Liquid Splash-Protective Ensembles and Clothing for Hazardous Material Emergencies),* 88
NFPA 1994 *(Standard on Protective Ensembles for Chemical/Biological Terrorism Incidents),* 88
NFPA 5000 *(Building Construction and Safety Code),* 136, 140, 144, 145, 146, 155
NFPA 12A *(Standard on Halon 1301 Fire Extinguishing Systems),* 199
NFPA 13A *(Recommended Practice for the Inspection, Testing and Maintenance of Sprinkler Systems),* 182
NFPA 12B *(Standard on Halon 1211 Fire Extinguishing Systems),* 199
NFPA 13D *(Standard for the Installation of Sprinkler Systems in One- and Two-Family Dwellings and Mobile Homes),* 182
NFPA 13E *(Recommendations for Fire Department Operations in Properties Protected by Sprinkler and Standpipe Systems),* 182
NFPA 13R *(Standard for the Installation of Sprinkler Systems in Residential Occupancies up to Four Stories in Height),* 182
Noncoded alarm systems, 193
Non–fire suppression activities, 66
Nonindicating valve, 168
Non-load-bearing walls, 126
Nonpressurized dry hydrant system, 174
Normal atmospheric pressure, 100
Nucleus, atom, 97

O

Occupancy
  defined, 120, 228
  and preincident planning, 230–31
Occupational health and safety specialists, 27
Occupational Safety and Health Administration (OSHA), 80
Occupation classification, 158
Offices, fire stations, 69
Oklahoma State University, 56
On-call fire departments, 210
Online education, 56–57
On-site testing, 159
Open-circuit SCBA, 89
Operations, as IMS functional area, 217
Ordinances. *See also* Fire prevention codes
  defined, 152
Ordinary dry chemical extinguishing systems, 201
Organizational structure
  fire departments *(See* Fire departments)
OSHA (Occupational Safety and Health Administration), 80
OSHA Code of Federal Regulations 29 CFR 1910.120, 80, 90
OSHA Code of Federal Regulations 29 CFR 1910.156, 89, 210
Outside Stem and Yoke (OS&Y) valves, 183
Overhaul, as strategy, 241–42
Overseas firefighter opportunities, 30
Oxidation, 97, 104

P

Paramedic, 249
Paramilitary organizations, 209
Parmelee, Henry, 180
Patriot Act, 13
Performance codes, 159
Permit model, 157

Personal alert safety systems (PASS), 82, 90
Personal history, selection process, 41–42
Personal protective equipment (PPE), 82–93
   clothing, 82–88
   PASS, 90
   SCBA, 88–90
Personnel, and preincident planning, 231–32
Photoelectric detectors, 197
Physical fitness, 43
Pictometry, 31
Pier and beam foundations, 124
Pipes, for water supply
   size requirements, NFPA, 168–70
Pitched roofs, 131–35
   conventional, 131
   prefabricated trusses, 131
   types of, 131–35 (See also specific types)
Plain concrete, 139
Planning
   preincident (See Preincident planning)
   responsibilities, IMS, 217
Platform-frame construction, 146–48
Platforms, 68
Plug uglies, 5
Point of strain, steel, 137
Portable fire extinguisher, 8
Post indicator valves, 183–84
   assembly, 184–85
Postsecondary education
   for emergency management, 256–57
   for EMT, 250
Post-tensioned concrete, 140
Preaction automatic sprinkler system, 190–91
Preemployment medical exams. See Medical
     exams
Prehospital care, 247
Preincident planning, 228–40
   apparatus and, 231–32
   auxiliary appliances and, 234
   building construction and, 230
   command structure and, 239–40
   communication and, 239
   defined, 228
   hazardous materials and, 238
   life-safety hazards and, 232
   location and extent of fire, 237
   occupancy and, 230–31
   personnel and, 231–32
   street conditions and, 234–35
   time of day, 237–38
   topography and, 236–37
   water supply and, 232–34
   weather and, 235–36
Preparedness, 253–54
Prerequisites, for firefighter position, 36
Prescribed fires, 22
Prescription/performance-based models, 159
Prescriptive codes, 159
Pressure
   flow, 175, 179
   residual, 170, 175
   soil, 123–24
   static, 170, 175
   trip, 190
   vapor, 99
Pressure regulators
   SCBA, 10
Pressure surges, 189
Pressurized dry hydrant system, 174
Prestressed/pretensioned concrete, 139–40
Primary feeder mains, distribution grid, 168
Primary power supply, signaling systems, 192
Private sector career opportunities, 30–31
Probationary period, 42
Pro Board, 60–61

Procurement unit, 218
Promotion, 51–54
Propane, vapor density of, 101
Proprietary signaling system, 194
Protected construction, Type III-211, 143–44
Protons, defined, 97
Proximity ensemble, 86–87
Public fire education specialist, 29
Public information officer (PIO), 29, 216–17
Public safety answering point (PSAP), 194
Public safety department, 210
Public safety officer (PSO), 210
Public water systems, 5
Pump-and-roll operations, 71
Pumpers, 70–72
Pyrolysis, 99
Pyrophoric materials, 107

R

Radiation, 104
Radiation shielding extinguishing systems, 202
Rankine scale, 102
Rank structure, fire departments, 211–12
Rapid fire progress event, 112
Rapid intervention team (RIT), 243
Rapid intervention vehicles (RIVs), 80
Rate of heat release, 105
Rate-of-rise detector, 195, 196
Rattle Watch, 4
Recovery phase, emergency management, 255
Red fire engines, 11
Regulations, FAA, 81
Relative humidity, and preincident plans, 235
Remote station system, 194
Rescue apparatus, 77–79
Rescue strategy, 240–41
Residential sprinkler systems, 192
Residual pressure, 170, 175
Resistance, defined, 107
Resistance heating, 107–8
Resource management, as NIMS principle, 221
Resources, fire department, 64–93
   fire apparatus (See Fire apparatus)
   fire department facilities (See Facilities, fire
     department)
   overview, 65
   personal protective equipment (See Personal
     protective equipment (PPE))
   specialized apparatus (See Specialized appa-
     ratus)
   tools and equipment, 91–92
Respiratory protective equipment, 10
Response phase, emergency management,
   254–55
Right of entry, 157–58
Risers
   sprinkler system, 186
   standpipe systems, 177
Risk Watch program, 29
Rome, 2–3
Roofs
   configurations, 131–35
   flat-roof, 131
   pitched, 131–35
   stick-framed, 131
Royer, Keith, 231

S

Safety officer, 217
Safety standards, fire stations, 67
Salvage, as strategy, 242
Sand lime mortar, 143
Sandy, Hurricane, 257

San Francisco Auxiliary Water Supply
   System, 170
Saponification, 115, 202
Sawtooth roof, 132, 134
Search and rescue (SAR) teams, 23–24
Secondary backup power supply, signaling
   systems, 192
Secondary feeder mains, distribution grid, 168
Sedimentation, 169
See v. The City of Seattle, 157
Selection process, 35–44
   hiring conditions, 42
   interviewing skills, 39–41
   overview, 35–36
   personal history, 41–42
   preparation, 36–38
   self-marketing, 42–44
Self-contained apparatus foam extinguishing
     systems, 203
Self-contained breathing apparatus (SCBA), 10,
     82, 88–90, 238
   closed-circuit, 89
   open-circuit, 89
Self-marketing, 42–44
Self-sustained combustion, chain reaction of, 99
Semiautomatic dry system, 179
Semi-fixed type A foam extinguishing systems,
   203
Semi-fixed type B foam extinguishing systems,
   203
Separate water distribution models, 170
Shank, Able, 8
Shape of the Area, 236
Shear forces, 139
Shed roof, 132, 134
Short-term recovery, 255
Shut-off valves. See Water flow control valves
Signaling systems, 192–95
   auxiliary, 194
   central station system, 195
   features, 192
   local alarm system, 193
   proprietary, 194
   remote station system, 194
Siphon, 6
Sleeping accommodations, 68, 69
Slope, 236
Smith, James P., 240
Smoke detectors, 197
Smoke jumpers, 22
Smokey Bear campaign, 29
Smothering dry chemical extinguishing systems,
   202
Social media, for hiring purposes, 44
Soda-acid extinguisher, 8
Soil pressure, 123–24
Sole plate, 147
Solid matter, 98–99
Southern Building Code (SBC), 135
Southern Building Code Congress International
   (SBCCI), 154
Spalling, 103, 140
Span of Control, 214, 222
Sparking heating, 109
Sparky the Fire Dog®, 29
Special extinguishing systems, 198–204
   carbon dioxide systems, 200–201
   dry chemical systems, 201–2
   foam systems, 203–4
   Halon systems, 199–200
   wet chemical systems, 202–3
Specialized apparatus, 76–82
   Aircraft Rescue and Firefighting Apparatus,
     80–81
   ambulances, 77

firefighting aircraft, 81–82
hazardous materials apparatus, 79–80
rescue apparatus, 77–79
Specialized fire brigade, 210
Special operations bureaus, 50
Specialty certifications, 52, 53
Specific gravity, 101
Specific heat, 102
Spontaneous heating, 105–6
Spot-type rate-of-rise detector, 196
Spot types, 195
Sprinkler head, 187–89
Sprinkler system. *See* Automatic sprinkler system
Squirts, 6
Staff operation, 211
Staff organization, 213
Stair climb test, 38
Standard operating procedures (SOP), 20, 227–28
Standards
    for base-valve dry-barrel hydrants, 172
    manufacturing, 65
    safety, fire stations, 67
    setting, for higher education, 57–59
    for wet-barrel hydrants, 172
Standpipe systems, 177–80
    categories/types of, 179
    Class I, 178
    Class II, 178
    Class III, 179
    defined, 177
    flow rate requirements, 179–80
    pressure-regulating devices, 180
Static electricity heating, 109
Static loads, 121
Static pressure, 170, 175
Station work uniforms, 82–83
Statistician, 30
Statute, 158
Steam engine, 7
Steam-powered fire pump, 7
Steel, in building construction, 136–39
    flexibility, 137
    point of strain, 137
    structural, methods of protecting, 138–39
    structural failure of, heat exposure and, 137
    unrestrained beams, 137
Steel-reinforced concrete, 139
Stick frame, 131
Stockton (CA) Fire Department, 148
Stoichiometric reactions, 104
Strategic action plan, 221, 240
Strategic and Tactical Considerations on the Fireground, 240
Strategy, 228, 240–42. *See also* Preincident planning
    defined, 240
Street conditions, and preincident planning, 234–35
Stress, in building construction
    compressive, 122, 123
    tensile, 122, 136
Structural components, in building construction, 123–35
    beams, 124–25
    ceiling joist, 130–31
    columns, 125–26
    foundations, 123–24
    I-joist, 129–30
    load-carrying capability of, 145
    roofs (*See* Roofs)
    walls, 126–28
    wood trusses, 128–29
Structural firefighting ensembles, 83–85
Subsidence, 164
Summons, 160

Surface water, 164, 165
Système International (SI), 102

T
Tactical training, 49, 55
Tactics, 228, 242–43
    decision making, 242–43
    defined, 240
Target areas for hydrants, 177
Task Force 1. *See* Texas Task Force 1
Telescoping boom, 74, 75–76
Temperature, and preincident plans, 235
Tendons, 139
Tensile stress, 122, 136
Tenth Amendment, U.S. Constitution, 156
Tests
    employment, 36–38
    hydrants, 175
Texas A&M Engineering Extension Service (TEEX), 23
Texas Intrastate Fire Mutual Aid System (TIFMAS), 22
Texas Task Force 1, 23–24
Thermal barrier, 83–84
Thermal imaging camera (TIC), 91–92
Thermal lag, 196
Time of day, and preincident planning, 237–38
Time unit, 218
Tools, 91–92
Topography, and preincident planning, 236–37
Top plate, 147
Torsional loads, 122
Total flooding systems, 200, 201
Traditions, firefighting service and, 11–12
Training, 48–54. *See also* Higher education
    advancement and promotion, 51–54
    defined, 48
    facilities, 66–67
    IFSAC, 59–60
    instructor, codes/standards for, 54
    manipulative, 49
    overview, 47–48
    Pro Board accreditation, 60–61
    records, 50
    special operations bureaus, 50
    tactical, 49
Training bureau/officers, 54
Transportation Security Administration (TSA), 9
Triple certification, 13–14
Trip pressure, 190
Trouble signal power supply, signaling systems, 192
Trusses
    component of, 132
    flat-roof and, 131
    prefabricated, 131
    wood, 128–29
Tuberculation, 169
Two-in/two-out rule, 243
Type I ambulance, 77
Type I building construction
    concrete, 139–40
    steel, 136–39
Type I combustion, 110
Type II ambulance, 77
Type II building construction, 140–41
Type II combustion, 110–11
Type III ambulance, 77
Type III building construction, 141–44
    danger indicators in, 142–43
    floors in, 141
    subdivisions, 143–44
Type II pumpers, 70

Type I pumper, 70
Type IV building construction, 144–45
Type V-000, 146
Type V-111, 146
Type V building construction, 146–48
    balloon-frame construction, 146, 147
    platform-frame construction, 146–48

U
Ultraviolet flame detectors, 197
Uncle Joe Ross, 8
Underwriters Laboratory (UL), 195
    Standard UL 246 (*Hydrants for Fire Protection Service*), 172
Unified command, as NIMS principle, 222
Uniform Building Code (UBC), 135, 154
Union Volunteer Fire Company of Philadelphia, 4
United States, fire problems in, 12–13
United States Fire Administration (USFA)
    Executive Fire Officer Program, 50–51
    fire danger report, 12
    training/education programs, 50–51
Unity of command, 214, 222
Unprotected construction, Type III-200, 144
Upper air seat clapper, 190

V
Valves
    alarm, 186
    check, 182
    distribution grids, 168
    water flow control, 183–85
van der Heiden, Jan, 7
Vapor density
    defined, 100
    of propane, 101
Vapor pressure, 99
Vehicles, early, 6–10
Ventilation, as strategy, 242
Vigiles (watchmen), 2
Volunteer career opportunities, 31
Volunteer experience, 43
Volunteer fire departments, 209–10
Volunteer Firefighter and EMS Personnel Job Protection Act, 32

W
Wall post indicator valves, 184
Walls, 126–28
    bracing in, 126
    cantilevered, 126
    load-bearing, 126
    non-load-bearing, 126
Warrantless entry, 156–57
Warrants, defined, 157
Water distribution models, 170–71
Water flow control valves, 183–85
Water gong, 186
Watersheds, 165
Water supplies, 164–70
    components, 163–68
    groundwater, 164–65
    main, as automatic sprinkler system component, 182
    pipe size requirements, 168–70
    and preincident planning, 232–34
    surface water, 165
    transport systems, 165–67
    treatment systems, 167
Water transport systems, 165–67

Weather, and preincident
 planning, 235–36
Wet-barrel hydrant, 172–73
Wet chemical agents, 202–3
Wet-pipe system, 189–90
Wildland firefighter, 21–22
Wildland firefighting

apparatus, 70, 71–72
 ensemble, 85–86
William the Conqueror, 3
Wind, and preincident
 plans, 235
Women firefighters, 44
Wood trusses, 128–29

Written tests, 36–38

Z

Zone-coded alarm systems, 193
Zone-noncoded alarm systems, 193